Graph Machine Learning

Second Edition

Learn about the latest advancements in graph data to build robust machine learning models

Aldo Marzullo

Enrico Deusebio

Claudio Stamile

‹packt›

Graph Machine Learning
Second Edition
Copyright © 2025 Packt Publishing

Portfolio Director: Sunith Shetty

Relationship Lead: Sanjana Gupta

Project Manager: Shashank Desai

Content Engineer: Tiksha Abhimanyu Lad

Technical Editor: Gaurav Gavas

Copy Editor: Safis Editing

Indexer: Pratik Shirodkar

Proofreader: Safis Editing and Tiksha Lad

Production Designer: Ganesh Bhadwalkar and Salma Patel

Growth Lead: Bhavesh Amin

First published: May 2021

Second edition: June 2025

Production reference: 2150725

Published by Packt Publishing Ltd.

Grosvenor House

11 St Paul's Square

Birmingham

B3 1RB, UK.

ISBN 978-1-80324-806-6

www.packtpub.com

Contributors

About the authors

Aldo Marzullo received an M.Sc. degree in computer science from the University of Calabria (Cosenza, Italy) in September 2016. During his studies, he developed a solid background in several areas, including algorithm design, graph theory, and machine learning. In January 2020, he received his joint Ph.D. from the University of Calabria and Université Claude Bernard Lyon 1 (Lyon, France), with a thesis titled *Deep Learning and Graph Theory for Brain Connectivity Analysis in Multiple Sclerosis*. He is currently a postdoctoral researcher and collaborates with several international institutions.

For those I hold close, and those I carry quietly with me.

Aldo Marzullo

Enrico Deusebio is currently working as engineering manager at Canonical, the publisher of Ubuntu, to promote open source technologies in the data and AI space and to make them more accessible to everyone. He has been working with data and distributed computing for over 15 years, both in an academic and industrial context, helping organizations implement data-driven strategies and build AI-powered solutions. He has collaborated and worked with top-tier universities, such as the University of Cambridge, University of Turin, and the Royal Institute of Technology (KTH) in Stockholm, where he obtained a Ph.D. in 2014. He holds a B.Sc. and an M.Sc. degree in aerospace engineering from Politecnico di Torino.

To Lili and Pepe, for always reminding me, with your learning process, how wonderful the human brain and life are.

Enrico Deusebio

Claudio Stamile received an M.Sc. degree in computer science from the University of Calabria (Cosenza, Italy) in September 2013 and, in September 2017, he received his joint Ph.D. from KU Leuven (Leuven, Belgium) and Université Claude Bernard Lyon 1 (Lyon, France). During his career, he developed a solid background in AI, graph theory and machine learning with a focus on the biomedical field.

A Enea, che ha dato senso a molti silenzi.

To Enea, who gave meaning to many silences.

Claudio Stamile

About the reviewer

Nathan Smith is a principal consultant in the professional services division of Neo4j. As a data scientist, he works with companies to apply graph algorithms for machine learning and analytics. He is an organizer of the Data Science Kansas City Meetup, and he enjoys engaging with the data science community through blog posts *in Towards Data Science, Medium, and the Neo4j Developer Blog.*

Table of Contents

Part II: Machine Learning on Graphs   105

Chapter 4: Unsupervised Graph Learning   107

Chapter 5: Supervised Graph Learning   149

Part III: Practical Applications of Graph Machine Learning 211

Chapter 7: Social Network Graphs   213

Chapter 8: Text Analytics and Natural Language Processing Using Graphs 237

Preface

This updated and expanded second edition brings several significant improvements to help you stay ahead in the evolving field of graph machine learning. Compared to the previous version, this edition features refined chapters for improved clarity and flow, new examples utilizing both legacy tools and modern frameworks such as PyTorch and DGL, and entirely new chapters covering cutting-edge topics such as temporal graph machine learning and the integration of **large language models (LLMs)**.

Graph Machine Learning provides a powerful toolkit for processing network-structured data and leveraging the relationships between entities for predictive modeling, analytics, and more. You'll begin with a concise introduction to graph theory, graph machine learning, and neural networks, building a foundational understanding of their principles and applications. As you progress, you'll dive into the core machine learning models for graph representation learning, exploring their goals, inner workings, and practical implementation across various supervised and unsupervised tasks. You'll develop an end-to-end machine learning pipeline, from data preprocessing to training and prediction, to fully harness the potential of graph data. Throughout the book, you'll find real-world scenarios such as social network analysis, natural language processing with graphs, and financial transaction systems. The later chapters take you through the creation of scalable, data-intensive applications for storing, querying, and processing graph data and introduce you to the recent breakthroughs and emerging trends in the domain, some of which are the interaction between graphs and LLMs used in the context of generative AI and **retrieval-augmented generation (RAG)** systems.

By the end of this book, you will have understood the key concepts of graph theory and machine learning algorithms, allowing you to develop impactful graph-based machine learning solutions.

Who this book is for

This book is for data analysts, graph developers, graph analysts, and graph professionals who want to leverage the information embedded in the connections and relations between data points, unravel hidden structures, and exploit topological information to boost their analysis and models' performance. The book will also be useful for data scientists and machine learning developers who want to build machine learning-driven graph databases.

What this book covers

Chapter 1, Getting Started with Graphs, introduces the basic concepts of graph theory using the NetworkX Python library.

Chapter 2, Graph Machine Learning, introduces the main concepts of graph machine learning and graph embedding techniques.

Chapter 3, Neural Networks and Graphs, introduces **Graph Neural Networks** (**GNNs**) and the leading libraries for graph-based deep learning.

Chapter 4, Unsupervised Graph Learning, covers recent unsupervised graph embedding methods.

Chapter 5, Supervised Graph Learning, covers recent supervised graph embedding methods.

Chapter 6, Solving Common Graph-Based Machine Learning Problems, introduces the most common machine learning tasks on graphs.

Chapter 7, Social Network Graphs, shows an application of machine learning algorithms on social network data.

Chapter 8, Text Analytics and Natural Language Processing Using Graphs, shows an application of machine learning algorithms on a natural language processing task.

Chapter 9, Graphs Analysis for Credit Card Transactions, shows an application of machine learning algorithms in credit card fraud detection.

Chapter 10, Building a Data-Driven Graph-Powered Application, introduces some technologies and techniques useful to deal with large graphs.

Chapter 11, Temporal Graph Machine Learning, focuses on techniques to model and learn from dynamic, time-evolving graph data.

Chapter 12, GraphML and LLMs, explores how graph structures can enhance LLMs and how LLMs can be used for graph-based tasks.

Chapter 13, Novel Trends on Graphs, introduces some novel trends (algorithms and applications) of graph machine learning.

To get the most out of this book

We recommend that you use Docker to have a reproducible environment and stable dependency sets. The provided Docker images – one for each chapter – ship with a Jupyter installation and a Python kernel with the dependencies pre-installed, which you can use to run all the examples. For some chapters, Neo4j, JanusGraph, and Gephi are also needed.

Software/hardware covered in the book	OS requirements
Python	Windows, macOS, and Linux (any)
Neo4j	Windows, macOS, and Linux (any)
Gephi	Windows, macOS, and Linux (any)
Docker	Windows, macOS, and Linux (any)

A beginner-level understanding of graph databases and graph data is required. Intermediate-level working knowledge of Python programming and machine learning is also expected to make the most of this book.

The authors acknowledge the use of cutting-edge AI, such as ChatGPT, with the sole aim of enhancing the language and clarity within the book, thereby ensuring a smooth reading experience for readers. It's important to note that the content itself has been crafted by the authors and edited by a professional publishing team.

Download the example code files

The code bundle for the book is hosted on GitHub at https://github.com/PacktPublishing/Graph-Machine-Learning. We also have other code bundles from our rich catalog of books and videos available at https://github.com/PacktPublishing. Check them out!

Conventions used

There are a number of text conventions used throughout this book.

CodeInText: Indicates code words in text, database table names, folder names, filenames, file extensions, pathnames, dummy URLs, user input, and X/Twitter handles. For example: "For this exercise, we will be using a GraphSAGE encoder with three layers of 32, 32, and 16 dimensions, respectively."

A block of code is set as follows:

```
TMF_model = TMF(num_nodes, hid_dim, win_size, num_epochs, alpha, beta,
theta, learn_rate, device)
adj_est = TMF_model.TMF_fun(adj_list)
```

Any command-line input or output is written as follows:

```
Precision: 0.9636952636282395
Recall: 0.9777853337866939
F1-Score: 0.9706891701828411
```

Warnings or important notes appear like this.

Tips and tricks appear like this.

Get in touch

Feedback from our readers is always welcome.

General feedback: Email feedback@packtpub.com and mention the book's title in the subject of your message. If you have questions about any aspect of this book, please email us at questions@packtpub.com.

Errata: Although we have taken every care to ensure the accuracy of our content, mistakes do happen. If you have found a mistake in this book, we would be grateful if you reported this to us. Please visit http://www.packtpub.com/submit-errata, click **Submit Errata**, and fill in the form.

Piracy: If you come across any illegal copies of our works in any form on the internet, we would be grateful if you would provide us with the location address or website name. Please contact us at copyright@packtpub.com with a link to the material.

If you are interested in becoming an author: If there is a topic that you have expertise in and you are interested in either writing or contributing to a book, please visit http://authors.packtpub.com/.

Share Your Thoughts

Once you've read *Graph Machine Learning, Second Edition,* we'd love to hear your thoughts! Scan the QR code below to go straight to the Amazon review page for this book and share your feedback.

https://packt.link/r/1803248068

Your review is important to us and the tech community and will help us make sure we're delivering excellent quality content.

Free Benefits with Your Book

This book comes with free benefits to support your learning. Activate them now for instant access (see the "*How to Unlock*" section for instructions).

Here's a quick overview of what you can instantly unlock with your purchase:

PDF and ePub Copies

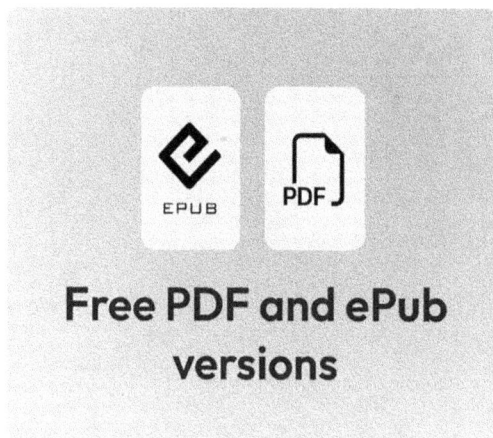

Free PDF and ePub versions

Next-Gen Web-Based Reader

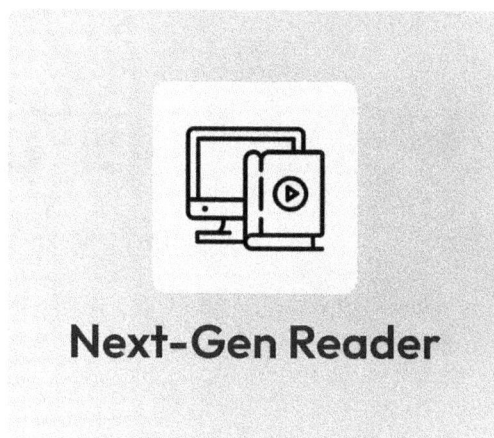

Next-Gen Reader

Access a DRM-free PDF copy of this book to read anywhere, on any device.

Use a DRM-free ePub version with your favorite e-reader.

Multi-device progress sync: Pick up where you left off, on any device.

Highlighting and notetaking: Capture ideas and turn reading into lasting knowledge.

Bookmarking: Save and revisit key sections whenever you need them.

Dark mode: Reduce eye strain by switching to dark or sepia themes.

How to Unlock

Scan the QR code (or go to packtpub.com/unlock).
Search for this book by name, confirm the edition,
and then follow the steps on the page.

*Note: Keep your invoice handy. Purchases made directly
from Packt don't require one.*

Part 1

Introduction to Graph Machine Learning

In this part, you will get a brief introduction to graph machine learning, showing the potential of graphs combined with the right machine learning algorithms. Moreover, a general overview of graph theory and Python libraries is provided in order to allow you to deal with (create, modify, and plot) graph data structures.

This part comprises the following chapters:

- *Chapter 1, Getting Started with Graphs*
- *Chapter 2, Graph Machine Learning*
- *Chapter 3, Neural Networks and Graphs*

1

Getting Started with Graphs

Graphs are mathematical structures that are used for describing relationships between entities, and they are used almost everywhere. They can be used for representing maps, where cities are linked through streets. Graphs can describe biological structures, web pages, and even the progression of neurodegenerative diseases. For example, social networks are graphs, where users are connected by links representing the "follow" relationship.

Graph theory, the study of graphs, has received major interest for years, leading people to develop algorithms, identify properties, and define mathematical models to better understand complex behaviors.

This chapter will review some of the concepts behind graph-structured data. Theoretical notions will be presented, together with examples to help you understand some of the more general concepts and put them into practice. In this chapter, we will introduce and use some of the most widely used Python libraries for the creation, manipulation, and study of the structure dynamics and functions of complex networks.

The following topics will be covered in this chapter:

- General information on the practical exercises and how to set up the Python environment to run them
- Introduction to graphs with `networkx`
- Plotting graphs
- Graph properties
- Benchmarks and repositories
- Dealing with large graphs

Practical exercises

For all of our exercises, we will be using Jupyter Notebook. Along with the book, we provide a GitHub repository at https://github.com/PacktPublishing/Graph-Machine-Learning, where all of the notebooks are provided and organized in different folders, one for each chapter of the book.

Each chapter is also based on a self-contained and separated environment, bundling all of the dependencies required to run the exercises of a given chapter. The Python version and the version of the dependencies may slightly vary depending on the set of libraries used in the chapter. Version management is implemented using **Poetry**, which allows us to resolve, manage, and update dependencies easily, making sure that the environments are fully reproducible.

Direct dependencies (including the Python version) are specified in each chapter/folder in the pyproject.toml file. If you are using Poetry, you can simply install the environment by using:

```
poetry install
```

Otherwise, if you don't have a Poetry installation on your local machine, you can also use pip. Along with the pyproject.toml and poetry.lock files, we also provide a requirements.txt file with the entire set of dependencies (also transitive) pinned to the exact version used to run the examples, which can be installed using:

```
pip install -r requirements.txt
```

Moreover, we also provide a Docker image with a Jupyter server installation integrated with the different Python environments. Each chapter's environment is loaded as a separated kernel and the different notebooks are already configured to use the respective environment. **Docker** can be installed on multiple operating systems (Linux, Windows, and macOS). Please refer to the website for guidance on how to set up Docker on your system. If you are a beginner, we also suggest you install **Docker Desktop** for an easy-to-use **graphical user interface** (**GUI**) to interact with the Docker Engine.

Once Docker is installed, you can start the containerized image either via the GUI or using the CLI:

```
docker run \
    -p 8888:8888 \
    --name graph-machine-learning-box \
    graph-machine-learning:latest
```

You can find more information on how to run and build the *Graph Machine Learning* book image in the `README.md` file at `https://github.com/PacktPublishing/Graph-Machine-Learning/blob/main/docker/README.md`.

The image will run a Jupyter server, available at `http://localhost:8888/`. The environments of the different chapters have already been configured and loaded in the Jupyter server, and they can be selected when creating a new notebook. The notebooks in the different chapters are already configured to bind to the correct kernel.

Conventions

In this book, the following Python commands will be referred to:

- `import networkx as nx`
- `import pandas as pd`
- `import numpy as np`

Technical requirements

All code files relevant to this chapter are available at `https://github.com/PacktPublishing/Graph-Machine-Learning/tree/main/Chapter01`. Please refer to the *Practical exercises* section for guidance on how to set up the environment to run the examples in this chapter, either using Poetry, pip, or Docker.

For more complex data visualization tasks provided in this chapter, Gephi (`https://gephi.org/`) may also be required. The installation manual is available here: `https://gephi.org/users/install/.https://gephi.org/users/install/.`

Introduction to graphs with networkx

In this section, we will give a general introduction to graph theory. Moreover, to link theoretical concepts to their practical application, we will use code snippets in Python. We will use Networkx, a powerful Python library for creating, manipulating, and studying complex networks and graphs. networkx is flexible and easy to use, which makes it an excellent didactic tool for beginners and a practical tool for advanced users. It can handle relatively large graphs, and features many built-in algorithms for analyzing networks.

A graph G is defined as a couple $G=(V,E)$, where $V=\{V_1,..., V_n\}$ is a set of nodes (also called **vertices**) and $E=\{\{V_k, V_w\}, ..., \{V_i, V_j\}\}$ is a set of two-sets (set of two elements) of edges (also called **links**), representing the connection between two nodes belonging to V.

It is important to underline that since each element of E is a two-set, there is no order between each edge. To provide more detail, $\{V_k, V_w\}$ and $\{V_w, V_k\}$ represent the same edge. We will call this kind of graph undirected.

We'll now provide definitions for some basic properties of graphs and nodes, as follows:

- The **order** of a graph is the number of its vertices $|V|$. The **size** of a graph is the number of its edges $|E|$.

- The **degree** of a vertex is the number of edges that are adjacent to it. The **neighbors** of a vertex v in a graph G are a subset of vertex V' induced by all vertices adjacent to v.

- The **neighborhood graph** (also known as an ego graph) of a vertex v in a graph G is a subgraph of G, composed of the vertices adjacent to v and all edges connecting vertices adjacent to v.

For example, imagine a graph representation of a road map, where nodes represent cities and edges represent roads connecting those cities. An example of what such a graph may look like is illustrated in the following figure:

V = [Paris, Milan, Dublin, Rome]

E = [{Milan, Dublin}, {Milan, Paris}, {Paris, Dublin}, {Milan, Rome}]

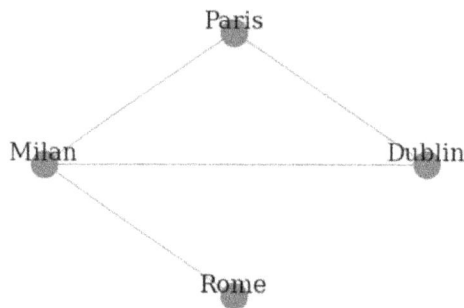

Figure 1.1: Example of a graph

According to this representation, since there is no direction, an edge from **Milan** to **Paris** is equal to an edge from **Paris** to **Milan**. Thus, it is possible to move in the two directions without any constraint. If we analyze the properties of the graph depicted in *Figure 1.1*, we can see that it has *order* and *size* equal to **4** (there are, in total, four vertices and four edges). The **Paris** and **Dublin** vertices have degree **2**, **Milan** has degree **3**, and **Rome** has degree **1**. The neighbors for each node are shown in the following list:

- **Paris = {Milan, Dublin}**
- **Milan = {Paris, Dublin, Rome}**
- **Dublin = {Paris, Milan}**
- **Rome = {Milan}**

The same graph can be represented in Networkx, as follows:

```
import networkx as nx
G = nx.Graph()
V = {'Dublin', 'Paris', 'Milan', 'Rome'}
E = [('Milan','Dublin'), ('Milan','Paris'), ('Paris','Dublin'),
('Milan','Rome')]
G.add_nodes_from(V)
G.add_edges_from(E)
```

Since, by default, the nx.Graph() command generates an undirected graph, we do not need to specify both directions of each edge. In Networkx, nodes can be any hashable object: strings, classes, or even other Networkx graphs. Let's now compute some properties of the graph we previously generated.

All the nodes and edges of the graph can be obtained by running the following code:

```
print(f"V = {G.nodes}")
print(f"E = {G.edges}")
```

Here is the output of the previous commands:

```
V = ['Rome', 'Dublin', 'Milan', 'Paris']
E = [('Rome', 'Milan'), ('Dublin', 'Milan'), ('Dublin', 'Paris'),
('Milan', 'Paris')]
```

We can also compute the graph order, the graph size, and the degree and neighbors for each of the nodes, using the following commands:

```
print(f"Graph Order: {G.number_of_nodes()}")
print(f"Graph Size: {G.number_of_edges()}")
print(f"Degree for nodes: { {v: G.degree(v) for v in G.nodes} }")
print(f"Neighbors for nodes: { {v: list(G.neighbors(v)) for v in G.nodes}
}")
```

The result will be the following:

```
Graph Order: 4
Graph Size: 4
Degree for nodes: {'Rome': 1, 'Paris': 2, 'Dublin':2, 'Milan': 3}
Neighbors for nodes: {'Rome': ['Milan'], 'Paris': ['Milan', 'Dublin'],
'Dublin': ['Milan', 'Paris'], 'Milan': ['Dublin', 'Paris', 'Rome']}
```

Finally, we can also compute an ego graph of a specific node for the graph G, as follows:

```
ego_graph_milan = nx.ego_graph(G, "Milan")
print(f"Nodes: {ego_graph_milan.nodes}")
print(f"Edges: {ego_graph_milan.edges}")
```

The result will be the following:

```
Nodes: ['Paris', 'Milan', 'Dublin', 'Rome']
Edges: [('Paris', 'Milan'), ('Paris', 'Dublin'), ('Milan', 'Dublin'),
('Milan', 'Rome')]
```

The original graph can be also modified by adding new nodes and/or edges, as follows:

```
# Add new nodes and edges
new_nodes = {'London', 'Madrid'}
new_edges = [('London','Rome'), ('Madrid','Paris')]
G.add_nodes_from(new_nodes)
G.add_edges_from(new_edges)
print(f"V = {G.nodes}")
print(f"E = {G.edges}")
```

This would output the following lines:

```
V = ['Rome', 'Dublin', 'Milan', 'Paris', 'London', 'Madrid']
E = [('Rome', 'Milan'), ('Rome', 'London'), ('Dublin', 'Milan'),
('Dublin', 'Paris'), ('Milan', 'Paris'), ('Paris', 'Madrid')]
```

Removal of nodes can be done by running the following code:

```
node_remove = {'London', 'Madrid'}
G.remove_nodes_from(node_remove)
print(f"V = {G.nodes}")
print(f"E = {G.edges}")
```

This is the result of the preceding commands:

```
V = ['Rome', 'Dublin', 'Milan', 'Paris']
E = [('Rome', 'Milan'), ('Dublin', 'Milan'), ('Dublin', 'Paris'),
('Milan', 'Paris')]
```

As expected, all the edges that contain the removed nodes are automatically deleted from the edge list.

Also, edges can be removed by running the following code:

```
node_edges = [('Milan','Dublin'), ('Milan','Paris')]
G.remove_edges_from(node_edges)
print(f"V = {G.nodes}")
print(f"E = {G.edges}")
```

The final result will be as follows:

```
V = ['Dublin', 'Paris', 'Milan', 'Rome']
E = [('Dublin', 'Paris'), ('Milan', 'Rome')]
```

The networkx library also allows us to remove a single node or a single edge from graph G by using the following commands: G.remove_node('Dublin') and G.remove_edge('Dublin', 'Paris').

Types of graphs

In the previous section, we discussed how to create and modify simple undirected graphs. However, there are other formalisms available for modeling graphs. In this section, we will explore how to extend graphs to capture more detailed information by introducing **directed graphs (digraphs)**, **weighted graphs**, and **multigraphs**.

Digraphs

A digraph G is defined as a couple $G=(V, E)$, where $V=\{V_1, ..., V_n\}$ is a set of nodes and $E=\{(V_k, V_w), ..., (V_i, V_j)\}$ is a set of ordered couples representing the connection between two nodes belonging to V.

Since each element of E is an ordered couple, it enforces the direction of the connection. The edge (V_k, V_w) means the node V_k goes into V_w. This is different from (V_w, V_k) since it means the node V_w goes into V_k. The starting node V_w is called the *head*, while the ending node is called the *tail*.

Due to the presence of edge direction, the definition of node degree needs to be extended.

In-degree and out-degree

For a vertex *v*, the number of head ends adjacent to *v* is called the **in-degree** (indicated by $deg^-(v)$) of *v*, while the number of tail ends adjacent to *v* is its **out-degree** (indicated by $deg^+(v)$).

For instance, imagine our road map where, this time, certain roads are one-way. For example, you can travel from Milan to Rome, but not back using the same road. We can use a digraph to represent such a situation, which will look like the following figure:

V = [Paris, Milan, Dublin, Rome]

E = [(Milan, Dublin), (Paris , Milan), (Paris, Dublin), (Milan, Rome)]

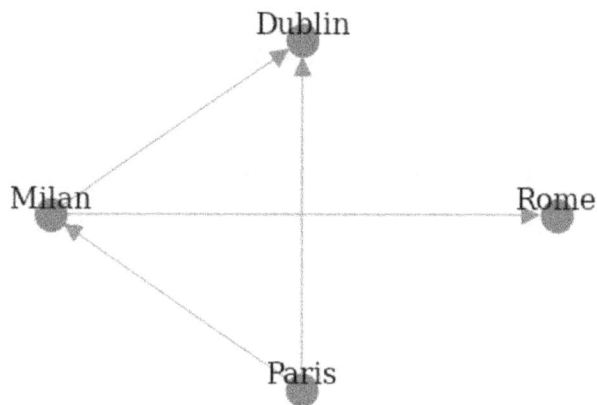

Figure 1.2: Example of a digraph

The direction of the edge is visible from the arrow—for example, **Milan -> Dublin** means from **Milan** to **Dublin**. **Dublin** has $deg^-(v) = 2$ and $deg^+(v) = 0$, **Paris** has $deg^-(v) = 0$ and $deg^+(v) = 2$, **Milan** has $deg^-(v) = 1$ and $deg^+(v) = 2$, and **Rome** has $deg^-(v) = 1$ and $deg^+(v) = 0$.

The same graph can be represented in **networkx**, as follows:

```
G = nx.DiGraph()
V = {'Dublin', 'Paris', 'Milan', 'Rome'}
E = [('Milan','Dublin'), ('Paris','Milan'), ('Paris','Dublin'),
```

```
    ('Milan','Rome')]
    G.add_nodes_from(V)
    G.add_edges_from(E)
```

The definition is the same as that used for simple undirected graphs; the only difference is in the **networkx** classes that are used to instantiate the object. For digraphs, the nx.DiGraph() class is used.

In-degree and **out-degree** can be computed using the following commands:

```
    print(f"Indegree for nodes: { {v: G.in_degree(v) for v in G.nodes} }")
    print(f"Outdegree for nodes: { {v: G.out_degree(v) for v in G.nodes} }")
```

The results will be as follows:

```
    Indegree for nodes: {'Rome': 1, 'Paris': 0, 'Dublin': 2, 'Milan': 1}
    Outdegree for nodes: {'Rome': 0, 'Paris': 2, 'Dublin': 0, 'Milan': 2}
```

As for the undirected graphs, the G.add_nodes_from(), G.add_edges_from(), G.remove_nodes_from(), and G.remove_edges_from() functions can be used to modify a given graph G.

Multigraph

We will now introduce the multigraph object, which is a generalization of the graph definition that allows multiple edges to have the same pair of start and end nodes.

A **multigraph G** is defined as *G=(V, E)*, where *V* is a set of nodes and *E* is a multi-set (a set allowing multiple instances for each of its elements) of edges.

A multigraph is called a **directed multigraph** if *E* is a multi-set of ordered couples; otherwise, if *E* is a multi-set of two-sets, then it is called an **undirected multigraph**.

To make this clearer, imagine our road map where some cities (nodes) are connected by multiple roads (edges). For example, there could be two highways between Milan and Dublin: one might be a direct toll road, while the other is a scenic route. These multiple connections between the same cities can be captured by a multigraph, where both roads are represented as distinct edges between the same pair of nodes. Similarly, if one of these roads is one-way, the graph becomes a directed multigraph, allowing us to represent complex road networks more accurately. An example of a directed multigraph is shown in the following figure:

V = [Paris, Milan, Dublin, Rome]

E = [(Milan, Dublin), (Milan, Dublin), (Paris , Milan), (Paris, Dublin), (Milan, Rome), (Milan, Rome)]

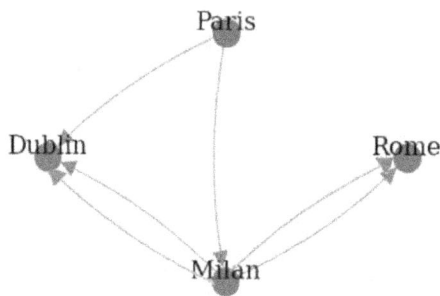

Figure 1.3: Example of a multigraph

In the following code snippet, we show how to use Networkx in order to create a directed or an undirected multigraph:

```
directed_multi_graph = nx.MultiDiGraph()
undirected_multi_graph = nx.MultiGraph()
V = {'Dublin', 'Paris', 'Milan', 'Rome'}
E = [('Milan','Dublin'), ('Milan','Dublin'), ('Paris','Milan'),
('Paris','Dublin'), ('Milan','Rome'), ('Milan','Rome')]
directed_multi_graph.add_nodes_from(V)
undirected_multi_graph.add_nodes_from(V)
directed_multi_graph.add_edges_from(E)
undirected_multi_graph.add_edges_from(E)
```

The only difference between a directed and an undirected multigraph is in the first two lines, where two different objects are created: nx.MultiDiGraph() is used to create a directed multigraph, while nx.MultiGraph() is used to build an undirected multigraph. The function used to add nodes and edges is the same for both objects.

Weighted graphs

We will now introduce directed, undirected, and multi-weighted graphs.

An **edge-weighted graph** (or simply a weighted graph) G is defined as $G=(V, E, w)$ where V is a set of nodes, E is a set of edges, and $w: E \rightarrow \mathbb{R}$ is the weighted function that assigns at each edge $e \in E$ a weight expressed as a real number.

A **node-weighted graph** G is defined as $G=(V, E, w)$, where V is a set of nodes, E is a set of edges, and $w: V \rightarrow \mathbb{R}$ is the weighted function that assigns at each node $v \in V$ a weight expressed as a real number.

Please keep in mind the following points:

- If E is a set of ordered couples, then we call it a **directed weighted graph**.
- If E is a set of two-sets, then we call it an **undirected weighted graph**.
- If E is a multi-set of ordered couples, we will call it a **directed weighted multigraph**.
- If E is a multi-set of two-sets, it is an **undirected weighted multigraph**.

An example of a directed edge-weighted graph is shown in the following figure:

V = [Paris, Milan, Dublin, Rome]

E = [(Milan, Dublin, 19), (Paris , Milan, 8), (Paris, Dublin, 11), (Milan, Rome, 5)]

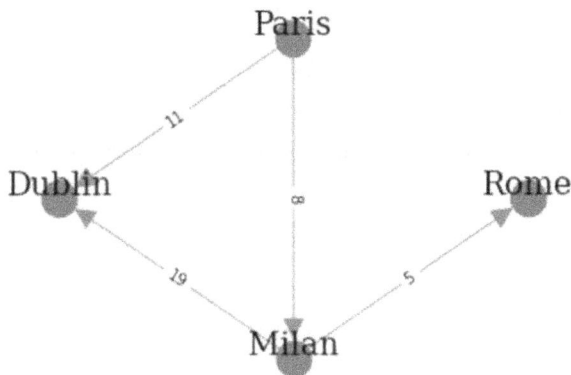

Figure 1.4: Example of a directed edge-weighted graph

From *Figure 1.4*, it is easy to see how the presence of weights on graphs helps to add useful information to the data structures. Indeed, we can imagine the edge weight as a "cost" to reach a node from another node. For example, reaching **Dublin** from **Milan** has a "cost" of **19**, while reaching **Dublin** from **Paris** has a "cost" of **11**.

In **networkx**, a directed weighted graph can be generated as follows:

```
G = nx.DiGraph()
V = {'Dublin', 'Paris', 'Milan', 'Rome'}
E = [('Milan','Dublin', 19), ('Paris','Milan', 8), ('Paris','Dublin', 11),
('Milan','Rome', 5)]
G.add_nodes_from(V)
G.add_weighted_edges_from(E)
```

Multipartite graphs

We will now introduce another type of graph that will be used in this section: multipartite graphs. Bi- and tri-partite graphs—and, more generally, k^{th}-partite graphs—are graphs whose vertices can be partitioned in two, three, or more k-th sets of nodes, respectively. Edges are only allowed across different sets and are not allowed within nodes belonging to the same set. In most cases, nodes belonging to different sets are also characterized by particular node types. To illustrate this with our road map example, imagine a scenario where we want to represent different types of entities: cities, highways, and rest stops. Here, we can model the system using a tripartite graph. One set of nodes could represent the cities, another set the highways, and a third set the rest stops. Edges would exist only between these sets—such as connecting a city to a highway or a highway to a rest stop—but not between cities directly or between rest stops.

In *Chapter 8*, *Text Analytics and Natural Language Processing Using Graphs*, and *Chapter 9*, *Graph Analysis for Credit Card Transactions*, we will deal with some practical examples of graph-based applications and you will see how multipartite graphs can indeed arise in several contexts—for example, in the following scenarios:

- When processing documents and structuring the information in a bipartite graph of documents and entities that appear in the documents

- When dealing with transactional data, in order to encode the relations between the buyers and the merchants

A bipartite graph can be easily created in **networkx** with the following code:

```
import pandas as pd
import numpy as np
n_nodes = 10
n_edges = 12
bottom_nodes = [ith for ith in range(n_nodes) if ith % 2 ==0]
top_nodes = [ith for ith in range(n_nodes) if ith % 2 ==1]
iter_edges = zip(
    np.random.choice(bottom_nodes, n_edges),
    np.random.choice(top_nodes, n_edges))
edges = pd.DataFrame([
    {"source": a, "target": b} for a, b in iter_edges])
B = nx.Graph()
B.add_nodes_from(bottom_nodes, bipartite=0)
B.add_nodes_from(top_nodes, bipartite=1)
B.add_edges_from([tuple(x) for x in edges.values])
```

The network can also be conveniently plotted using the bipartite_layout utility function of Networkx, as illustrated in the following code snippet:

```
from networkx.drawing.layout import bipartite_layout
pos = bipartite_layout(B, bottom_nodes)
nx.draw_networkx(B, pos=pos)
```

The `bipartite_layout` function produces a graph, as shown in the following figure. Intuitively, we can see this graph is bipartite because there are no "vertical" edges connecting left nodes with left nodes or right nodes with right nodes. Notice that the nodes in the `bottom_nodes` parameter appear on one side of the layout, while all the remaining nodes appear on the other side. This arrangement helps visualize the two sets and the connections between them clearly.

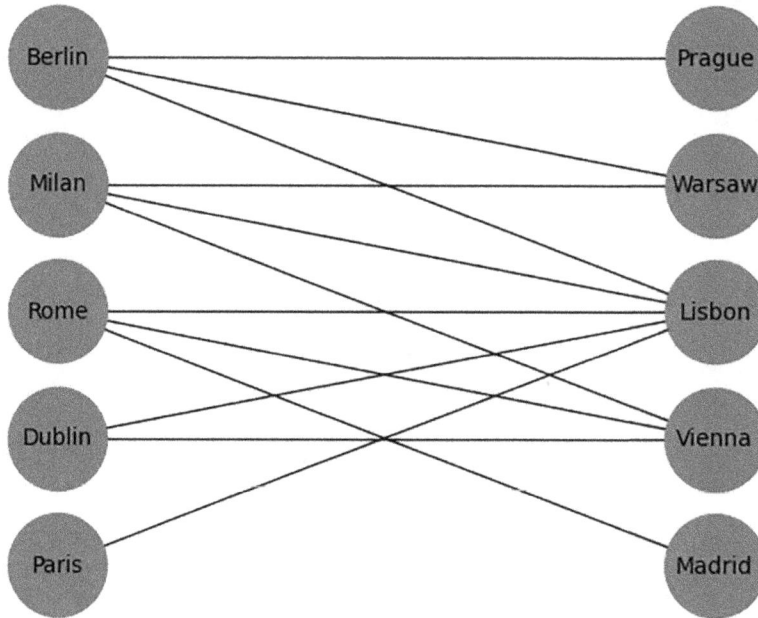

Figure 1.5: Example of a bipartite graph

Connected graphs

Finally, it's important to note that not all parts of a graph are always connected. In some cases, a set of connected nodes can exist independently from another set within the same graph. We define **connected graphs** as graphs in which every node is reachable from any other node.

Disconnected graphs

Conversely, a **disconnected graph** contains at least one pair of nodes that cannot be reached from each other. For example, consider a road map where one cluster of cities is connected by roads (like Dublin, Paris, and Milan), while another cluster (such as Rome, Naples, and Moscow) is completely separate, with no direct roads linking them.

Complete graphs

We define a **complete graph** as a graph in which all nodes are directly reachable from each other, leading to a highly interconnected structure.

Graph representations

As described earlier, with **networkx**, we can define and manipulate a graph by using node and edge objects. However, in certain cases, this representation may become cumbersome to work with. For instance, if you have a large, densely connected graph (such as a network of thousands of interconnected cities), visualizing and managing individual node and edge objects can be overwhelming and inefficient. In this section, we will introduce two more compact ways to represent a graph: the adjacency matrix and the edge list. These methods allow us to represent the same graph data in a more structured and manageable form, especially for large or complex networks. For example, if your application requires no or minor modification to the graph structure and needs to check for the presence of an edge as fast as possible, an adjacency matrix is what you are looking for because accessing a cell in a matrix is a very fast operation from a computational point of view.

Adjacency matrix

The **adjacency matrix** M of a graph $G=(V, E)$ is a square matrix $(|V| \times |V|)$ such that its element M_{ij} is 1 when there is an edge from node i to node j, and 0 when there is no edge. In the following figure, we show a simple example of the adjacency matrix of different types of graphs:

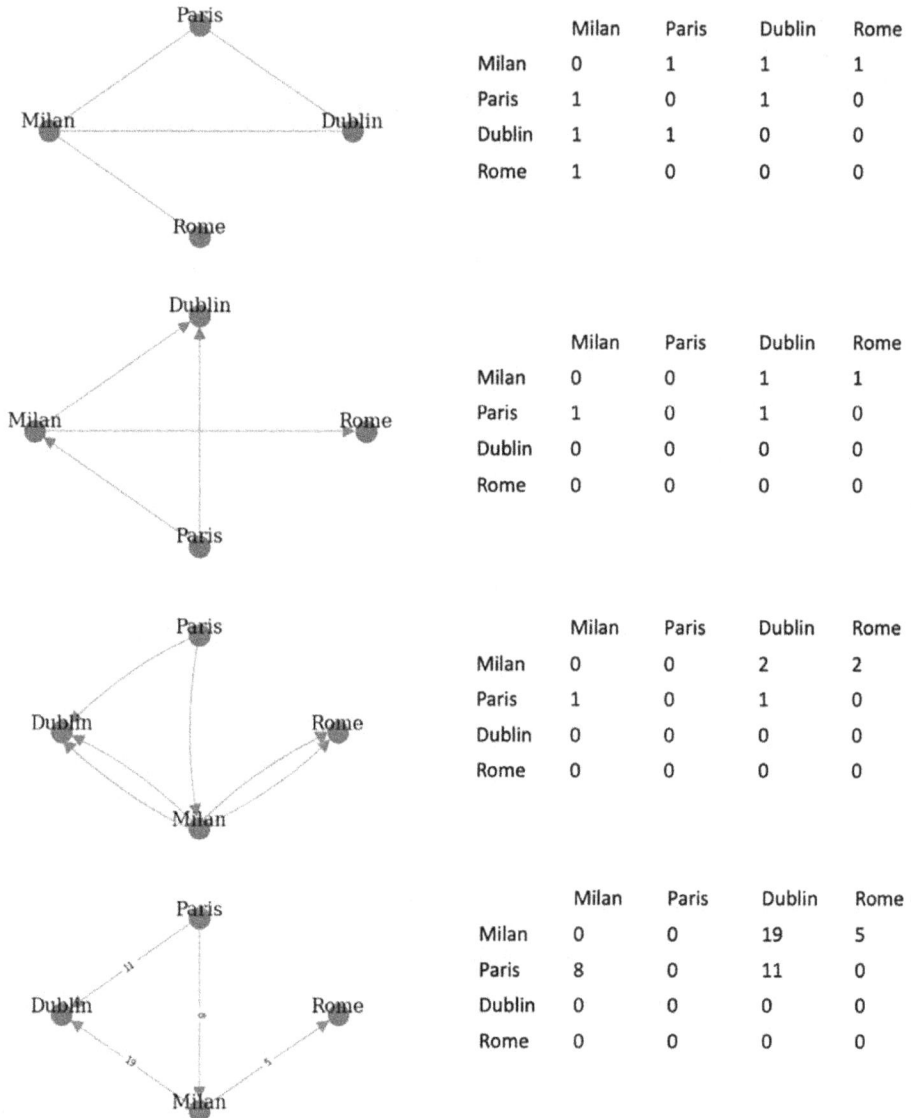

	Milan	Paris	Dublin	Rome
Milan	0	1	1	1
Paris	1	0	1	0
Dublin	1	1	0	0
Rome	1	0	0	0

	Milan	Paris	Dublin	Rome
Milan	0	0	1	1
Paris	1	0	1	0
Dublin	0	0	0	0
Rome	0	0	0	0

	Milan	Paris	Dublin	Rome
Milan	0	0	2	2
Paris	1	0	1	0
Dublin	0	0	0	0
Rome	0	0	0	0

	Milan	Paris	Dublin	Rome
Milan	0	0	19	5
Paris	8	0	11	0
Dublin	0	0	0	0
Rome	0	0	0	0

Figure 1.6: Adjacency matrix for an undirected graph, a digraph, a multigraph, and a weighted graph

It is easy to see that adjacency matrices for undirected graphs are always symmetric since no direction is defined for the edge. However, the symmetry is not guaranteed for the adjacency matrix of a digraph due to the presence of constraints in the direction of the edges. For a multigraph, we can instead have values greater than 1 since multiple edges can be used to connect the same couple of nodes. For a weighted graph, the value in a specific cell is equal to the weight of the edge connecting the two nodes.

In **networkx**, the adjacency matrix for a given graph can be computed using a pandas **DataFrame** or **numpy** matrix. If G is the graph shown in *Figure 1.6*, we can compute its adjacency matrix as follows:

```
nx.to_pandas_adjacency(G) #adjacency matrix as pd DataFrame
nt.to_numpy_matrix(G) #adjacency matrix as numpy matrix
```

For the first and second lines, we get the following results, respectively:

```
        Rome   Dublin Milan  Paris
Rome    0.0    0.0    0.0    0.0
Dublin  0.0    0.0    0.0    0.0
Milan   1.0    1.0    0.0    0.0
Paris   0.0    1.0    1.0    0.0
[[0. 0. 0. 0.]
 [0. 0. 0. 0.]
 [1. 1. 0. 0.]
 [0. 1. 1. 0.]]
```

Since a **numpy** matrix cannot represent the name of the nodes, the order of the element in the adjacency matrix is the one defined in the G.nodes list.

In general, you can choose a pandas DataFrame for better readability and data manipulation, or a **numpy** matrix for efficient numerical operations.

Edge list

As well as an adjacency matrix, an edge list is another compact way to represent graphs. The idea behind this format is to represent a graph as a list of edges.

The **edge list** *L* of a graph *G=(V, E)* is a list of size *|E|* matrix such that its element L_i is a couple representing the tail and the end node of the edge *i*. An example of the edge list for each type of graph is shown in the following figure:

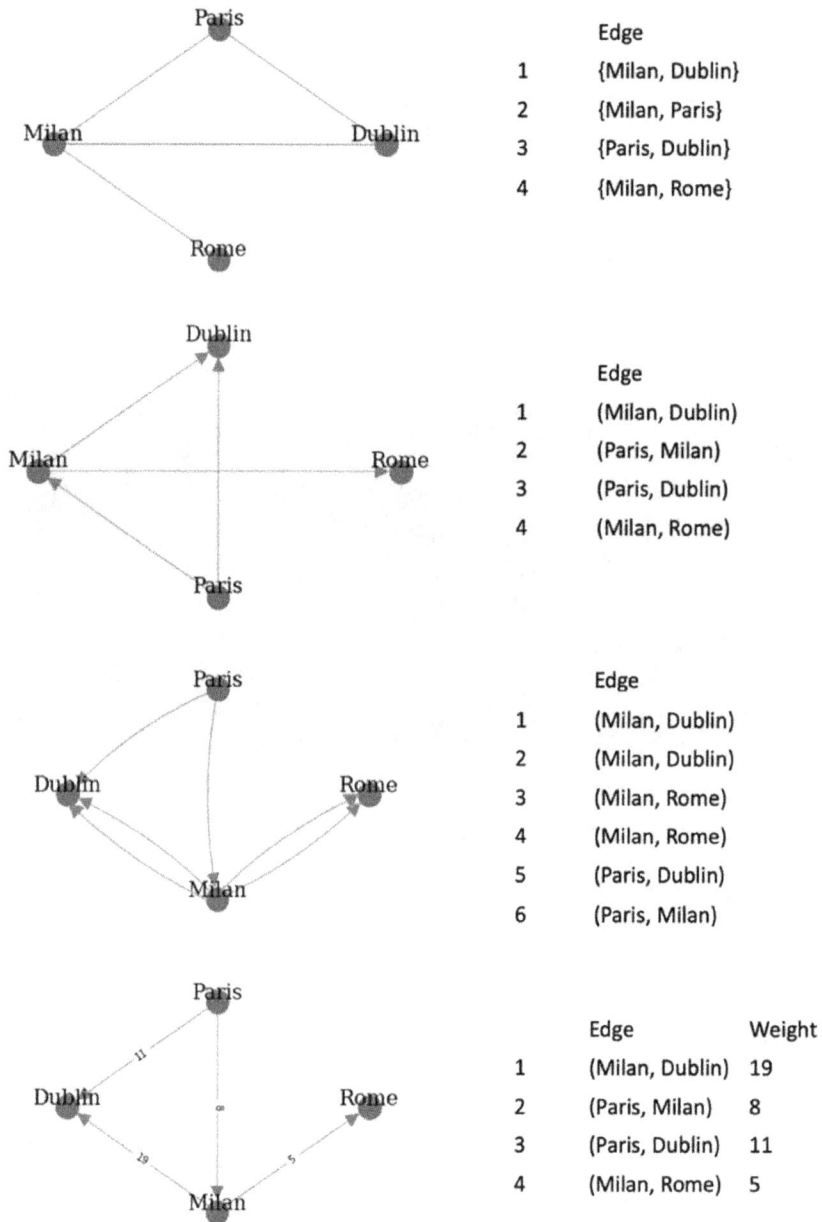

	Edge
1	{Milan, Dublin}
2	{Milan, Paris}
3	{Paris, Dublin}
4	{Milan, Rome}

	Edge
1	(Milan, Dublin)
2	(Paris, Milan)
3	(Paris, Dublin)
4	(Milan, Rome)

	Edge
1	(Milan, Dublin)
2	(Milan, Dublin)
3	(Milan, Rome)
4	(Milan, Rome)
5	(Paris, Dublin)
6	(Paris, Milan)

	Edge	Weight
1	(Milan, Dublin)	19
2	(Paris, Milan)	8
3	(Paris, Dublin)	11
4	(Milan, Rome)	5

Figure 1.7: Edge list for an undirected graph, a digraph, a multigraph, and a weighted graph

In the following code snippet, we show how to compute in **networkx** the edge list of the simple undirected graph G shown in *Figure 1.7*:

```
print(nx.to_pandas_edgelist(G))
```

By running the preceding command, we get the following result:

```
    source  target
0   Milan   Dublin
1   Milan     Rome
2   Paris    Milan
3   Paris   Dublin
```

It is noteworthy that nodes with zero degrees may never appear in the list.

Adjacency matrices and edge lists are two of the most common graph representation methods. However, other representation methods, which we will not discuss in detail, are also available in **networkx**. Some examples are nx.to_dict_of_dicts(G) and nx.to_numpy_array(G), among others.

Plotting graphs

As we have seen in previous sections, graphs are intuitive data structures represented graphically. Nodes can be plotted as simple circles, while edges are lines connecting two nodes.

Despite their simplicity, it could be quite difficult to make a clear representation when the number of edges and nodes increases. The source of this complexity is mainly related to the position (space/Cartesian coordinates) assigned to each node in the final plot. Indeed, it could be unfeasible to manually assign to a graph with hundreds of nodes the specific position of each node in the final plot.

In this section, we will see how we can plot graphs without specifying coordinates for each node. We will exploit two different solutions: Networkx and Gephi.

NetworkX

NetworkX offers a simple interface to plot graph objects through the nx.draw library. In the following code snippet, we show how to use the library in order to plot graphs:

```
def draw_graph(G, nodes_position, weight):
    nx.draw(G, nodes_position,
        with_labels=True,
```

```
        font_size=15,
        node_size=400,
        edge_color='gray',
        arrowsize=30)
    if plot_weight:
        edge_labels=nx.get_edge_attributes(G,'weight')
        nx.draw_networkx_edge_labels(G,
                                     node_position,
                                     edge_labels=edge_labels)
```

Here, nodes_position is a dictionary where the keys are the nodes and the value assigned to each key is an array of length 2, with the Cartesian coordinates used for plotting the specific node.

The nx.draw function will plot the whole graph by putting its nodes in the given positions. The with_labels option will plot its name on top of each node with the specific font_size value. node_size and edge_color will respectively specify the size of the circle, representing the node and the color of the edges. Finally, arrowsize will define the size of the arrow for directed edges (notice that arrowsize is meaningful only when plotting graphs in which edges are drawn as arrows, such as digraphs).

In the following code example, we show how to use the draw_graph function previously defined in order to plot a graph:

```
G = nx.Graph()
V = {'Paris', 'Dublin','Milan', 'Rome'}
E = [('Paris','Dublin', 11), ('Paris','Milan', 8),
    ('Milan','Rome', 5), ('Milan','Dublin', 19)]
G.add_nodes_from(V)
G.add_weighted_edges_from(E)
node_position = {"Paris": [0,0], "Dublin": [0,1], "Milan": [1,0], "Rome":
[1,1]}
draw_graph(G, node_position, True)
```

The result of the plot is shown in the following figure:

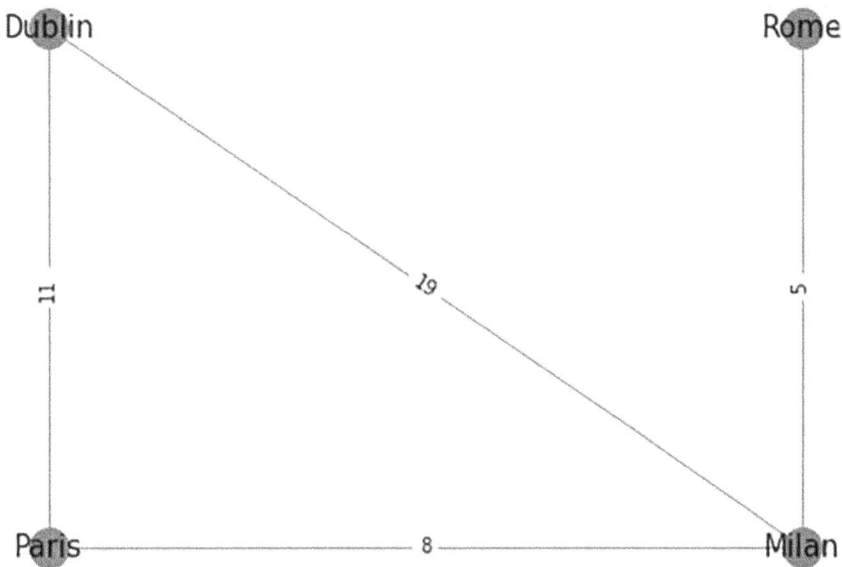

Figure 1.8: Result of the plotting function

The method previously described is simple but unfeasible to use in a real scenario since the node_position value could be difficult to decide. In order to solve this issue, **networkx** offers a different function to automatically compute the position of each node according to different layouts. In *Figure 1.9*, we show a series of plots of an undirected graph, obtained using the different layouts available in NetworkX. In order to use them in the function we proposed, we simply need to assign node_position to the result of the layout we want to use—for example, node_position = nx.circular_layout(G). The plots can be seen in the following figure:

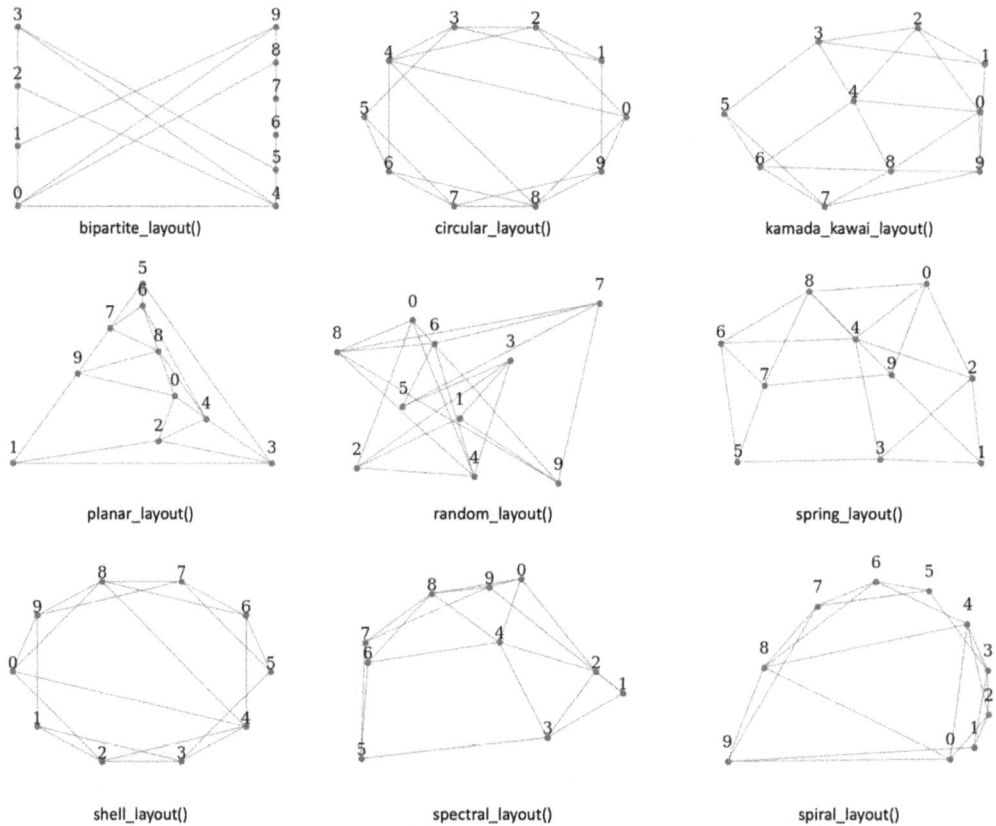

Figure 1.9: Plots of the same undirected graph with different layouts

networkx is a great tool for easily manipulating and analyzing graphs, but it has limitations when it comes to creating complex and visually appealing plots of large graphs. For instance, when visualizing a social network with thousands of users and their interactions, overlapping nodes and edges can make graph interpretation difficult.

In the next section, we will investigate another tool to perform complex graph visualization: Gephi.

Gephi

In this section, we will show how **Gephi** (open source network analysis and visualization software) can be used for performing complex, fancy plots of graphs. For all the examples shown in this section, we will use the Les Miserables.gexf sample (a weighted undirected graph), which can be selected in the **Welcome** window when the application starts.

The main interface of Gephi is shown in *Figure 1.10*. It can be divided into four main areas, as follows:

- **Graph**: This section shows the final plot of the graph. The image is automatically updated each time a filter or a specific layout is applied.

- **Appearance**: Here, it is possible to specify the appearance of nodes and edges.

- **Layout:** In this section, it is possible to select the layout (as in NetworkX) to adjust the node position in the graph. Different algorithms, from a simple random position generator to a more complex Yifan Hu algorithm, are available.

- **Filters & Statistics**: In this set area, two main functions are available, outlined as follows:

    a. **Filters**: In this tab, it is possible to filter and visualize specific subregions of the graph according to a set of properties computed using the **Statistics** tab.

    b. **Statistics**: This tab contains a list of available graph metrics that can be computed on the graph using the **Run** button. Once metrics are computed, they can be used as properties to specify the edges' and nodes' appearance (such as node and edge size and color) or to filter a specific subregion of the graph.

You can see the main interface of Gephi in the following screenshot:

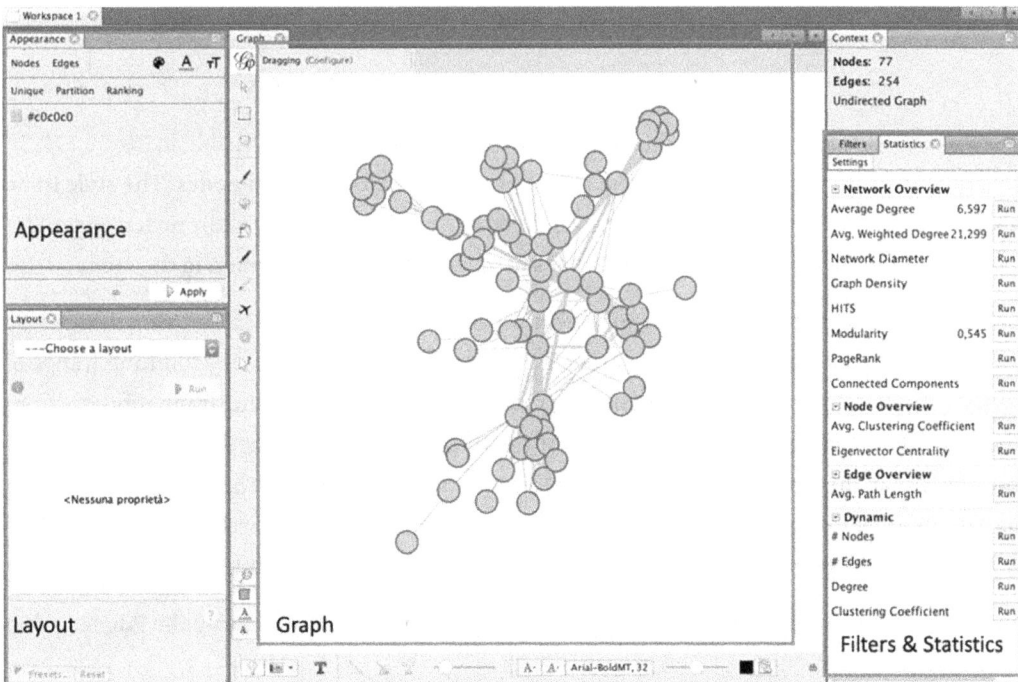

Figure 1.10: Gephi main window

Our exploration of Gephi starts with the application of different layouts to the graph. As previously described, in NetworkX, the layouts allow us to assign to each node a specific position in the final plot. In Gephi 1.2, different layouts are available. In order to apply a specific layout, we have to select one of the available layouts from the **Layout** area, and then click on the **Run** button that appears after the selection.

The graph representation, visible in the **Graph** area, will be automatically updated according to the new coordinates defined by the layout. It should be noted that some layouts are parametric, hence the final graph plot can significantly change according to the parameters used. In the following figure, we propose several examples for the application of three different layouts:

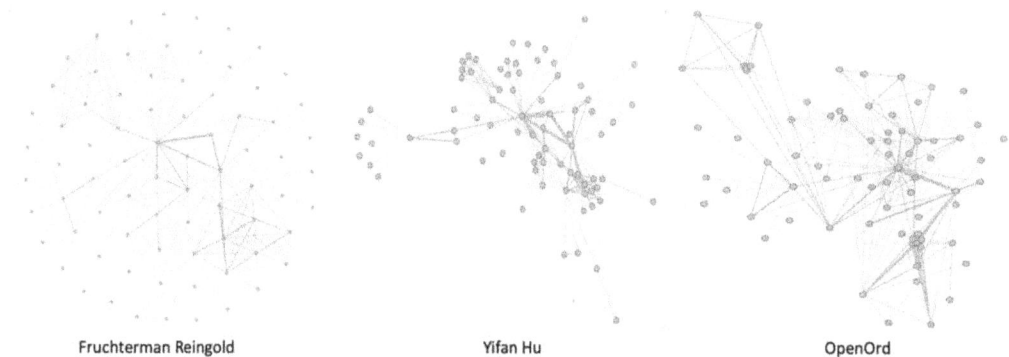

Fruchterman Reingold   Yifan Hu   OpenOrd

Figure 1.11: Plot of the same graph with different layouts

We will now introduce the available options in the **Appearance** menu, visible in *Figure 1.10*. In this section, it is possible to specify the style to be applied to edges and nodes. The style to be applied can be static or can be dynamically defined by specific properties of the nodes/edges. We can change the color and the size of the nodes by selecting the **Nodes** option in the menu.

In order to change the color, we have to select the color palette icon and decide, using the specific button, if we want to assign a **Unique** color, a **Partition** (discrete values), or a **Ranking** (range of values) of colors. For **Partition** and **Ranking**, it is possible to select a specific **Graph** property from the drop-down menu to use as a reference for the color range. Only the properties computed by clicking **Run** in the **Statistics** area are available in the drop-down menu. The same procedure can be used in order to set the size of the nodes. By selecting the concentric circles icon, it is possible to set a **Unique** size to all the nodes or to specify a **Ranking** of size according to a specific property.

As for the nodes, it is also possible to change the style of the edges by selecting the **Edges** option in the menu.

We can then select to assign a **Unique** color, a **Partition** (discrete values), or a **Ranking** (range of values) of colors. For **Partition** and **Ranking**, the reference value to build the color scale is defined by a specific **Graph** property that can be selected from the drop-down menu.

It is important to remember that in order to apply a specific style to the graph, the **Apply** button should be clicked. As a result, the graph plot will be updated according to the style defined. In the following figure, we show an example where the color of the nodes is given by the **Modularity Class** value, the size of each node is given by its degree, and the color of each edge is defined by the edge weight:

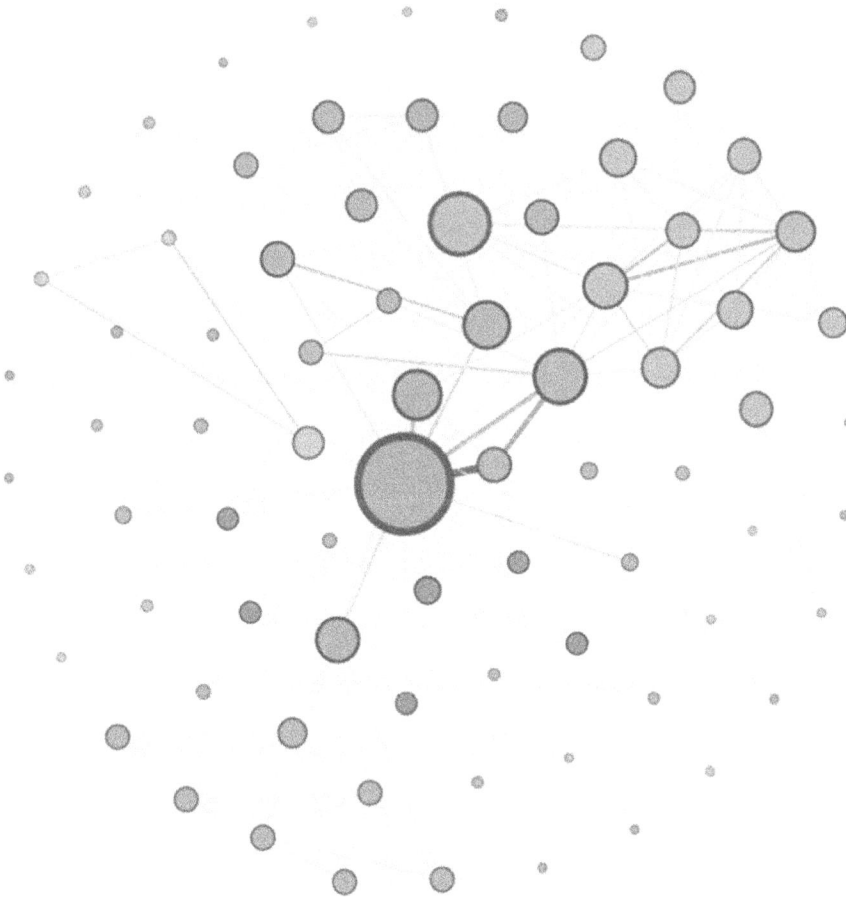

Figure 1.12: Example of graph plot changing nodes' and edges' appearance

Another important section that needs to be described is **Filters & Statistics**. In this menu, it is possible to compute some statistics based on graph metrics.

Finally, we conclude our discussion on Gephi by introducing the functionalities available in the **Statistics** menu, visible in the right panel in *Figure 1.10*. Through this menu, it is possible to compute different statistics on the input graph. Those statistics can be easily used to set some properties of the final plot, such as nodes'/edges' color and size, or to filter the original graph to plot just a specific subset of it. In order to compute a specific statistic, the user then needs to explicitly select one of the metrics available in the menu and click on the **Run** button (*Figure 1.10*, right panel).

Moreover, the user can select a subregion of the graph, using the options available in the **Filters** tab of the **Statistics** menu, visible in the right panel in *Figure 1.10*. An example of filtering a graph can be seen in *Figure 1.13*. To provide more details of this, we build and apply to the graph a filter, using the **Degree** property. The result of the filters is a subset of the original graph, where only the nodes (and their edges) having the specific range of values for the **Degree** property are visible.

This is illustrated in the following screenshot:

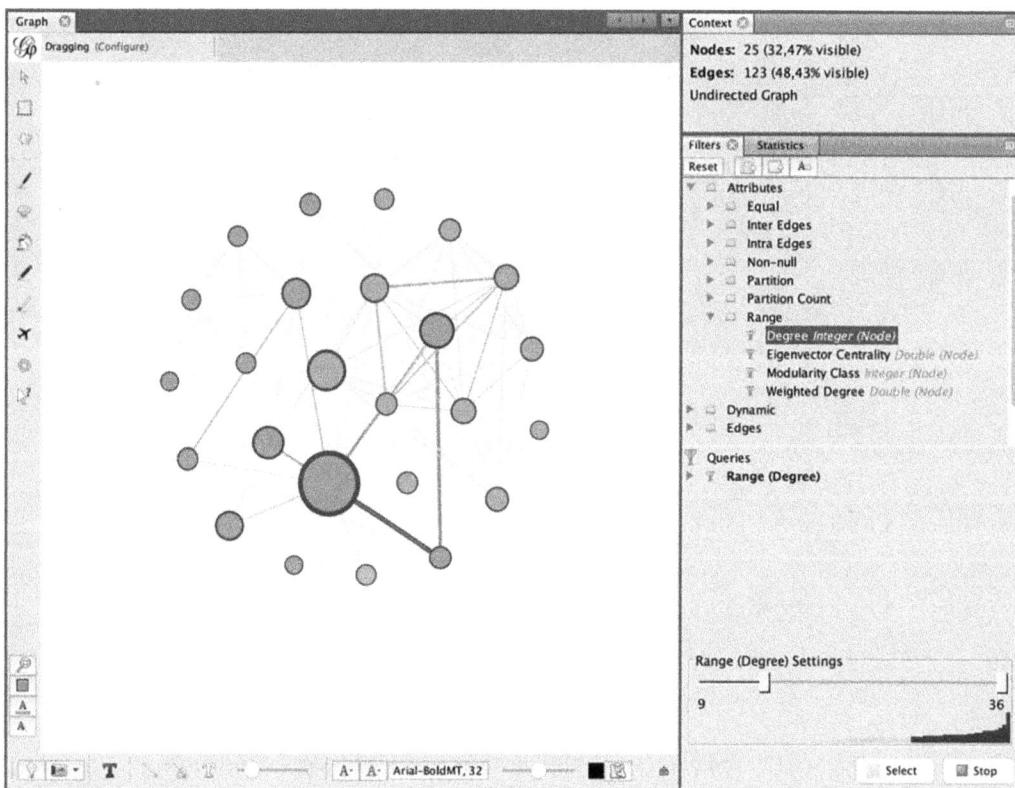

Figure 1.13: Example of a graph filtered according to a range of values for Degree

Of course, Gephi allows us to perform more complex visualization tasks and contains a lot of functionalities that cannot be fully covered in this book. Some good references to better investigate all the features available in Gephi are the official Gephi guide (`https://gephi.org/users/`) and the *Gephi Cookbook* book by Packt Publishing.

Graph properties

As we have already learned, a *graph* is a mathematical model that is used for describing relationships between entities. However, each complex network presents intrinsic properties. Such properties can be measured by particular metrics, and each measure may characterize one or several local and global aspects of the graph.

In a graph for a social network such as X (formerly known as Twitter), for example, users (represented by the *nodes* of the graph) are connected to each other. However, there are users who are more connected than others (influencers). On the Reddit social graph, users with similar characteristics tend to group into communities.

We have already mentioned some of the *basic features* of graphs, such as the *number of nodes and edges* in a graph, which constitute the size of the graph itself. Those properties already provide a good description of the structure of a network. Think about the Facebook graph, for example: it can be described in terms of the number of nodes and edges. Such numbers easily allow it to be distinguished from a much smaller network (for example, the social structure of an office) but fail to characterize more complex dynamics (for example, how *similar* nodes are connected). To this end, more advanced graph-derived **metrics** can be considered, which can be grouped into four main categories, outlined as follows:

- **Integration metrics**: These measure how nodes tend to be interconnected with each other.
- **Segregation metrics**: These quantify the presence of groups of interconnected nodes, known as communities or modules, within a network.
- **Centrality metrics**: These assess the importance of individual nodes inside a network.
- **Resilience metrics**: These can be thought of as a measure of how much a network can maintain and adapt its operational performance when facing failures or other adverse conditions.

Those metrics are defined as **global** when expressing a measure of an overall network. On the other hand, **local** metrics measure values of individual network elements (nodes or edges). In weighted graphs, each property may or may not account for the *edge weights*, leading to **weighted and unweighted metrics**.

In the following section, we describe some of the most used metrics that measure global and local properties. For simplicity, unless specified differently in the text, we illustrate the global unweighted version of the metric. In several cases, this is obtained by averaging the local unweighted properties of the node.

Integration metrics

In this section, some of the most frequently used integration metrics will be described.

Distance, path, and shortest path

The concept of distance in a graph is often related to the number of edges to traverse to reach a target node from a given source node.

Consider a source node i and a target node j. The set of edges connecting node i to node j is called a **path**. When studying complex networks, we are often interested in finding the **shortest path** between two nodes. The shortest path between a source node i and a target node j is the path having the lowest number of edges compared to all the possible paths between i and j. The **diameter** of a network is the number of edges contained in the longest shortest path among all possible shortest paths.

Look at the following figure. There are different paths to reach **Tokyo** from **Dublin**. However, one of them is the shortest (the edges on the shortest path are highlighted):

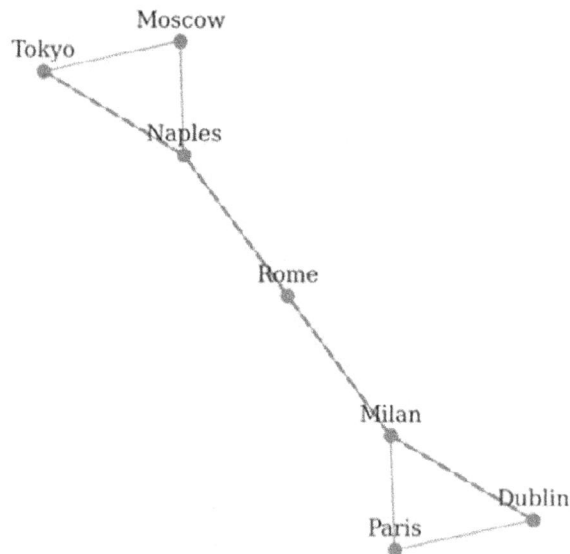

Figure 1.14: The shortest path between two nodes

The shortest_path function of the NetworkX Python library enables users to quickly compute the shortest path between two nodes in a graph. Consider the following code, in which a seven-node graph is created using **networkx.** For clarity and simplicity in creating the graph structure, we will use numerical identifiers for the nodes, even though they represent cities:

```
G = nx.Graph()
nodes = {1:'Dublin',2:'Paris',3:'Milan',4:'Rome',5:'Naples',
         6:'Moscow',7:'Tokyo'}
G.add_nodes_from(nodes.keys())
G.add_edges_from([(1,2),(1,3),(2,3),(3,4),(4,5),(5,6),(6,7),(7,5)])
```

The shortest path between a source node (for example, 'Dublin', identified by key 1) and a target node (for example, 'Tokyo', identified by key 7) can be obtained as follows:

```
path = nx.shortest_path(G,source=1,target=7)
```

This should output the following:

```
[1,3,4,5,7]
```

Here, [1,3,4,5,7] are the nodes contained in the shortest path between 'Tokyo' and 'Dublin'.

Characteristic path length

Let's assume we have a fully connected graph. The **characteristic path length** is defined as the average of all the shortest path lengths between all possible pairs of nodes. If l_i is the average path length between node i and all the other nodes, the characteristic path length is computed as follows:

$$\frac{1}{q(q-1)} \sum_{i,j \in V; i \neq j} l_{ij}$$

Here, V is the set of nodes in the graph, and $q = |V|$ represents its *order*. This equation calculates the average distance across the entire network by summing the shortest path lengths from each node to every other node and normalizing it by the total number of pairs, $q(q-1)$. The characteristic path length is a crucial measure of how efficiently information spreads across a network. Networks with shorter characteristic path lengths facilitate quick information transfer, thereby reducing communication costs. This concept is particularly important in fields such as social network analysis, where understanding the speed of information dissemination can provide insights into network dynamics. Characteristic path length can be computed through NetworkX using the following function:

```
nx.average_shortest_path_length(G)
```

This should give us the following number, quantifying the average shortest path length:

```
2.1904761904761907
```

In the following figure, two examples of graphs are depicted. As observable, fully connected graphs have a lower average shortest path length compared to circular graphs. Indeed, in a fully connected graph, the number of edges to traverse to reach a node from another is, on average, less than the one in a circular graph, where multiple edges need to be traversed.

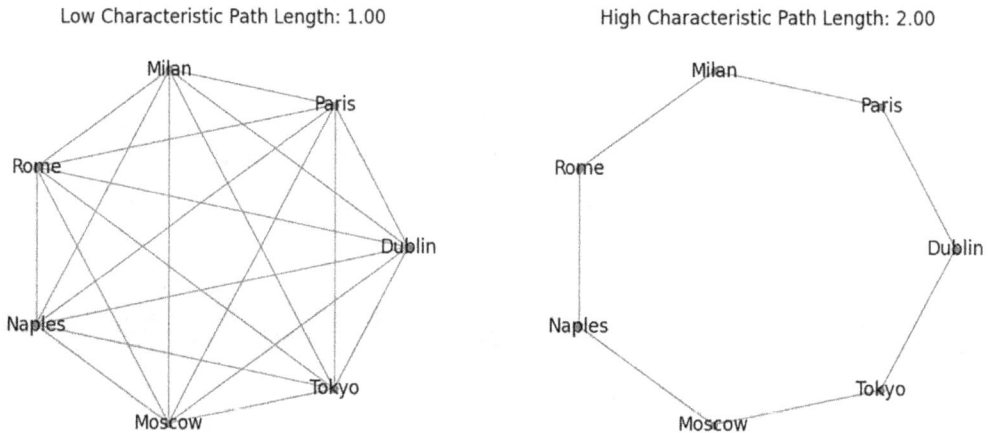

Figure 1.15: Characteristic path length of a fully connected graph (left) and a circular graph (right)

Notice that this metric cannot always be defined since it is not possible to compute a path among all the nodes in *disconnected graphs*. For this reason, **network efficiency** is also widely used.

Global and local efficiency

Global efficiency is the average of the inverse shortest path length for all pairs of nodes. Such a metric can be seen as a measure of how efficiently information is exchanged across a network. Consider that l_{ij} is the shortest path between a node i and a node j. The network efficiency is defined as follows:

$$Eff_g = \frac{1}{q(q-1)} \sum_{i,j \in V; i \neq j} \frac{1}{l_{ij}}$$

The contribution to the efficiency for pairs of disconnected nodes is 0, such that disconnected pairs can be dropped from the summation above.

Efficiency has a maximum value of 1 when a graph is fully connected, while it has a minimum value of 0 for completely disconnected graphs. Intuitively, the shorter the path, the lower the measure.

The **local efficiency** of a node can be computed by considering only the neighborhood of the node in the calculation, without the node itself. In the formula, $N(i)$ is the neighborhood of the node i and $q_i = |N(i)|$ represents the number of neighbors of node i. The local coefficient is computed as:

$$Eff_i = \frac{1}{q_i(q_i - 1)} \sum_{i,j \in N(i); i \neq j} \frac{1}{l_{ij}}$$

Global efficiency is computed in NetworkX using the following command:

```
nx.global_efficiency(G)
```

The output should be as follows:

```
0.6111111111111109
```

Average local efficiency is computed in NetworkX using the following command:

```
nx.local_efficiency(G)
```

The output should be as follows:

```
0.6666666666666667
```

In the following figure, two examples of graphs are depicted. As observed, a fully connected graph on the left presents a higher level of efficiency compared to a circular graph on the right. In a fully connected graph, each node can be reached from any other node in the graph, and information is exchanged rapidly across the network. However, in a circular graph, several nodes should instead be traversed to reach the target node, making it less efficient:

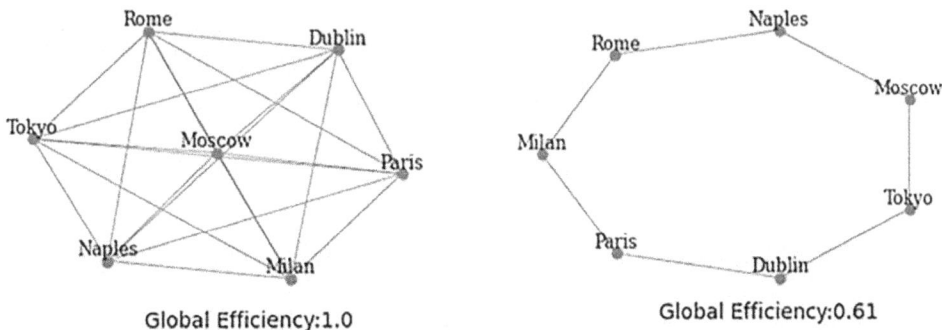

Figure 1.16: Global efficiency of a fully connected graph (left) and a circular graph (right)

Integration metrics describe the connection among nodes. However, more information about the presence of groups can be extracted by considering segregation metrics.

Segregation metrics

In this section, some of the most common segregation metrics will be described.

Clustering coefficient

The **clustering coefficient** is a measure of how closely nodes are grouped together. It is defined as the fraction of **triangles** (complete subgraph of three nodes and three edges) around a node and is equivalent to the fraction of the node's *neighbors* that are neighbors of each other. In the formula, let k_i be the number of neighbors of a node i and let E_i be the number of edges that exist between these k_i neighbors. So $k_i(k_i - 1)$ will be the maximum possible number of edges that could exist among the neighbors of node i. The local clustering coefficient can then be calculated as follows:

$$C_i = \frac{2E_i}{k_i(k_i - 1)}$$

Therefore, the global clustering coefficient is computed by averaging the clustering coefficients for all nodes:

$$C = \frac{1}{q} \sum_{i=1}^{q} C_i$$

The global clustering coefficient is computed in **networkx** using the following command:

```
nx.average_clustering(G)
```

This should output the following:

```
0.6666666666666667
```

The local clustering coefficient is computed in **networkx** using the following command:

```
nx.clustering(G)
```

This should output the following:

```
{1: 1.0,
2: 1.0,
3: 0.3333333333333333,
4: 0,
```

```
5: 0.3333333333333333,
6: 1.0,
7: 1.0}
```

The output is a Python dictionary containing, for each node (identified by the respective key), the corresponding value. In the graph shown in *Figure 1.17*, two clusters of nodes can be easily identified. By computing the clustering coefficient for each single node, it can be observed that **Rome** has the lowest value. **Tokyo** and **Moscow**, as well as **Paris** and **Dublin**, are instead very well connected within their respective groups (notice the size of each node is drawn proportionally to each node's clustering coefficient). The graph can be seen in the following figure:

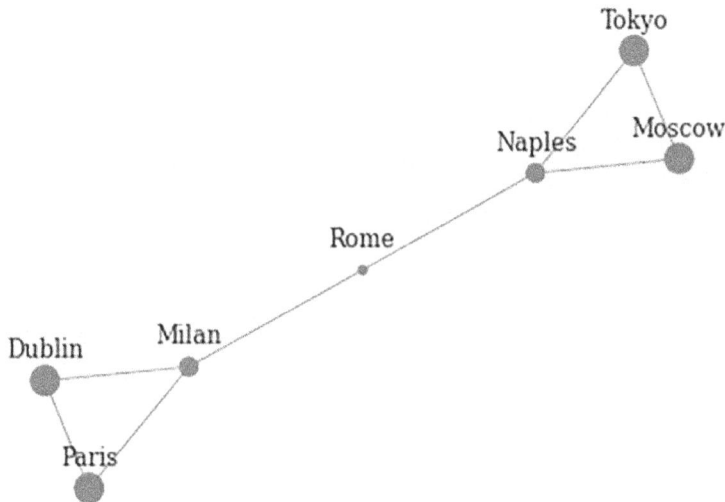

Figure 1.17: Local clustering coefficient representation

Modularity

Modularity was designed to quantify the division of a network into aggregated sets of highly interconnected nodes, commonly known as **modules**, **communities**, **groups**, or **clusters**. The main idea is that networks having high modularity will show dense connections within the module and sparse connections between modules.

Consider a social network such as Reddit: members of communities related to video games tend to interact much more with other users in the same community, talking about recent news, favorite consoles, and so on. However, they will probably interact less with users talking about fashion. Differently from many other graph metrics, modularity is often computed by means of optimization algorithms.

We will discuss this metric in more detail in *Chapter 6, Solving Common Graph-Based Machine Learning Problems,* when discussing the algorithms used for community extractions and identifications in more depth. For now, it is sufficient to understand that high modularity indicates a strong community structure, where many connections exist within communities and fewer connections exist between them. Low modularity, instead, suggests that the network does not have a strong community structure (we can say that the distribution of the edges is more random).

Modularity in NetworkX is computed using the modularity function of the networkx.algorithms. community module, as follows:

```
import networkx.algorithms.community as nx_comm
nx_comm.modularity(G, communities=[{1,2,3}, {4,5,6,7}])
```

Here, the second argument—**communities**—is a list of sets, each representing a partition of the graph. The output should be as follows:

```
0.3671875
```

Segregation metrics help to understand the presence of groups. However, each node in a graph has its own *importance*. To quantify this, we can use centrality metrics, which are discussed in the next sections.

Centrality metrics

In this section, some of the most common centrality metrics will be described. Centrality metrics are extremely useful for identifying the most relevant nodes in a network. As a result, these quantities may be the most used metrics when filtering and targeting nodes and edges (e.g., finding influencers, critical points of failures, etc.).

Degree centrality

One of the most common and simple centrality metrics is the **degree centrality** metric. This is directly connected with the *degree* of a node, measuring the number of *incident* edges on a certain node i.

Intuitively, the more a node is connected to another node, the more its degree centrality will assume high values. Note that if a graph is *directed*, the **in-degree centrality** and **out-degree centrality** should be considered for each node, related to the number of *incoming* and *outcoming* edges, respectively. This number is then normalized by the graph's size to obtain a number between 0 and 1. Degree centrality is computed in NetworkX by using the following command:

```
nx.degree_centrality(G)
```

The output should be as follows:

```
{1: 0.3333333333333333, 2: 0.3333333333333333, 3: 0.5, 4:
0.3333333333333333, 5: 0.5, 6: 0.3333333333333333, 7: 0.3333333333333333}
```

Closeness centrality

The **closeness centrality** metric attempts to quantify how close a node is (well connected) to other nodes. More formally, it refers to the average distance of a node i to all other nodes in the network. If l_{ij} is the shortest path between node i and node j, the closeness centrality c_j is defined as follows:

$$c_j = \frac{N - 1}{\sum l_{ij}}$$

Here, V is the set of nodes in the graph. Closeness centrality can be computed in NetworkX using the following command:

```
nx.closeness_centrality(G)
```

The output should be as follows:

```
{1: 0.4, 2: 0.4, 3: 0.5454545454545454, 4: 0.6, 5: 0.5454545454545454, 6:
0.4, 7: 0.4}
```

Betweenness centrality

The **betweenness centrality** metric evaluates how much a node acts as a **bridge** between other nodes. Even if a node has a low degree or closeness centrality, it can still be strategically connected because of a high betweenness centrality, if it helps to keep the whole network connected.

If L_{wj}^{tot} is the total number of shortest paths between node w and node j and $L_{wj}(i)$ is the total number of shortest paths between w and j passing through node i, then the betweenness centrality is defined as follows:

$$b_i = \sum_{w, j \in V; w \neq j \neq i} \frac{L_{wj}(i)}{L_{wj}^{tot}}$$

If we observe the formula, we can notice that the higher the number of shortest paths passing through node i, the higher the value of the betweenness centrality. Betweenness centrality is computed in **networkx** by using the following command:

```
nx.betweenness_centrality(G)
```

The output should be as follows:

```
{1: 0.0, 2: 0.0, 3: 0.5333333333333333, 4: 0.6, 5: 0.5333333333333333, 6:
0.0, 7: 0.0}
```

In *Figure 1.18*, we illustrate the difference between *degree centrality*, *closeness centrality*, and *betweenness centrality*. **Milan** and **Naples** have the highest degree centrality. **Rome** has the highest closeness centrality since it is the closest to any other node. It also shows the highest betweenness centrality because of its crucial role in connecting the two visible clusters and keeping the whole network connected.

You can see the differences here:

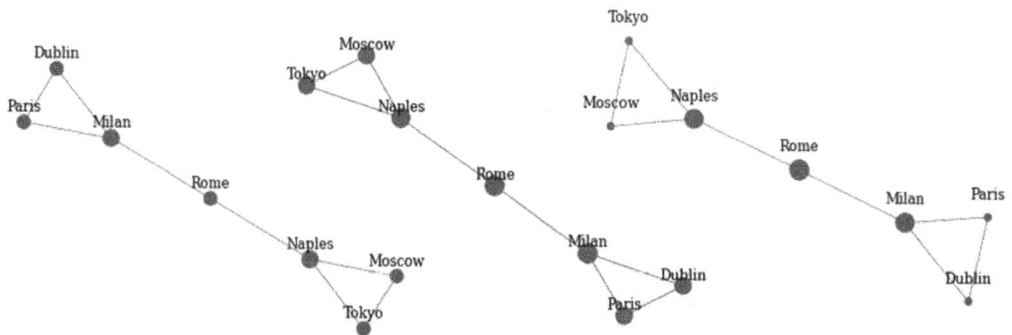

Figure 1.18: Degree centrality (left), closeness centrality (center), and betweenness centrality (right)

Finally, we will mention resilience metrics, which enable us to measure the vulnerability of a graph—that is, how susceptible a network is to disconnection or functional failure when certain nodes are removed.

Resilience metrics

There are several metrics that measure a network's resilience. Assortativity is one of the most used.

Assortativity coefficient

Assortativity is used to quantify the tendency of nodes being connected to similar nodes, which can impact the network's ability to withstand failures or "attacks." High assortativity indicates that nodes of similar degrees are more likely to be connected, leading to a resilient structure where the failure of some nodes does not significantly disrupt overall connectivity. Conversely, networks with low assortativity tend to have nodes connecting with dissimilar degrees, making them more vulnerable to targeted attacks on high-degree nodes, as shown in the figure below:

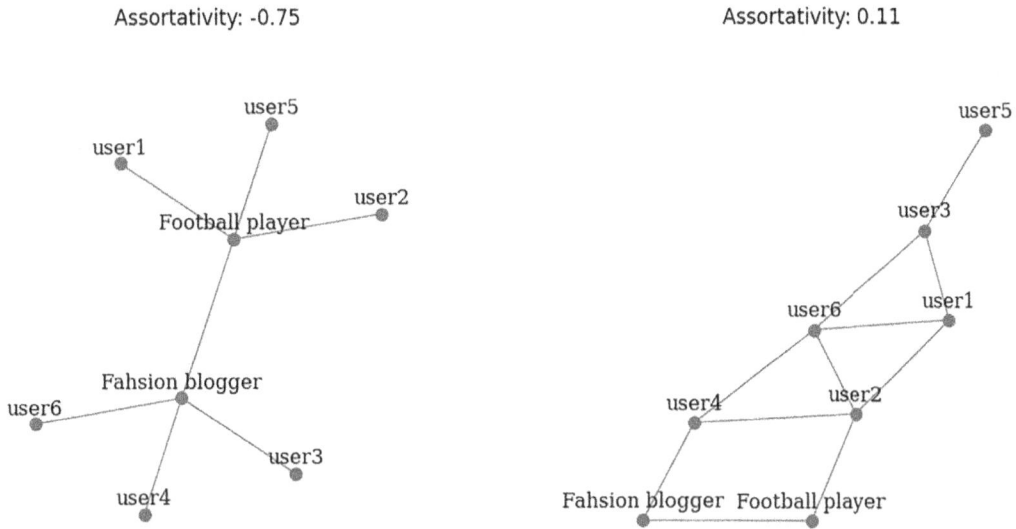

Figure 1.19: Disassortative (left) and assortative (right) graphs

There are several ways to measure such correlations. One of the most used methods is the **Pearson correlation coefficient** between the degrees of directly connected nodes (nodes on two opposite ends of a link). The coefficient assumes positive values when there is a correlation between nodes of a similar degree, while it assumes negative values when there is a correlation between nodes of a different degree. Assortativity using the Pearson correlation coefficient is computed in NetworkX by using the following command:

```
nx.degree_pearson_correlation_coefficient(G)
```

The output should be as follows:

```
-0.6
```

Social networks are mostly assortative. However, the so-called *influencers* (famous singers, football players, fashion bloggers, etc.) tend to be *followed* (incoming edges) by several standard users, while tending to not be connected with each other and showing disassortative behavior.

It is important to note that the properties previously presented are just a subset of the many metrics available for describing graphs. We have chosen to focus on these specific metrics because they offer foundational insights into graph theory and are frequently used in practical applications. A wider set of metrics and algorithms can be found at https://networkx.org/documentation/stable/reference/algorithms/.

Hands-on examples

Now that we understand the basic concepts and notions about graphs and network analysis, it is time to dive into some practical examples that will help us start to put into practice the general concepts we have learned so far. In this section, we will present some examples and toy problems that are generally used to study the properties of networks, as well as benchmark performances and the effectiveness of networks' algorithms.

Simple graphs

We will start by looking at some very simple examples of networks. Fortunately, **networkx** already comes with a number of graphs already implemented, ready to be used and played with. Let's start by creating a **fully connected undirected graph** with *n* nodes, as follows:

```
complete = nx.complete_graph(n=7)
```

This has $\frac{n(n-1)}{2} = 21$ edges and a clustering coefficient *C=1*. Although fully connected graphs are not very interesting on their own, they represent a fundamental building block that may arise within larger graphs. A fully connected subgraph of *n* nodes within a larger graph is generally referred to as a **clique** of size *n*.

A **clique**, *C*, in an undirected graph is defined as the subset of its vertices, $C \subseteq V$, such that every two distinct vertices in the subset are adjacent. This is equivalent to the condition that the induced subgraph of *G* induced by *C* is a fully connected graph.

Cliques represent one of the basic concepts in graph theory and are often also used in mathematical problems where relationships need to be encoded. Besides, they also represent the simplest unit when constructing more complex graphs. On the other hand, the task of finding cliques of a given size *n* in larger graphs (clique problem) is of great interest and it can be shown that it is a **nondeterministic polynomial-time complete (NP-complete)** problem, often studied in computer science. In other words, finding a solution is computationally difficult, and the required time increases exponentially as the size of the graph increases.

Some simple examples of **networkx** graphs can be seen in the following figure:

Complete   Lollipop   Barbell

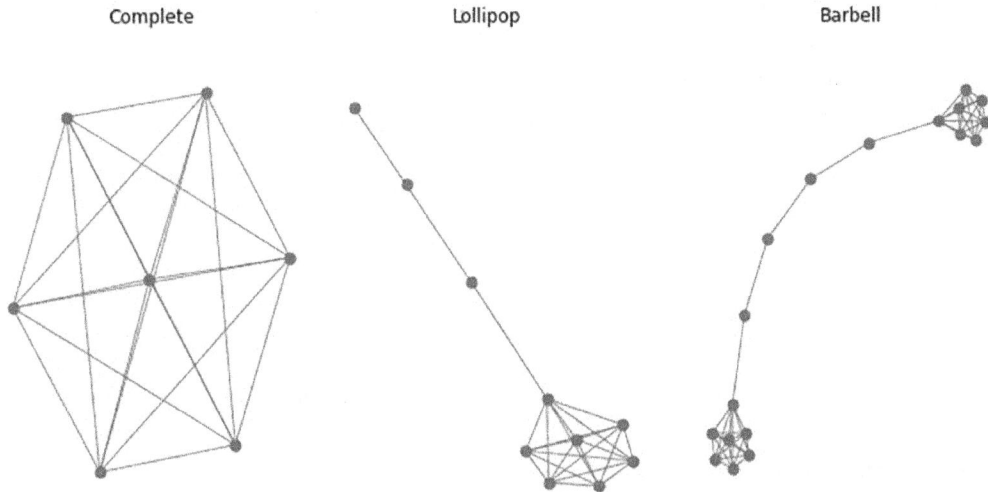

Figure 1.20: Simple examples of graphs with networkx: (left) fully connected graph; (center) lollipop graph; (right) barbell graph

In *Figure 1.20*, we show a fully connected graph (left). These graphs represent well-integrated groups, such as small teams in an organization, where every member interacts with every other member. This graph can help analyze communication patterns. It also contains two other simple examples containing cliques that can be easily generated with networkx, outlined as follows:

- A **lollipop graph** formed by a clique of size *m* and a branch of *n* nodes, as shown in the following code snippet:

```
lollipop = nx.lollipop_graph(m=7, n=3)
```

- A **barbell graph** formed by two cliques of size *m1* connected by a path of size *m2*, which resembles the sample graph we used previously to characterize some of the global and local properties. The code to generate this is shown in the following snippet:

```
barbell = nx.barbell_graph(m1=7, m2=4)
```

Such simple graphs are basic building blocks that can be used to generate more complex networks by combining them. Merging subgraphs is very easy with NetworkX and can be done with just a few lines of code, as shown in the following code snippet, where the three graphs are merged into a single graph and some random edges are placed to connect them:

```
def get_random_node(graph):
    return np.random.choice(graph.nodes)
allGraphs = nx.compose_all([complete, barbell, lollipop])
allGraphs.add_edge(get_random_node(lollipop), get_random_node(lollipop))
allGraphs.add_edge(get_random_node(complete), get_random_node(barbell))
```

Other very simple graphs (that can then be merged and played around with) can be found at `https://networkx.org/documentation/stable/reference/generators.html#module-networkx.generators.classic`.

Generative graph models

Although creating simple subgraphs and merging them is a way to generate new graphs of increasing complexity, networks may also be generated by means of **probabilistic models** and/or **generative models** that let a graph grow by itself. Such graphs usually share interesting properties with real networks and have long been used to create benchmarks and synthetic graphs, especially in times when the amount of data available was not as overwhelming as today. Here, we present some examples of random generated graphs, briefly describing the models that underlie them.

Watts and Strogatz (1998)

This model was introduced by the authors to study the behavior of small-world networks. A small-world network is characterized by a high clustering coefficient and a short average path length, meaning that most nodes can be reached from any other node through a small number of intermediate connections. This structure often mirrors real-world social networks, where individuals are typically connected through a few mutual acquaintances, allowing for rapid information dissemination. The graph is generated by first displacing n nodes in a ring and connecting each node with its k neighbors. Each edge of such a graph then has a probability p of being rewired to a randomly chosen node. By ranging p, the Watts and Strogatz model allows a shift from a regular network ($p=0$) to a completely random network ($p=1$). In between, graphs exhibit small-world features; that is, they tend to bring this model closer to social network graphs. These kinds of graphs can be easily created with the following command:

```
graph = nx.watts_strogatz_graph(n=20, k=5, p=0.2)
```

An illustration of a graph before and after rewiring can be found below:

Regular Ring Network (p=0)   Watts-Strogatz Network with Rewiring (p=0.2)

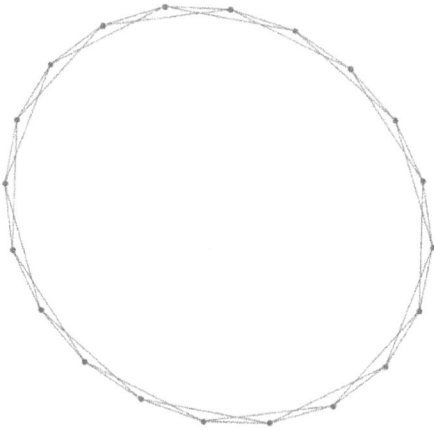 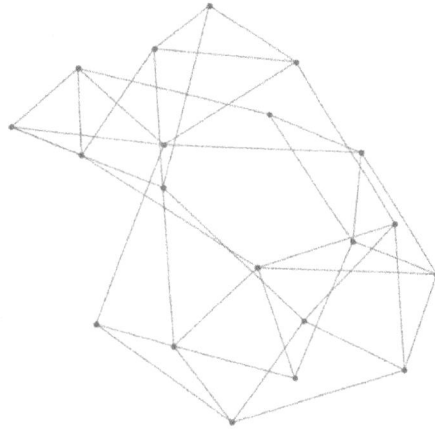

Figure 1.21: A sample graph before (left) and after (right) rewiring with p=0.2

Barabási-Albert (1999)

The model proposed by Albert and Barabási is based on a generative model that allows the creation of random scale-free networks by using a **preferential attachment** schema, where a network is created by progressively adding new nodes and attaching them to already existing nodes, with a preference for nodes that have more neighbors. Mathematically speaking, the underlying idea of this model is that the probability for a new node to be attached to an existing node i depends on the degree of the i-th node, k_i, and the sum of the degrees of all existing nodes in the network, $\sum k_j$, according to the following formula:

$$p_i = \frac{k_i}{\sum k_j}$$

Thus, nodes with a large number of edges (hubs) tend to develop even more edges, whereas nodes with few links are unlikely to develop other links (periphery). Networks generated by this model exhibit a *power-law distribution* for the connectivity (that is, degree) between nodes. Such a behavior is also found in real networks, such as the **World Wide Web** (**WWW**) network and the actor collaboration network (connections between actors based on their collaborations in films and television shows).

Interestingly, this model illustrates that it is the popularity of a node (how many edges it already has) rather than its intrinsic node properties that influences the creation of new connections. The initial model has then been extended (and this is the version that is available on NetworkX) to also allow the preferential attachment of new edges or the rewiring of existing edges.

The Barabási-Albert model is illustrated in the following figure:

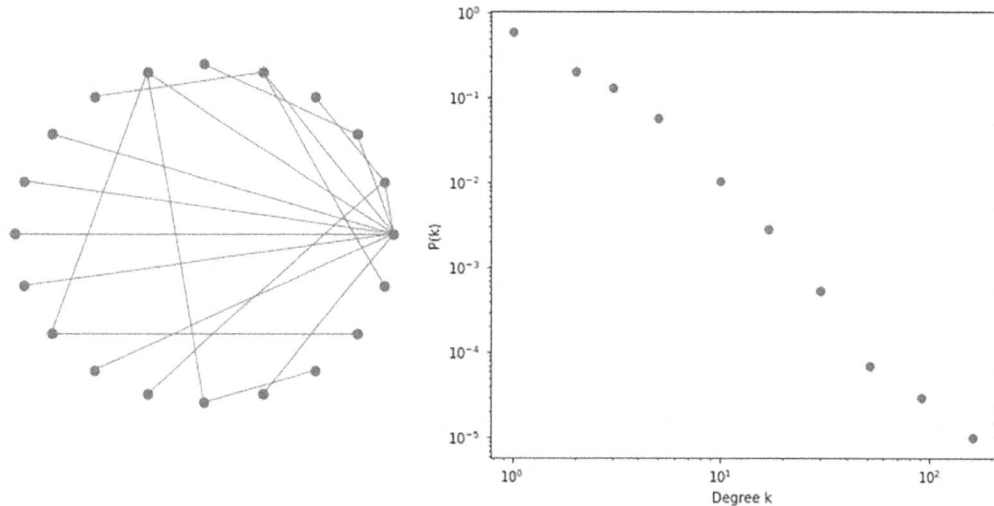

Figure 1.22: Barabási-Albert model (left) with 20 nodes; distribution of connectivity with
n=100.000 nodes (right), showing the scale-free power law distribution

In *Figure 1.22*, we showed an example of the Barabási-Albert model for a small network, where you can already observe the emergence of hubs (on the left), as well as the probability distribution of the degree of the nodes, which exhibits a scale-free power-law behavior (on the right). The preceding distribution can easily be replicated in **networkx**, as in the following code snippet, where n and m are the number of nodes and edges, respectively, p is the probability value for adding an edge between existing nodes, and q is the probability value of the rewiring of existing edges (with $p + q < 1$):

```
ba_model = nx.extended_barabasi_albert_graph(n=100,m=1,p=0,q=0)
degree = dict(nx.degree(ba_model)).values()
bins = np.round(np.logspace(np.log10(min(degree)), np.log10(max(degree)),
10))
cnt = Counter(np.digitize(np.array(list(degree)), bins))
```

Data resources for network analysis

Digitalization has profoundly changed our lives, and today, any activity, person, or process generates data, providing a huge amount of information to be drilled into, analyzed, and used to promote data-driven decision-making. A few decades ago, it was hard to find datasets ready to be used to develop or test new algorithms. On the other hand, there exist today plenty of repositories that provide us with datasets, even of fairly large dimensions, to be downloaded and analyzed. These repositories, where people can share datasets, also provide a benchmark where algorithms can be applied, validated, and compared with each other.

In this section, we will briefly go through some of the main repositories and file formats used in network science, in order to provide you with all the tools needed to import datasets—of different sizes—to analyze and play around with.

In such repositories, you will find network datasets coming from some of the common areas of network science, such as social networks, biochemistry, dynamic networks, documents, co-authoring and citation networks, and networks arising from financial transactions. In *Part 3, Advanced Applications of Graph Machine Learning*, we will discuss some of the most common types of networks (social networks, graphs arising when processing corpus documents, and financial networks) and analyze them more thoroughly by applying the techniques and algorithms described in *Part 2, Machine Learning on Graphs*.

Also, **networkx** already comes with some basic (and very small) networks that are generally used to explain algorithms and basic measures, which can be found at https://networkx.org/documentation/stable/reference/generators.html#module-networkx.generators.social. These datasets are, however, generally quite small. For larger datasets, refer to the repositories we present next.

Network Repository

Network Repository is surely one of the largest repositories of network data (http://networkrepository.com/) with several thousand different networks, featuring users and donations from all over the world and top-tier academic institutions. If a network dataset is freely available, chances are that you will find it there. Datasets are classified into about *30 domains*, including biology, economics, citations, social network data, industrial applications (energy, road), and many others. Besides providing the data, the website also provides a tool for interactive visualization, exploration, and comparison of datasets, and we suggest you check it out and explore it.

The data in Network Repository is generally available under the **Matrix Market Exchange Format (MTX)** file format. The MTX file format is basically a file format for specifying dense or sparse matrices, real or complex, via readable text files (**American Standard Code for Information Interchange**, or **ASCII**). For more details, please refer to http://math.nist.gov/MatrixMarket/formats.html#MMformat.

A file in MTX format can be easily read in Python using **SciPy**. Some of the files we downloaded from Network Repository seemed slightly corrupted and required a minimal fix on a 10.15.2 macOS system. In order to fix them, just make sure the header of the file is compliant with the format specifications—that is, with a double **%** and no spaces at the beginning of the line, as in the following line:

```
%%MatrixMarket matrix coordinate pattern symmetric
```

Matrices should be in coordinate format. In this case, the specification points also to an unweighted, undirected graph (as understood by **pattern** and **symmetric**). Some of the files have some comments after the first header line, which are preceded by a single **%**.

As an example, we consider the **Astro Physics (ASTRO-PH)** collaboration network. The graph is generated using all the scientific papers available from the e-print *arXiv* repository published in the *Astrophysics* category in the period from January 1993 to April 2003. The network is built by connecting (via undirected edges) all the authors that co-authored a publication, thus resulting in a clique that includes all authors of a given paper. The code to generate the graph can be seen here:

```
from scipy.io import mmread
adj_matrix = mmread("ca-AstroPh.mtx")
graph = nx.from_scipy_sparse_matrix(adj_matrix)
```

The dataset has 17,903 nodes, connected by 196,072 edges. Visualizing so many nodes cannot be done easily, and even if we were to do it, it might not be very informative, as understanding the underlying structure would not be very easy with so much information. However, we can get some insights by looking at specific subgraphs, as we will do next.

First, we can start by computing some basic properties we described earlier and put them into a pandas **DataFrame** for our convenience to later use, sort, and analyze. The code to accomplish this is illustrated in the following snippet (it may require several minutes to complete):

```
stats = pd.DataFrame({
    "centrality": nx.centrality.betweenness_centrality(graph),
    "C_i": nx.clustering(graph),
```

```
    "degree": nx.degree(graph)
})
```

We can easily find out that the node with the largest **degree centrality** is the one with ID **6933**, which has 503 neighbors (surely a very popular and important scientist in astrophysics!), as illustrated in the following code snippet:

```
neighbors = [n for n in nx.neighbors(graph, 6933)]
```

Of course, also plotting its **ego network** (the node with all its neighbors) would still be a bit messy. One way to produce some subgraphs that can be plotted is by sampling (for example, with a 0.1 ratio) its neighbors in three different ways: random (sorting by index is a sort of random sorting), selecting the most central neighbors, or selecting the neighbors with the largest C_i values. The code to accomplish this is shown in the following code snippet:

```
sampling = 0.1 # this represents 10% of the neighbors
nTop = round(len(neighbors)*sampling)
idx = {
    "random": stats.loc[neighbors].sort_index().index[:nTop],
    "centrality": stats.loc[neighbors]\
        .sort_values("centrality", ascending=False)\
        .index[:nTop],
    "C_i": stats.loc[neighbors]\
        .sort_values("C_i", ascending=False)\
        .index[:nTop]
}
```

We can then define a simple function for extracting and plotting a subgraph that includes only the nodes related to certain indices, as shown in the following code snippet:

```
def plotSubgraph(graph, indices, center = 6933):
    nx.draw_kamada_kawai(
        nx.subgraph(graph, list(indices) + [center])
    )
```

Using the function above, we can plot the different subgraphs. Each subgraph will be obtained by filtering the ego network using three different criteria, based on random sampling, centrality, and the clustering coefficient. An example is provided here:

```
plotSubgraph(graph, idx["random"])
```

In *Figure 1.23*, we compare these results where the other networks have been obtained by changing the key value to **centrality** and C_i. The random representation seems to show some emerging structure with separated communities. The graph with the most central nodes clearly shows an almost fully connected network, possibly made up of all full professors and influential figures in astrophysics science, publishing on multiple topics and collaborating frequently with each other. Finally, the last representation, on the other hand, highlights some specific communities, possibly connected with a specific topic, by selecting the nodes that have a higher clustering coefficient. These nodes might not have a large degree of centrality, but they represent specific topics very well. You can see examples of the ego subgraph here:

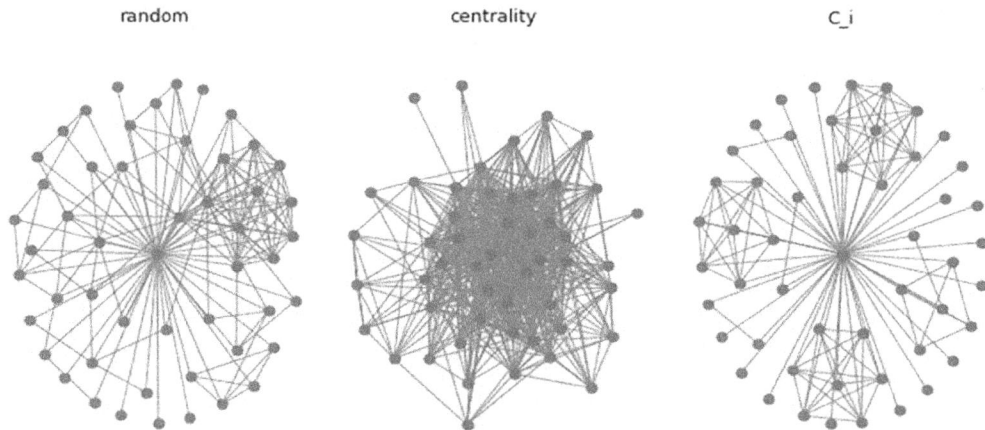

Figure 1.23: Examples of the ego subgraph for the node that has the largest degree in the ASTRO-PH dataset. Neighbors are sampled with a ratio=0.1 random sampling (left); nodes with largest betweenness centrality (center); nodes with largest clustering coefficient (right)

Another option to visualize this in NetworkX could also be to use the *Gephi* software, which allows for fast filtering and visualizations of graphs. In order to do so, we need to first export the data in **Graph Exchange XML Format (GEXF)** (which is a file format that can be imported in Gephi), as follows:

```
nx.write_gext(graph, "ca-AstroPh.gext")
```

Once data is imported in Gephi, with a few filters (by centrality or degree) and some computations (modularity), you can easily do plots as nice as the one shown in *Figure 1.24*, where nodes have been colored using modularity in order to highlight clusters. Coloring also allows us to easily spot nodes that connect the different communities and that therefore have large betweenness.

Some of the datasets in Network Repository may also be available in the **EDGE file format** (for instance, the citation networks). The EDGE file format slightly differs from the MTX file format, although it represents the same information. Probably the easiest way to import such files into NetworkX is to convert them by simply rewriting its header. Take, for instance, the **Digital Bibliography and Library Project (DBLP)** citation network.

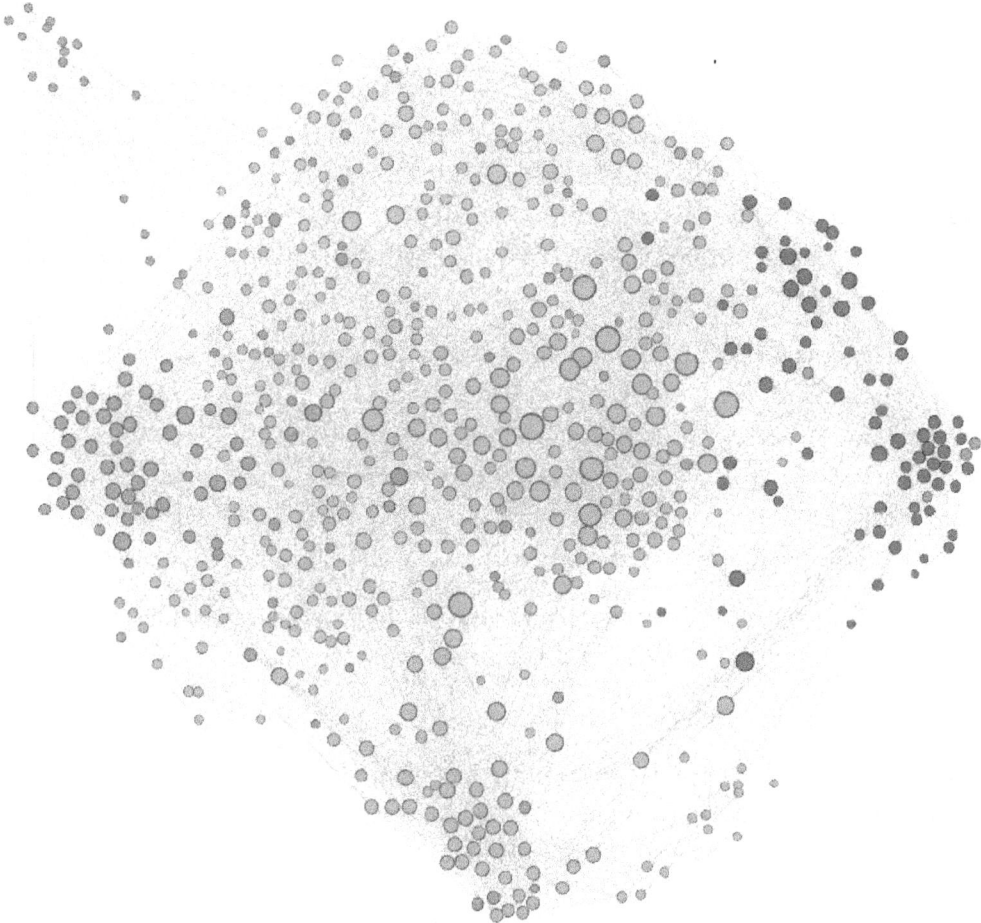

Figure 1.24: Example of the visualization ASTRO-PH dataset with Gephi. Nodes are filtered by degree centrality and colored by modularity class; node sizes are proportional to the value of the degree

The header of the file in this case reads:

```
% asym unweighted
% 49743 12591 12591
```

This can be easily converted to comply with the MTX file format by replacing these lines with the following code:

```
%%MatrixMarket matrix coordinate pattern general
12591 12591 49743
```

Then, you can use the import functions described previously.

Stanford Large Network Dataset Collection

Another valuable source of network datasets is the website of the **Stanford Network Analysis Platform (SNAP)** (https://snap.stanford.edu/index.html), which is a general-purpose network analysis library that was written in order to handle even fairly large graphs, with hundreds of millions of nodes and billions of edges. It is written in *C++* to achieve top computational performance, but it also features interfaces with Python in order to be imported and used in native Python applications.

Although **networkx** is currently the main library to study **networkx in Python**, SNAP or other libraries (more on this shortly) can be orders of magnitude faster than **networkx**, and they may be used in place of **networkx** for tasks that require higher performance. On the SNAP website, you will find a specific web page for **Biomedical Network Datasets** (https://snap.stanford.edu/biodata/index.html), besides other more general networks (https://snap.stanford.edu/data/index.html), covering similar domains and datasets as Network Repository, described previously.

Data is generally provided in a **text file format** containing a list of edges. Reading such files can be done with **networkx** in one code line, using the following command:

```
g = nx.read_edgelist("amazon0302.txt")
```

Some graphs might have extra information, other than about edges. Extra information is included in the archive of the dataset as a separated file—for example, where some metadata of the nodes is provided and is related to the graph via the *id* node.

Graphs can also be read directly using the SNAP library and its interface via Python. If you have a working version of SNAP on your local machine, you can easily read the data as follows:

```
from snap import LoadEdgeList, PNGraph
graph = LoadEdgeList(PNGraph, "amazon0302.txt", 0, 1, '\t')
```

Keep in mind that, at this point, you will have an instance of a PNGraph object of the SNAP library, and you can't directly use NetworkX functionalities on this object. If you want to use some NetworkX functions, you first need to convert the PNGraph object to a **networkx** object.

You can do that by creating a new graph and adding nodes and edges from `PNGraph` by using the networkx functionalities we have seen before.

Open Graph Benchmark

This is the most recent update (dated May 2020) in the graph benchmark landscape, and this repository is expected to gain increasing importance and support in the coming years. The **Open Graph Benchmark (OGB)** has been created to address one specific issue: current benchmarks are actually too small compared to real applications to be useful for ML advances. On the one hand, some of the models developed on small datasets turn out to not be able to scale to large datasets, proving them unsuitable in real-world applications. On the other hand, large datasets also allow us to increase the capacity (complexity) of the models used in ML tasks and explore new algorithmic solutions (such as neural networks) that can benefit from a large sample size to be efficiently trained, allowing us to achieve very high performance. The datasets belong to diverse domains and they have been ranked on three different dataset sizes (small, medium, and large), where the small graphs, despite their name, already have more than 100,000 nodes and/or more than 1 million edges. Conversely, large graphs feature networks with more than 100 million nodes and more than 1 billion edges, facilitating the development of scalable models.

Besides the datasets, the OGB also provides, in a *Kaggle fashion*, an end-to-end ML pipeline that standardizes the data loading, experimental setup, and model evaluation. OGB creates a platform to compare and evaluate models against each other, publishing a *leaderboard* that allows tracking of the performance evolution and advancements on specific tasks of node, edge, and graph property prediction. For more details on the datasets and the OGB project, please refer to the following paper by Hu et al. (2021): `https://arxiv.org/pdf/2005.00687.pdf`.

Dealing with large graphs

When approaching a use case or an analysis, it is very important to understand how large the data we focus on is or will be in the future, as the dimension of the datasets may very well impact both the technologies we use and the analysis that we can do. As already mentioned, some of the approaches that have been developed on small datasets hardly scale to real-world applications and larger datasets, making them useless in practice.

When dealing with (possibly) large graphs, it is crucial to understand potential bottlenecks and limitations of the tools, technologies, and/or algorithms we use, assessing which part of our application/analysis may not scale when increasing the number of nodes or edges. Even more importantly, it is crucial to structure a data-driven application, however simple or at what early stage of the **proof of concept (POC)**, in a way that would allow its scaling out in the future when data/users would increase, without rewriting the whole application.

Creating a data-driven application that resorts to graphical representation/modeling is a challenging task that requires a design and implementation that is a lot more complicated than simply importing NetworkX. In particular, it is often useful to decouple the component that processes the graph—the **graph processing engine**—from the one that allows querying and traversing the graph—the **graph storage layer**. We will further discuss these concepts in *Chapter 10, Building a Data-Driven Graph-Powered Application*. Nevertheless, given the focus of the book on ML and analytical techniques, it makes sense to focus more on graph processing engines than on graph storage layers. We, therefore, find it useful at this stage to provide you with some of the technologies that are used for graph processing engines to deal with large graphs, crucial when scaling out an application.

In this respect, it is important to classify graph processing engines into two categories (that impact the tools/libraries/algorithms to be used), depending on whether the graph can fit a *shared memory machine* or requires *distributed architectures* to be processed and analyzed.

Note that there is no absolute definition of large and small graphs, but it also depends on the chosen architecture. Nowadays, thanks to the vertical scaling of infrastructures, you can find servers with **random-access memory (RAM)** larger than 1 **terabyte (TB)** (usually called *fat nodes*), and with tens of thousands of **central processing units (CPUs)** for multithreading in most cloud-provider offerings, although these infrastructures might not be economically viable. Even without scaling out to such extreme architectures, graphs with millions of nodes and tens of millions of edges can nevertheless be easily handled in single servers with ~100 **gigabytes (GB)** of RAM and ~50 CPUs.

Although **networkx** is a very popular, user-friendly, and intuitive library, when scaling out to such reasonably large graphs, it may not be the best available choice. NetworkX, being natively written in pure Python, which is an interpreted language, can be substantially outperformed by other graph engines fully or partly written in more performant programming languages (such as C++ and Julia) and that make use of multithreading, such as the following:

- **SNAP** (http://snap.stanford.edu/), which we have already seen in the previous section, is a graph engine developed at Stanford and is written in C++ with available bindings in Python.

- **igraph** (https://igraph.org/) is a C library and features bindings in Python, R, and Mathematica.

- **graph-tool** (https://graph-tool.skewed.de/), despite being a Python module, has core algorithms and data structures written in C++ and uses OpenMP parallelization to scale on multi-core architectures.

- **NetworKit** (`https://networkit.github.io/`) is also written in C++ with OpenMP boost for parallelization for its core functionalities, integrated into a Python module.
- **LightGraphs** (`https://juliapackages.com/p/lightgraphs`) is a library written in Julia that aims to mirror **networkx** functionalities in a more performant and robust library.

All the preceding libraries are valid alternatives to NetworkX when achieving better performance becomes an issue. Improvements can be very substantial, with speed-ups varying from 30 to 300 times faster, with the best performance generally achieved by LightGraphs.

In the forthcoming chapters, we will mostly focus on NetworkX in order to provide a consistent presentation and provide you with basic concepts on network analysis. We want you to be aware that other options are available, as this becomes extremely relevant when pushing the edge from a performance standpoint.

Summary

In this chapter, we covered concepts such as graphs, nodes, and edges. We reviewed graph *representation* methods and explored how to *visualize* graphs. We also defined *properties* that are used to characterize networks or parts of them.

We went through a well-known Python library to deal with graphs, NetworkX, and learned how to use it to apply theoretical concepts in practice. We then ran examples and toy problems that are generally used to study the properties of networks, as well as benchmark performance and the effectiveness of network algorithms. We also provided you with some useful links to repositories where network datasets can be found and downloaded, together with some tips on how to parse and process them.

In the next chapter, we will go beyond defining notions of ML on graphs. We will learn how more advanced and latent properties can be automatically found by specific ML algorithms.

Get This Book's PDF Version and Exclusive Extras

Scan the QR code (or go to packtpub.com/unlock).
Search for this book by name, confirm the edition,
and then follow the steps on the page.

*Note: Keep your invoice handy. Purchases made
directly from Packt don't require one.*

2

Graph Machine Learning

Machine learning is a subset of artificial intelligence that aims to provide systems with the ability to *learn* and improve from data. It has achieved impressive results in many different applications, especially where it is difficult or unfeasible to explicitly define rules to solve a specific task. For instance, we can train algorithms to recognize spam emails, translate sentences into other languages, recognize objects in an image, and so on.

In recent years, there has been an increasing interest in applying machine learning to *graph-structured data*. Graphs, composed of nodes and edges, naturally represent relationships and interactions in many real-world systems, making them a better choice in many scenarios where "traditional" machine learning models may overlook these important dependencies. For example, graph machine learning has found wide applications in recommendation systems, where the relationships between users and products (e.g., who bought or liked what) can be modeled as a graph, improving prediction accuracy. Similarly, graphs excel in areas like social network analysis, where the connections between individuals are vital for tasks such as community detection or predicting user behavior.

This chapter will first review some of the basic machine learning concepts. Then, an introduction to graph machine learning will be provided, with a particular focus on **representation learning**.

The following topics will be covered in this chapter:

- A refresher on machine learning
- What is machine learning on graphs and why is it important?
- A general taxonomy to navigate graph machine learning algorithms

Technical requirements

All code files relevant to this chapter are available at `https://github.com/PacktPublishing/` `Graph-Machine-Learning/tree/main/Chapter02`. Please refer to the *Practical exercises* section of *Chapter 1, Getting Started with Graphs,* for guidance on how to set up the environment to run the examples in this chapter, either using Poetry, `pip`, or Docker.

For more complex data visualization tasks provided in this chapter, Gephi (`https://gephi.org/`) may also be required. The installation manual is available at `https://gephi.org/users/install/`.

Understanding machine learning on graphs

Out of the branches of artificial intelligence, **machine learning** is the one that has attracted the most attention in recent years. It refers to a class of computer algorithms that automatically learn and improve their skills through experience *without being explicitly programmed*. Such an approach takes inspiration from nature. Imagine an athlete who faces a novel movement for the first time: they start slowly, carefully imitating the gesture of a coach, trying, making mistakes, and trying again. Eventually, they will improve, becoming more and more confident.

Basic principles of machine learning

How does this concept translate to machines? It is essentially an optimization problem. The goal is to find a mathematical model that is able to achieve the best possible performance on a particular task. Performance can be measured using a specific performance metric (also known as a **loss function** or **cost function**). In a common learning task, the algorithm is provided with data, possibly lots of it. The algorithm uses this data to iteratively make decisions or predictions for the specific task. At each iteration, decisions are evaluated using the loss function. The resulting *error* is used to update the model parameters in a way that, hopefully, means the model will perform better. This process is commonly called **training**.

More formally, let's consider a particular task, T, and a performance metric, P, which allows us to quantify how well an algorithm is performing on T. According to Mitchell et al. (1997), an algorithm is said to learn from experience, E, if its performance at task T, measured by P, improves with experience E.

Machine learning algorithms fall into four main categories, known as *supervised*, *unsupervised*, *semi-supervised*, and *reinforcement* learning. These learning paradigms depend on the way data is provided to the algorithm and how performance is evaluated.

Supervised learning is the learning paradigm used when we know the answer to the problem. In this scenario, the dataset is composed of samples of pairs of the form <x,y>, where x is the input (for example, an image or voice signal) and y is the corresponding desired output (for example, what the image represents or what the voice is saying). The input variables are also known as *features*, while the output is usually referred to as *labels, targets,* or *annotations*. In supervised settings, performance is often evaluated using a *distance function*. This function measures the differences between the prediction and the expected target. According to the type of labels, supervised learning can be further divided as follows:

- **Classification**: Here, the labels are discrete and refer to the "class" the input belongs to. Examples of classification include determining what the object in a photo is and predicting whether an email is spam or not.

- **Regression**: The target is continuous here. Examples of regression problems include predicting the temperature in a building or predicting the selling price of any particular product.

Unsupervised learning differs from supervised learning since the answer to the problem is not known. In this context, we do not have any labels and only the inputs, <x>, are provided. The goal is thus deducing structures and patterns, attempting to find similarities or anomalies.

Discovering groups of similar examples (clustering) is one of these problems, as well as giving new representations of the data in a more compact and meaningful vector space.

In **semi-supervised learning**, the algorithm is trained using a combination of labeled and unlabeled data. Usually, to direct the research of structures present in the unlabeled input data, a limited amount of labeled data is used.

Reinforcement learning is used for training machine learning agents to make a sequence of decisions. The artificial intelligence algorithm faces a game-like situation, where the agent gets *penalties* or *rewards* based on the actions performed. The goal of the agent is to understand how to act in order to maximize rewards and minimize penalties.

Minimizing the error on the training data is, however, not enough. The keyword in machine learning is *learning*. It means that algorithms must be able to achieve the same level of performance even on unseen data. The most common way of evaluating the generalization capabilities of machine learning algorithms is to divide the dataset into two parts: the **training set** and the **test set**. The model is trained on the training set, where the loss function is computed and used to update the parameters. After training, the model's performance is evaluated on the test set.

Moreover, when more data is available, the test set can be further divided into **validation** and **test** sets. The validation set is commonly used for assessing the model's performance during training and is generally required when selecting the best model among several models, or families of models.

When training a machine learning algorithm, three situations can be observed:

- In the first situation, the model reaches a low level of performance over the training set. This situation is commonly known as **underfitting**, meaning that the model is not powerful enough to address the task.

- In the second situation, the model achieves a high level of performance over the training set but struggles with generalizing to testing data. This situation is known as **overfitting**. In this case, the model is simply memorizing the training data without actually understanding the true relations among the data instances.

- Finally, the ideal situation is when the model is able to achieve (possibly) the highest level of performance over both training and testing data.

A graphical representation of overfitting and underfitting is given by the risk curve shown in *Figure 2.1*. From the figure, it is possible to see how the performances on the training and test sets change according to the complexity of the model (the number of parameters to be fitted):

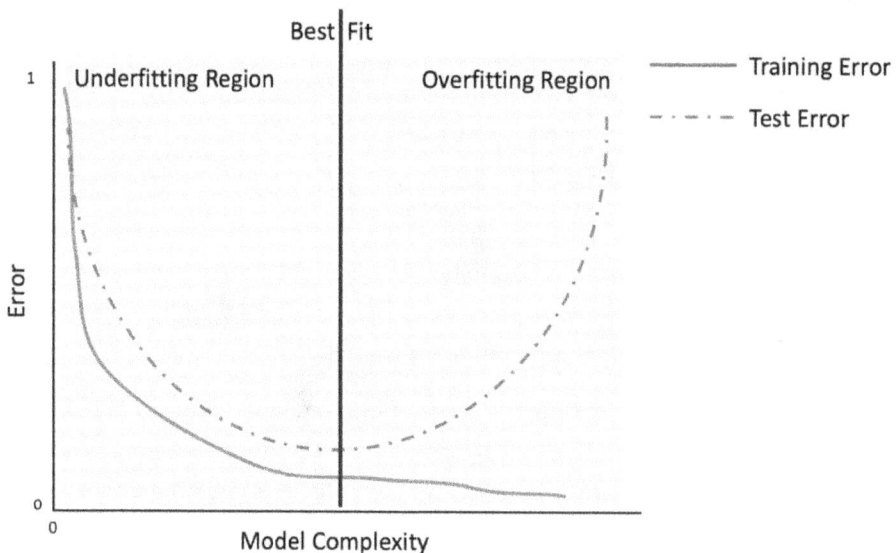

Figure 2.1: Risk curve describing the prediction error on training and test set error in the function of the model complexity (number of parameters of the model)

Interestingly, a study by Belkin et al. (2019) introduced the concept of a "double descent" curve, which suggests that beyond the shown region of overfitting, there exists another regime with extremely high model complexity. In this regime, the model can perfectly fit the training data (achieving near-zero error) but counterintuitively, the test error begins to decrease again after the initial overfitting phase. This phenomenon challenges the traditional view of model complexity and generalization. However, for the purposes of this book and the examples we will cover, we will not reach this double descent regime. Thus, the classical behavior of the risk curve, as illustrated in *Figure 2.1*, remains valid for our discussion.

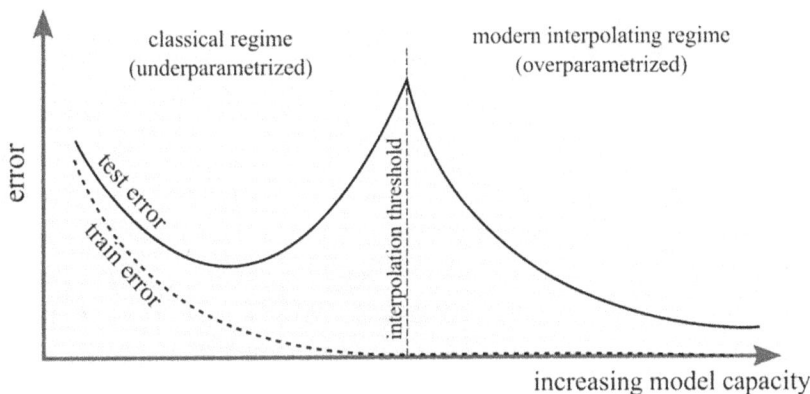

Figure 2.2: Double descent curve suggested by Belkin et al. (2019)

Overfitting is one of the main problems that affect machine learning practitioners. It can occur due to several reasons. Some of the reasons can be as follows:

- The dataset can be ill-defined or not sufficiently representative of the task. In this case, adding more data could help to mitigate the problem.
- The mathematical model used for addressing the problem is too powerful for the task. In this case, proper constraints can be added to the loss function in order to reduce the model's "power." Such constraints are called **regularization** terms.

The benefit of machine learning on graphs

Machine learning has achieved impressive results in many fields, becoming one of the most diffused and effective approaches in computer vision, pattern recognition, and natural language processing, among others. Several machine learning algorithms have been developed, each with its own advantages and limitations. Among those, it is worth mentioning regression algorithms (for example, linear and logistic regression), instance-based algorithms (for example, k-nearest neighbor or support vector machines), decision tree algorithms, Bayesian algorithms (for example, naïve Bayes), clustering algorithms (for example, k-means), and artificial neural networks.

Despite their success, these traditional algorithms often work best when the data can be represented in a structured, grid-like form, such as images or tabular data, where each sample has no encoded relation to other samples. However, many real-world systems involve data that is inherently structured as a graph. Traditional algorithms may struggle to capture the complex interactions in such data, which can result in suboptimal performance.

This is where graph machine learning comes into play. By leveraging the inherent structure of graph data, these algorithms can model relationships between data points more effectively.

So, what is the reason behind the growing success of graph machine learning?

Graph machine learning enables us to automatically detect and interpret complex, latent patterns in graph-structured data—patterns that are often too intricate for traditional machine learning models to capture.

In particular, there has been an increasing interest in *learning representations* for graph-structured data and many machine learning algorithms have been developed for handling graphs. For example, we might be interested in determining the role of a protein in a biological interaction graph, predicting the evolution of a collaboration network, recommending new products to a user in a social network, and much more (we will discuss applications later in the book).

Due to their nature, graphs can be analyzed at different levels of granularity: at the node, edge, and graph level (the whole graph). Machine learning algorithms can extract and process features coming from nodes, edges, and graphs, as depicted in *Figure 2.3*.

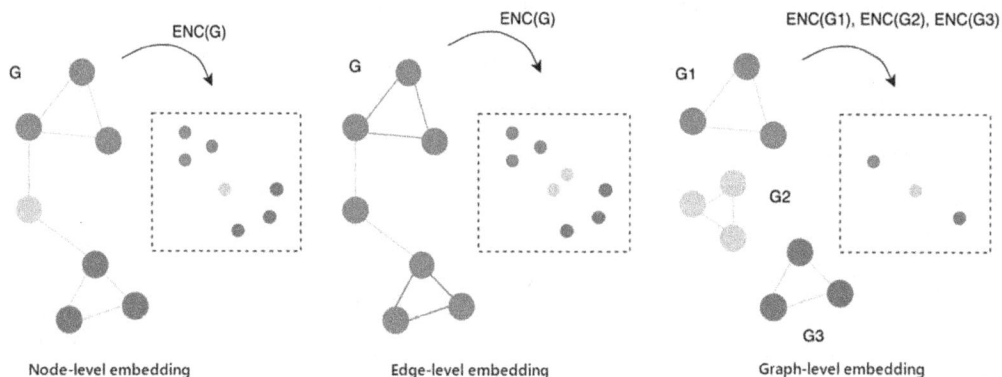

Figure 2.3: Visual representation of the three different levels of granularity in graphs

For each of those levels, several types of machine learning tasks may be addressed and, as a consequence, specific algorithms may be used in each case. In the following list, we give some examples of machine learning problems that correspond to each of those levels:

- **Node level:** Given a (possibly large) graph, $G=(V, E)$, the goal is to classify each vertex, $v \in V$, into the right class. In this setting, the dataset includes G and a list of pairs, $<v_i, y_i>$, where v_i is a node of graph G and y_i is the class to which the node belongs.

- **Edge level:** Given a (possibly large) graph, $G=(V, E)$, the goal is to classify each edge, $e \in E$, into the right class. In this setting, the dataset includes G and a list of pairs, $<e_i, y_i>$, where e_i is an edge of graph G and y_i is the class to which the edge belongs. Another typical task for this level of granularity is **link prediction**, the problem of predicting the existence of a link between two existing nodes in a graph. The task of link prediction can in fact be seen as a particular case of edge classification, where the target y_i is 1 if the edge e_i exists, and 0 otherwise.

- **Graph level:** Given a dataset with m different graphs, the task is to build a machine learning algorithm capable of classifying a graph into the right class. We can then see this problem as a classification problem, where the dataset is defined by a list of pairs, $<G_i, y_i>$, where G_i is a graph and y_i is the class the graph belongs to.

In this section, we have discussed some basic concepts of machine learning. Moreover, we have enriched our description by introducing some of the common machine learning problems when dealing with graphs. Having those theoretical principles as a basis, we will now introduce some more complex concepts relating to graph machine learning.

The generalized graph embedding problem

In classical machine learning applications, a common way to process the input data is to build from a set of features, in a process called **feature engineering**, which is capable of giving a compact and meaningful representation of each instance in the dataset.

The dataset obtained from the feature engineering step will then be used as input for the machine learning algorithm. If this process usually works well for a large range of problems, it may not be the optimal solution when we are dealing with graphs. Indeed, due to their well-defined structure, finding a suitable representation capable of incorporating all the useful information might not be an easy task.

The first, and most straightforward, way of creating features capable of representing structural information from graphs is the *extraction of certain statistics*. For instance, a graph could be represented by its degree distribution, efficiency, and all the metrics we described in the previous chapter.

A more complex procedure consists of applying specific kernel functions or, in other cases, engineering-specific features that are capable of incorporating the desired properties into the final machine learning model. However, as you can imagine, this process could be really time-consuming and, in certain cases, the features used in the model could just represent a subset of the information that is really needed to get the best performance for the final model.

In the last decade, a lot of work has been done in order to define new approaches for creating meaningful and compact representations of graphs. The general idea behind all these approaches is to create algorithms capable of *learning* a good representation of the original dataset made of geometric relationships into a new vector space that reflects the structure of the original graph. We usually call the process of learning a good representation of a given graph **representation learning** or **network embedding**. We will provide a more formal definition as follows.

Representation learning (network embedding) is the task that aims to learn a mapping function, $f: G \rightarrow \mathbb{R}^n$, that transforms a discrete graph into n-dimensional vector space. The function f will be capable of performing a low-dimensional vector representation such that the properties (local and global) of graph G are preserved. The vectorial representation is also often referred to as *embedding*.

Once mapping f is learned, it could be applied to the graph and the resulting mapping could be used as a feature set for a machine learning algorithm. A graphical example of this process is visible in *Figure 2.4*:

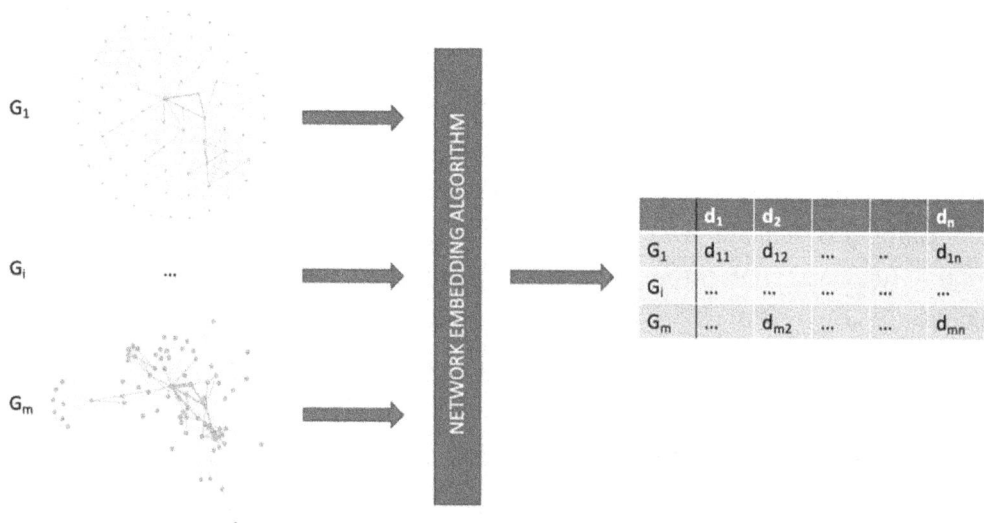

	d_1	d_2			d_n
G_1	d_{11}	d_{12}	...	..	d_{1n}
G_i	...	...	...	...	...
G_m	...	d_{m2}	...	...	d_{mn}

Figure 2.4: Example of a workflow for a network embedding algorithm

Mapping function *f* can also be applied in order to learn the vector representation for nodes and edges. As we already mentioned, machine learning problems on graphs could occur at different levels of granularity. As a consequence, different embedding algorithms have been developed in order to learn functions to generate the vectorial representation of nodes $f: V \rightarrow \mathbb{R}^n$ (also known as **node embedding**) or edges ($f: E \rightarrow \mathbb{R}^n$) (also known as **edge embedding**). Those mapping functions try to build a vector space such that the geometric relationships in the new space reflect the structure of the original graph, node, or edges. As a result, we will see that graphs, nodes, or edges that are similar in the original space will also be similar in the new space.

Thus, in the space generated by the embedding function, similar structures will have *a small Euclidean distance*, while dissimilar structures will have *a large Euclidean distance*. In other words, similar structures will have representations that lie close to each other, while dissimilar structures will have representations that are far apart. It is important to highlight that while most embedding algorithms generate a mapping in Euclidean vector spaces, there has recently been an interest in non-Euclidean mapping functions.

Let's now see a practical example of what an embedding space looks like, and how similarity can be seen in the new space. In the following code block, we show an example using a particular embedding algorithm known as **Node to Vector (Node2Vec)**. We will describe how it works in *Chapter 4, Unsupervised Graph Learning*. At the moment, we will just say that the algorithm will map each node of graph *G* in a vector:

```
import networkx as nx
from node2vec import Node2Vec
import matplotlib.pyplot as plt
G = nx.barbell_graph(m1=7, m2=4)
node2vec = Node2Vec(G, dimensions=2)
model = node2vec.fit(window=10)
fig, ax = plt.subplots()
for x in G.nodes():
    v = model.wv.get_vector(str(x))
    ax.scatter(v[0],v[1], s=1000)
    ax.annotate(str(x), (v[0],v[1]), fontsize=12)
```

In the preceding code, we have done the following:

1. We generated a barbell graph (this was described in the previous chapter; see *Figure 1.20* in *Chapter 1, Getting Started with Graphs*).

2. The Node2Vec embedding algorithm is then used in order to map each node of the graph in a vector of two dimensions.

3. Finally, the two-dimensional vectors generated by the embedding algorithm, representing the nodes of the original graph, are plotted.

The result is shown in *Figure 2.5*:

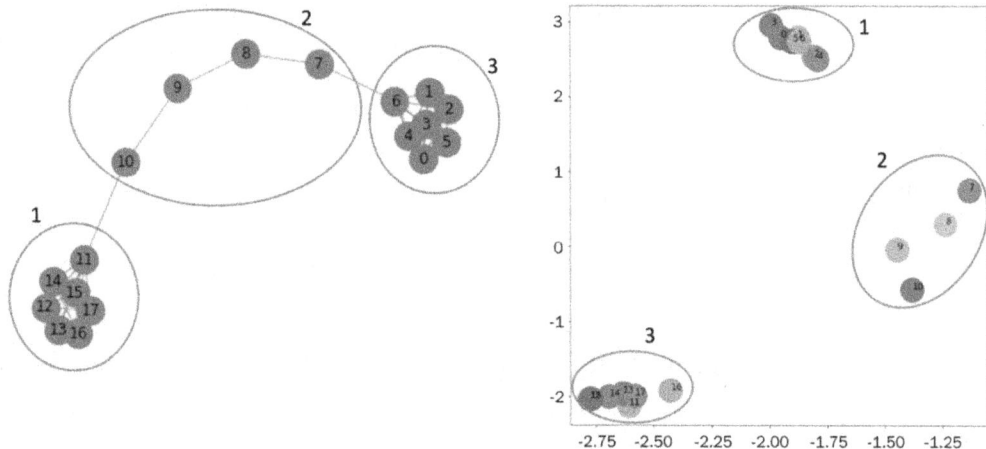

Figure 2.5: Application of the Node2Vec algorithm to a graph (left) to generate the embedding vector of its nodes (right)

From *Figure 2.5*, it is easy to see that nodes that have similar structures are close to each other and distant from nodes that have dissimilar structures. It is also interesting to observe how good Node2Vec is at discriminating group 1 from group 3. Since the algorithm uses neighboring information of each node to generate the representation, the clear discrimination of those two groups is possible. In a real-world example, we might think of the barbell graph as a social network, where the algorithm wants to recommend friends to a user. By generating node embeddings that capture the structural similarity between users (i.e., the patterns of friendships), the algorithm can identify users with similar network structures and suggest friend recommendations based on nearest neighbors' information. For instance, in a platform like LinkedIn, Node2Vec could be used to suggest professional connections who have similar career trajectories or connections.

Another example on the same graph can be performed using the **Edge to Vector (Edge2Vec)** algorithm in order to generate a mapping for the edges for the same graph, *G*:

```
from node2vec.edges import HadamardEmbedder
edges_embs = HadamardEmbedder(keyed_vectors=model.wv)
fig, ax = plt.subplots()
```

```
for x in G.edges():
    v = edges_embs[(str(x[0]), str(x[1]))]
    ax.scatter(v[0],v[1], s=1000)
    ax.annotate(str(x), (v[0],v[1]), fontsize=12)
```

In the preceding code, we have reused the same barbell graph used in the previous code snippet as well as the output of the Node2Vec algorithm to do the following:

1. The **HadamardEmbedder** embedding algorithm is applied to the result of the Node2Vec algorithm (keyed_vectors=model.wv) used in order to map each edge of the graph in a vector of two dimensions.

2. Finally, the two-dimensional vectors generated by the embedding algorithm, representing the nodes of the original graph, are plotted.

The results are shown in *Figure 2.6*:

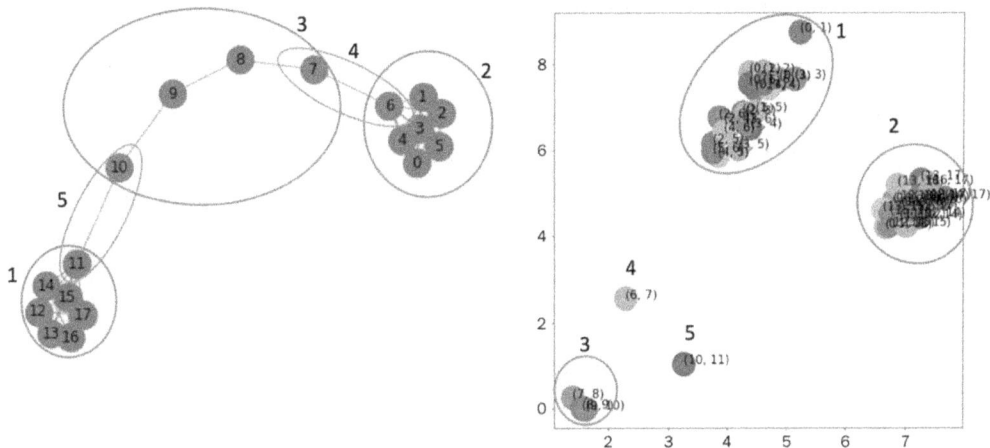

Figure 2.6: Application of the Hadamard algorithm to a graph (left) to generate the embedding vector of its edges (right)

As for node embedding, in *Figure 2.6*, we reported the results of the edge embedding algorithm. From the figure, it is easy to see that the edge embedding algorithm clearly identifies similar edges. As expected, edges belonging to groups 1, 2, and 3 are clustered in well-defined and well-grouped regions. Moreover, the (6,7) and (10,11) edges, belonging to groups 4 and 5, respectively, are well clustered in specific groups. Referring back to the social network example, we might think of edge embedding as a useful way to predict future connections between users.

For instance, on platforms like Facebook, such an algorithm may be used to suggest new friend recommendations by identifying potential links between users who are likely to connect based on their current network structures.

Finally, we will provide an example of a **Graph to Vector (Grap2Vec)** embedding algorithm. This algorithm maps a single graph in a vector. As for the other examples, we will discuss this algorithm in more detail in *Chapter 4, Unsupervised Graph Learning*. In the following code block, we provide a Python example showing how to use the Graph2Vec algorithm in order to generate the embedding representation on a set of graphs:

```python
import random
import matplotlib.pyplot as plt
from karateclub import Graph2Vec
n_graphs = 20
def generate_random():
    n = random.randint(6, 20)
    k = random.randint(5, n)
    p = random.uniform(0, 1)
    return nx.watts_strogatz_graph(n,k,p)
Gs = [generate_random() for x in range(n_graphs)]
model = Graph2Vec(dimensions=2)
model.fit(Gs)
embeddings = model.get_embedding()
fig, ax = plt.subplots(figsize=(10,10))
for i,vec in enumerate(embeddings):
    ax.scatter(vec[0],vec[1], s=1000)
    ax.annotate(str(i), (vec[0],vec[1]), fontsize=16)
```

In this example, the following has been done:

1. 20 Watts-Strogatz graphs (described in the previous chapter) have been generated with random parameters.

2. We have then executed the graph embedding algorithm in order to generate a two-dimensional vector representation of each graph.

3. Finally, the generated vectors are plotted in their Euclidean space.

The results of this example are shown in *Figure 2.7*:

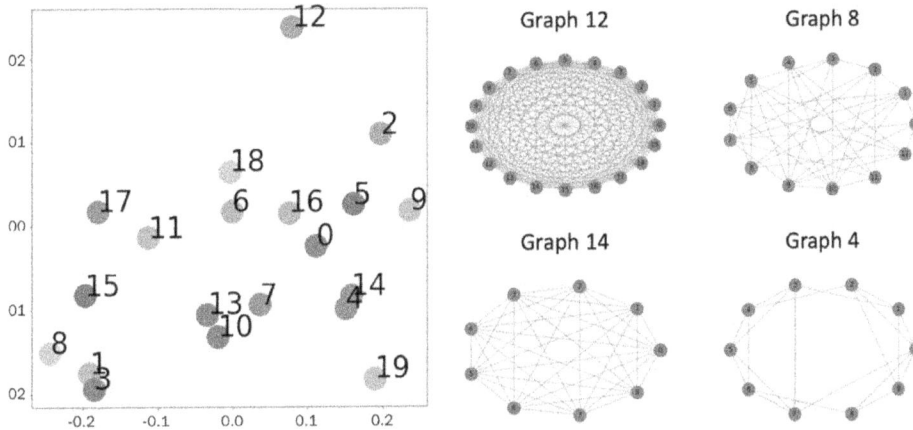

Figure 2.7: Plot of two embedding vectors generated by the Graph2Vec algorithm applied to 20 randomly generated Watts-Strogatz graphs (left). Extraction of two graphs with a large Euclidean distance (Graph 12 and Graph 8 at the top right) and two graphs with a low Euclidean distance (Graph 14 and Graph 4 at the bottom right) is shown

As we can see from *Figure 2.7*, graphs with a large Euclidean distance, such as Graphs 12 and 8, have a different structure. The former is generated with the nx.watts_strogatz_graph(20,20,0.2857) parameter and the latter with the nx.watts_strogatz_graph(13,6,0.8621) parameter. In contrast, a graph with a low Euclidean distance, such as Graphs 14 and 8, has a similar structure. Graph 14 is generated with the nx.watts_strogatz_graph(9,9,0.5091) command, while Graph 4 is generated with nx.watts_strogatz_graph(10,5,0.5659).

In this case, imagine decomposing a social network into unconnected subgraphs. This could be used to identify similar subnetworks, allowing the same advertising strategy to be applied across all of them.

In the scientific literature, a plethora of embedding methods have been developed, and many of them will be described in this book. These methods are usually classified into two main types: *transductive* and *inductive*, depending on the update procedure of the function when new samples are added. If new nodes are provided, transductive methods update the model (for example, re-train) to infer information about the nodes, while in inductive methods, models are expected to generalize to new nodes, edges, or graphs that were not observed during training. Which algorithm to choose strictly depends on the problem and use case you are aiming to address.

The taxonomy of graph embedding machine learning algorithms

A wide variety of methods to generate a compact space for graph representation has been developed. In recent years, a trend has been observed of researchers and machine learning practitioners converging toward a unified notation to provide a common definition to describe such algorithms. In this section, we will be introduced to a simplified version of the taxonomy defined in the *Machine Learning on Graphs: A Model and Comprehensive Taxonomy* paper (https://arxiv.org/abs/2005.03675; Chami et al., 2022).

In this formal representation, every graph, node, or edge embedding method can be described by two fundamental components, named the **encoder (ENC)** and the **decoder (DEC)**. The **ENC** maps the input into the embedding space, while the **DEC** decodes structural information about the graph from the learned embedding (*Figure 2.8*).

The framework described in the paper follows an intuitive idea: if we are able to encode a graph such that the DEC is able to retrieve all the necessary information, then the embedding must contain a compressed version of all this information and can be used to downstream machine learning tasks:

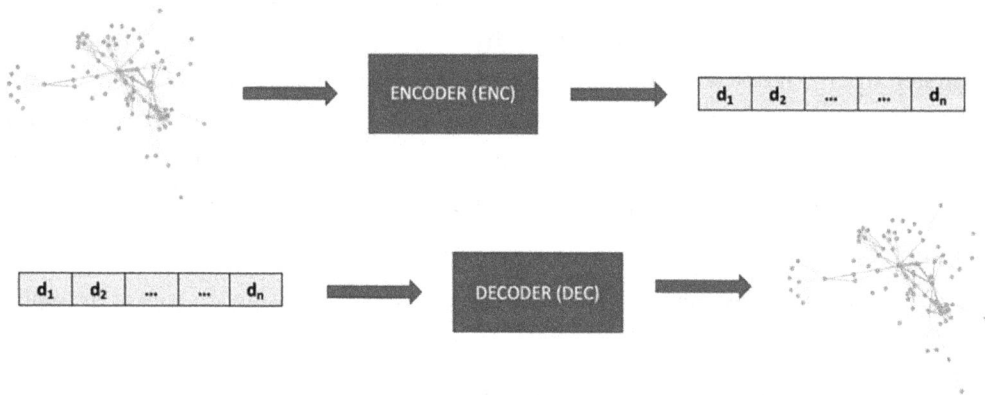

Figure 2.8: Generalized ENC and DEC architecture for embedding algorithms

In many graph-based machine learning algorithms for representation learning, the decoder is usually designed to map pairs of node embeddings to a real value, usually representing the proximity (distance) of the nodes in the original graphs. For example, the decoder can be implemented such that, given the embedding representation of two nodes, $z_i = ENC(V_i)$ and $z_j = ENC(V_j)$, $DEC(z_i, z_j) = 1$ if there is an edge connecting the two nodes z_i and z_j in the input graph. In practice, more effective *proximity functions* can be used to measure the similarity between nodes.

Inspired by the general framework depicted in *Figure 2.8*, we will now provide a categorization of the various embedding algorithms into four main groups. Moreover, in order to help you to better understand this categorization, we shall provide simple code snapshots in pseudo-code. In our pseudo-code formalism, we denote G as a generic NetworkX graph, with `graphs_list` as a list of NetworkX graphs and **model** as a generic embedding algorithm:

- **Shallow embedding methods**: These methods are able to learn and return only the embedding values for the learned input data. *Node2Vec*, *Edge2Vec*, and *Graph2Vec*, which we previously discussed, are examples of shallow embedding methods. These methods are therefore transductive, and indeed, they can only return a vectorial representation of the data they learned during the *fit* procedure. It is not possible to obtain the embedding vector for unseen data. A typical way to use these methods is as follows:

```
model.fit(graphs_list)
embedding = model.get_embedding()[i]
```

  In the code, a generic shallow embedding method is trained on a list of graphs (line 1). Once the model is fitted, we can only get the embedding vector of the ith graph belonging to `graphs_list` (line 2). Unsupervised and supervised shallow embedding methods will be described in *Chapter 4*, *Unsupervised Graph Learning*, and *Chapter 5*, *Supervised Graph Learning*, respectively.

- **Graph autoencoding methods**: These methods do not simply learn how to map the input graphs in vectors; they learn a more general mapping function, *f(G)*, capable of also generating the embedding vector for unseen instances. Due to this, these methods are generally inductive and a typical way to use them is as follows:

```
model.fit(graphs_list)
embedding = model.get_embedding(G)
```

  The model is trained on `graphs_list` (line 1). Once the model is fitted on the input training set, it is possible to use it to generate the embedding vector of a new unseen graph, G. Graph autoencoding methods will be described in *Chapter 4*, *Unsupervised Graph Learning*.

- **Neighborhood aggregation methods**: These algorithms can be used to extract embeddings at the graph level, where nodes are labeled with some properties. Moreover, as for the graph autoencoding methods, the algorithms belonging to this class are able to learn a general mapping function, *f(G)*, and are capable of generating the embedding vector for unseen instances. This generally makes them inductive.

- A nice property of those algorithms is the possibility to build an embedding space where not only the internal structure of the graph but also some external information, defined as properties of its nodes, is taken into account. For instance, with this method, we can have an embedding space capable of simultaneously identifying graphs with similar structures and different properties on nodes. Unsupervised and supervised neighborhood aggregation methods will be described in *Chapter 4*, *Unsupervised Graph Learning*, and *Chapter 5*, *Supervised Graph Learning*, respectively.

- **Graph regularization methods**: Methods based on graph regularization are slightly different from the ones listed in the preceding points. Here, we do not have a graph as input. Instead, the objective is to learn from a set of features by exploiting their "interactions" to regularize the process, either for inductive or transductive functions. In more detail, a graph can be constructed from the features by considering feature similarities. The main idea is based on the assumption that nearby nodes in a graph are likely to have the same labels. Therefore, the loss function is designed to constrain the labels to be consistent with the graph structure. For example, regularization might constrain neighboring nodes to share similar embeddings, in terms of their distance in the L2 norm. For this reason, the encoder only uses X-node features as input.

The algorithms belonging to this family learn a function, $f(X)$, that maps a specific set of features (X) to an embedding vector. As for the graph autoencoding and neighborhood aggregation methods, this algorithm is also able to apply the learned function to new, unseen features. Graph regularization methods will be described in *Chapter 5*, *Supervised Graph Learning*.

For algorithms belonging to the group of shallow embedding methods and neighborhood aggregation methods, it is possible to define an *unsupervised* and *supervised* version. The ones belonging to graph autoencoding methods are suitable for unsupervised tasks, while the algorithms belonging to graph regularization methods are used in semi-supervised or supervised settings.

For unsupervised algorithms, the embedding of a specific dataset is performed only using the information contained in the input dataset, such as nodes, edges, or graphs. For the supervised setting, external information is used to guide the embedding process. That information is usually classed as a label, such as a pair, $<G_i, y_i>$, that assigns a specific class to each graph. This process is more complex than the unsupervised one since the model tries to find the best vectorial representation in order to find the best assignment of a label to an instance. In order to clarify this concept, we can think, for instance, of the *convolutional neural networks* for image classification. During their training process, neural networks try to classify each image into the right class by performing the fitting of various convolutional filters at the same time.

The goal of those convolutional filters is to find a compact representation of the input data in order to maximize the prediction performances. The same concept is also valid for supervised graph embedding, where the algorithm tries to find the best graph representation in order to maximize the performance of a class assignment task.

From a more mathematical perspective, all these models are trained with a proper loss function. This function can be generalized using two terms:

- The first is used in supervised settings to minimize the difference between the prediction and the target
- The second is used to evaluate the similarity between the input graph and the one reconstructed after the ENC and DEC steps (which is the structure reconstruction error)

Formally, it can be defined as follows:

$$\text{Loss} = \alpha L_{sup}(y, \overline{y}) + L_{rec}(G, \overline{G})$$

Here, $\alpha L_{sup}(y, \overline{y})$ is the loss function in the supervised settings, and α represents a regularization coefficient to tune the weight of the supervised loss. The model is optimized to minimize, for each instance, the error between the true y class and the predicted $\overline{y}$. $L_{rec}(G, \overline{G})$ class is the loss function representing the reconstruction error between the input graph (G) and the one obtained after the ENC + DEC process $(\overline{G})$. For unsupervised settings, we have the same loss but $\alpha = 0$ since we do not have a target variable to use.

It is important to highlight the main role that these algorithms play when we try to solve a machine learning problem on a graph. They can be used *passively* in order to transform a graph into a feature vector suitable for a classical machine learning algorithm or for data visualization tasks. However, they can also be used *actively* during the learning process, where the machine learning algorithm finds a compact and meaningful solution to a specific problem.

Summary

In this chapter, we refreshed our knowledge of some basic *machine learning* concepts and discovered how they can be applied to graphs. We defined basic *graph machine learning* terminology with a particular focus on *graph representation learning*. A taxonomy of the main graph machine learning algorithms was presented in order to clarify what differentiates the various ranges of solutions developed over the years. Finally, practical examples were provided to begin understanding how the theory can be applied to practical problems.

In the next chapter, we will look at the concepts of neural networks and neural networks applied to graphs. We will also present the principal frameworks for deep learning and deep learning for graphs in order to better understand the examples throughout the rest of this book.

Get This Book's PDF Version and Exclusive Extras

UNLOCK NOW

Scan the QR code (or go to packtpub.com/unlock).
Search for this book by name, confirm the edition, and then follow the steps on the page.

Note: Keep your invoice handy. Purchases made directly from Packt don't require one.

3

Neural Networks and Graphs

The machine learning landscape of the last decade has seen the rise and explosion of a particular type of model that is extremely popular nowadays, and whose name is becoming very familiar even to non-technical people and practitioners: **artificial neural networks (ANNs)**. Their versatility and potency have resulted in widespread adoption globally, including in the graph domain. Several frameworks have been developed to support their study, use, and development.

Although the first attempts to train ANNs date back to the early 1980s (with the seminal work of Paul Werbos and Geoffrey Hinton), their rise and success has come around only recently, thanks to the advances in computing power (via CPUs but mostly thanks to the highly efficient parallelization of computation enabled by GPUs) as well as the availability of large datasets. ANNs are in fact very general models, able to virtually learn any function, but as such, they need to be trained on vast amounts of data with a large computational cost. In return, however, they can achieve extremely high performances (even super-human) on very complex tasks.

Given such capabilities, recent research and advances in graph machine learning have therefore gone through the development of ANN-based algorithms. The integration between neural networks and graphs has generated a new class of algorithms that are commonly referred to as **graph neural networks (GNNs)**, which – as you will see in the next chapters – will constitute an important backbone of this book.

In this chapter, we will introduce the basic concepts around neural networks and the libraries/frameworks commonly used to deal with ANNs and GNNs.

The following topics will be covered in this chapter:

- A refresher on ANNs and deep learning

- An overview of the most widely used deep learning frameworks

- An introduction to GNNs and the most widely used frameworks for deep learning on graphs

Technical requirements

All code files relevant to this chapter are available at `https://github.com/PacktPublishing/` `Graph-Machine-Learning/tree/main/Chapter03`. Please refer to the *Practical exercises* section of *Chapter 1, Getting Started with Graphs* for guidance on how to set up the environment to run the examples of this chapter, either using Poetry, `pip`, or Docker.

Introduction to ANNs

Neural networks are formed by the integration of several simpler units, called **neurons**. A neuron is just a representation of the following relation between some inputs x_i and one output y of the form

$$ y = f_a \left(b + \sum_i W_i \, x_i \right) $$

Here:

- b is called the bias

- W_i is called the weights

- f_a is the activation functions

The earliest biological models of the neuron in animals and humans historically use a mathematical formulation similar to the one above, in which dendrites transmit incoming information from synapses, propagating and combining it in the neuron's axons, which is then connected to the synapses of nearby, connected neurons. However, it is worth stressing that the formula above is an extremely simplified, naive, and (to some extent) poor representation of what really happens in our brains, lacking the more complex time-dependence behaviors produced by the propagation of electric signals. Over the years, researchers have come up with more realistic and complex models (i.e., generalized integrate-and-fire models) that can more closely capture the neuronal firing patterns observed in living creatures.

Nevertheless, it is extremely interesting to see what a combination of simple relations such as the one above can provide. Indeed, neural networks generally build upon this simplified neuron mode where the output of one neuron becomes the input of the following one.

This is represented using a diagram like the one shown in *Figure 3.1* (this representation will also be used throughout this book), where you can see that neurons can be stacked in progressive and subsequent **layers**. An ANN is therefore specified by the set of biases, activation functions, and weights (also defining the connectivity) of its neurons.

As you can see from *Figure 3.1*, we use the notation $W_{ij}^{(l)}$ to define the weight connecting the i neuron of the l layer to the j neuron of the $l + 1$ layer. Similarly, $b_i^{(l)}$ represents the bias of the i neuron of layer l. Although the activation functions could vary from neuron to neuron, it is common to choose a single function for each layer. Some common choices are sigmoid, **rectified linear unit (ReLU)**, and softmax.

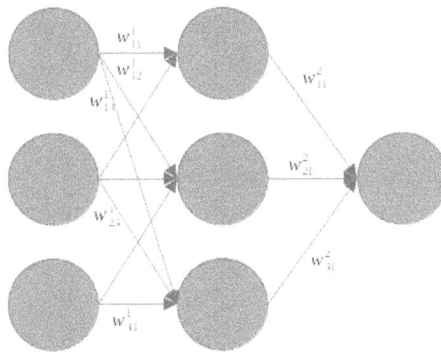

Figure 3.1: Simple schematic representation of a neural network

The various combinations and choices for the network's parameters give rise to the most common neural network architectures nowadays. One can also pose some sort of constraints among the parameters, like imposing that some weights must have the same values (in other words, are shared between neural connections). This is quite common for **kernels** largely used in **Convolutional Neural Networks (CNNs)**, where a given mask (of generally small size) carrying the same set of weights is applied to nearby neurons and rolled throughout the entire layer.

It is possible to demonstrate that such simple combinations of smaller units are – quite surprisingly – extremely powerful and general, as stated by the **universal approximation theorem**: a simple neural network made of *just* one hidden layer (besides the input and output layers) can represent any continuous function, provided it has a large enough number of units on its hidden layer. For some time, this theorem has provided support to people for building **wide neural networks**, thus characterized by a small number of large-neuron layers. However, experience and practice over the years have shown that **deep neural networks**, characterized by a very large number of layers stacked and connected on top of each other, are able to "learn" high-level features that allow the network to generalize better.

Training neural networks

So far, we have just described the way that some inputs can be propagated through the neural network to produce one or multiple outputs. This would be of little use if we did not have a way to find a combination of the biases b and weight $W_{ij}^{(l)}$ to produce meaningful outputs. Training a neural network refers to the process that allows us to optimize the values of the parameters to make the neural network able to carry out a specific task. And to do so, we need data. Often, a lot of data.

For the training data, in fact, we know exactly what the input and the output of the neural network should be, and the training process can be seen as the optimization problem to identify the set of values of the biases b and weights $W_{ij}^{(l)}$ that makes the predicted outputs the closest to the expected ones. Therefore, like in any optimization problem, when training a neural network, we define a loss function that we would like to minimize:

$$ Loss = f\left(y_i, \widehat{y}_i \mid \overline{b}, \overline{\overline{W}}\right) + g\left(\overline{b}, \overline{\overline{W}}\right) $$

Here, y_i and $\widehat{y}_i$ represent the predicted and expected output, $f\left(. \mid \overline{b}, \overline{\overline{W}}\right)$ is the loss function quantifying the dissimilarity between predicted and expected outputs, based on the value of $\overline{\overline{W}}$, and $g\left(\overline{b}, \overline{\overline{W}}\right)$ is a regularization term. Note that the single bar on top of b and the double bar on top of W indicate that we are dealing with arrays and matrices, respectively. The activation functions of the neural network are not "trained" but rather encoded in the neural network architecture, although there have also been attempts to use families of activation functions and/or a combination of activation functions of some sort to be identified during the learning process.

If all the functions used in the formulation above – namely, the activation function f_a, the dissimilarity function f, and the regularization function g are continuous and differentiable, the loss to be minimized is also differentiable; therefore, minimization gradient-based techniques can generally be applied. The very famous **backpropagation algorithm** is one implementation of such methods where the gradients of the loss with respect to the weights are computed very efficiently by iterating backward from the last layers up to the first, making use of the chain rule and avoiding redundant calculation. There exist several variants of the general approach, depending on how the direction and the step updates are handled across the learning. Unfortunately, you must also note that the shape of the loss function can be extremely complex, with multiple local minima as well as saddle points. Finding the absolute minima is often very hard, but finding local minima can be good enough for most cases. For more information, please refer to *Deep Learning*, a book by Goodfellow *et al.*

Computational frameworks for ANNs

Over the years, many frameworks and libraries have been developed to deal with the implementation, training, and computation of ANNs, also following the different trends in the programming languages traditionally used by researchers, like C, C++, Lua, Python, and Java. Generally, all these libraries provide a framework to define:

- The representation of the network, which specifies the relationship between inputs, outputs, and the parameter that this representation depends on
- The loss function relating input, outputs, and parameters, which we want to minimize
- The optimizer algorithm to be used to minimize the loss function

Besides this basic information needed to train ANNs, the different libraries also commonly offer (and this is where the key differences between libraries often emerge from a developer's standpoint) some predefined types of neurons, kernels, and layers. These predefined components make it easier to define new neural network architectures – even very complex ones – resulting in an extremely easy and lean coding process.

Of course, besides the high-level functionalities, the different frameworks and libraries may very well achieve significantly different performances, in terms of CPU/GPU usage, memory usage, and training time, as well as support (or lack thereof) for distributed training.

These capabilities can determine the choice of one framework over another, and these **key performance indicators** (**KPIs**) are tightly connected to the backbone and low-level implementation of the underlying computational engine.

All the neural network frameworks depend – in one way or another – on a symbolic representation of the graph, allowing you to build a structure that describes how inputs are combined to produce intermediate results and the final outputs. This is often referred to as a **computational graph** or a **stateful dataflow graph,** where nodes are the variables (scalar, arrays, and tensors) – eventually transformed using activation functions – and edges represent the combinations of variables (edge source) to produce the output (edge target) of a single neuron. Each edge is also described by a weight that is multiplied to the input. You can therefore understand how such a graph can generically represent a composition of equations of the form shown in the *Training neural networks* section above.

The loss function can equally be represented symbolically, and the computational graph representation is extremely useful for computing the differentiation of the loss function (needed by the backpropagation algorithm) with respect to the various parameters (i.e., the weights and biases) in a modular way.

However, two classes of neural networks emerged based on how the differentiation is done: **symbol-to-number** and **symbol-to-symbol** differentiation.

Some neural network frameworks implement differentiation by taking the computational graph and a set of inputs and then returning a set of numerical values representing the gradients of the loss function with respect to their parameters. This approach is commonly referred to as **symbol-to-number** differentiation, and it is used by PyTorch and Caffe. On the other hand, other frameworks take the input computation graph and the loss function, and they generate additional nodes and edges that provide a symbolic description of the derivatives, therefore effectively creating a computational graph for the differentiation computations. This is referred to as **symbol-to-symbol** differentiation, and – compared to the **symbol-to-number** approach – it provides the advantage of easily extending to higher derivatives very naturally since the same logic can be recursively applied to the resulting graph of the differentiation. This is the approach that is taken by *Theano* and *TensorFlow*.

But this advantage comes with a cost. In fact, symbol-to-symbol frameworks are characterized by a static graph, where the computational graph (and its derivatives) is usually defined upfront, and – in this immutable structure – the data is then fed into it, flowing through the operations, providing either predictions or derivatives. The idea of the data *flowing* through the graph is also the reason behind the *Flow* suffix in *TensorFlow*. On the other hand, frameworks like Torch build the computational graph at execution time (when data is fed to it), therefore easily allowing you to change its topology and deal with dynamic graphs.

In the next subsections, we provide a more detailed description of two popular neural network frameworks that will be used in this book – TensorFlow and PyTorch – giving you tools that will hopefully allow you to tackle different use cases. As we will see in the *Frameworks for deep learning on graphs* section, this is also true when dealing with GNNs: some algorithms and frameworks will be better suited for your tasks, depending on their requirements.

TensorFlow

Released as open source by Google in 2017, TensorFlow is now the standard, de facto framework that allows symbolic computations and differential programming, especially in production and industrial use cases. By abstracting its computation, TensorFlow is a tool that can run on multiple backends: on machines powered by CPUs, GPUs, or even ad hoc, specifically-designed processing units such as the **tensor processing units** (**TPUs**) designed by Google. TensorFlow-powered applications can also be deployed on different devices, ranging from single and distributed servers to mobile devices.

As previously mentioned, TensorFlow provides a framework to perform symbol-to-symbol differentiation, which allows you to symbolically define a computational graph to be differentiated with respect to any of its variables, resulting in a new computational graph that can also be differentiated to produce higher-order derivatives. Historically, TensorFlow only supported static graphs. However, with the release of version 2, TensorFlow now also provides support for dynamic graphs, with the so-called **eager execution** in which the graph is built at execution time.

TensorFlow is organized around the concept of **tensors**. These abstractions represent any kind of data, be it inputs, outputs, intermediate results, and/or variables to be optimized during training. Although one could implement the different operations between tensors (e.g., matrix multiplication between inputs and weights as well as the application of activation functions), since its last major release, 2.x, the standard way of building a model with TensorFlow is by using the Keras API.

Keras was natively a side external project with respect to TensorFlow, aimed at providing a common and simple API to use several differential programming frameworks, such as TensorFlow, Theano, and CNTK, for implementing a neural network model. It abstracts the low-level implementation of the computation graph and provides you with the most common layers used when building neural networks (although custom layers can also be easily implemented), such as the following:

- Convolutional layers
- Recurrent layers
- Regularization layers
- Loss functions

Keras also exposes APIs that are very similar to scikit-learn, the most popular library for machine learning in the Python ecosystem, making it very easy for data scientists to build, train, and integrate neural network-based models in their applications.

Moreover, TensorFlow comes with a set of modules and tools that makes it a very mature technology, oriented for industrial and production use cases. One such tool worth mentioning is TensorBoard, which is a **graphical user interface** (GUI) that allows you to monitor model training and debug/understand models. It is extremely helpful to data scientists and seamlessly integrated into the framework.

A simple classification example

To showcase how easy it is to train a neural network in TensorFlow and Keras, we will apply this framework to a simple case of image classification using the Fashion-MNIST dataset. This example, based on a well-known task, will serve as preparation for later understanding graph-based applications.

The Fashion-MNIST dataset is similar to the famous MNIST dataset, a collection of hand-written numbers on a black-and-white image. In fact, Fashion-MNIST has 10 categories and consists of 60k and 10k (training dataset and test dataset) 28x28 pixel grayscale images that represent a piece of clothing (*T-shirt, Trouser, Pullover, Dress, Coat, Sandal, Shirt, Sneaker, Bag*, and *Ankle boot*). However, the Fashion-MNIST dataset image classification is a harder task than the original MNIST dataset and it has historically been used for benchmarking algorithms, with recent algorithms achieving top performances on this dataset.

The dataset is already integrated into the Keras library and can be easily imported using the following code:

```
from tensorflow.keras.datasets import fashion_mnist
(x_train, y_train), (x_test, y_test) = fashion_mnist.load_data()
```

It is usually good practice to rescale the inputs with an order of magnitude of around 1 (for which activation functions are most efficient) and make sure that the numerical data is in single precision (32 bits) instead of double precision (64 bits). This is because it is often desirable to promote speed rather than precision when training a neural network, which is a computationally expensive process. In certain cases, the precision could even be lowered to half-precision (16 bits).

We transform the input with the following:

```
x_train = x_train.astype('float32') / 255.
x_test = x_test.astype('float32') / 255.
```

We can grasp the type of inputs we are dealing with by plotting some of the samples from the training set using the following code:

```
n = 6
plt.figure(figsize=(20, 4))
for i in range(n):
    ax = plt.subplot(1, n, i + 1)
    plt.imshow(x_train[i])
    plt.title(classes[y_train[i]])
```

```
    plt.gray()
    ax.get_xaxis().set_visible(False)
    ax.get_yaxis().set_visible(False)
plt.show()
```

In the preceding code, `classes` represents the mapping between integers and class names, for example, *T-shirt, Trouser, Pullover, Dress, Coat, Sandal, Shirt, Sneaker, Bag,* and *Ankle boot*:

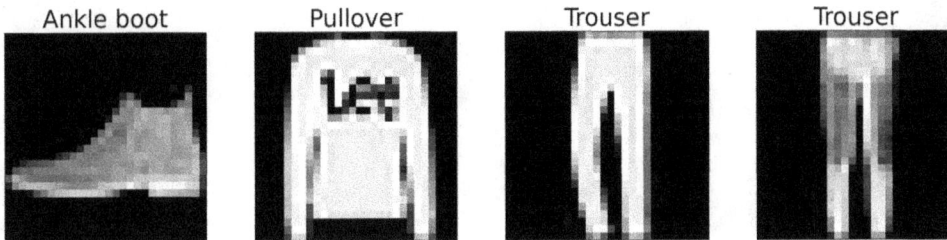

Figure 3.2 – Some samples taken from the training set of the Fashion-MNIST dataset

> Note that images in this dataset have 28x28 pixels, which is why they appear low-resolution and blurry in the preceding figure to accurately reflect the original raw data.

Let's now build the first model. Using the Keras API, we create a simple model that unrolls the images into a flat structure (1D vector of 784 elements) and adds one hidden layer, fully connected, of 128 units:

```
model = tf.keras.models.Sequential([
  tf.keras.layers.Flatten(input_shape=(28, 28)),
  tf.keras.layers.Dense(128, activation='relu'),
  tf.keras.layers.Dropout(0.2),
  tf.keras.layers.Dense(10)
])
```

Note that the last layer represents a fully connected layer that maps out the number of classes of our task (e.g., 10), representing the probabilities for each of the 10 classes. To prevent overfitting, before the output layer, we use a `Dropout` layer, which randomly removes (with a probability of 0.2 – i.e., 20%) the activation of a neuron during training. The idea is that the neural network should be resilient and able to make good predictions, even when some of its single units are removed.

Once the model is defined, it can be helpful to inspect it with the following:

```
model.summary()
```

This should provide the following output:

```
Model: "sequential"

_____
 Layer (type)                Output Shape              Param #
=================================================================
 flatten (Flatten)           (None, 784)               0

 dense (Dense)               (None, 128)               100480

 dropout (Dropout)           (None, 128)               0

 dense_1 (Dense)             (None, 10)                1290


=================================================================
Total params: 101,770
Trainable params: 101,770
Non-trainable params: 0
```

Even this very simple model has more than 100k parameters.

Once the topology/architecture of our neural network is defined, we need to "compile" the model, which effectively means building the **computational graph** associated with its training. However, to fully specify the computational plan of the training, we also need to define the loss function (with respect to which we need to compute the gradients of the weights) as well as the optimization algorithm (that defines how the gradients are to be used):

```
loss_fn = tf.keras.losses.SparseCategoricalCrossentropy(from_logits=True)
optimizer = tf.keras.optimizers.Adam()
```

We can now compile the model with the following code:

```
model.compile(optimizer=optimizer,
              loss=loss_fn,
              metrics=['accuracy'])
```

As you can see, when compiling the model, we can also provide extra metrics that will be monitored during training and evaluation. In fact, the loss function we would like to minimize may be – in some cases, such as this one – somewhat obscure, based on the difference between probability distributions. This allows efficient training, but it is not readily meaningful to us. It is indeed a lot more natural to refer to the "accuracy," which represents how many correct predictions (taking the class with the highest probability) the model is able to make.

We are now fully ready to train the neural network, which can be easily done with:

```
model.fit(
    x_train,
    y_train,
    validation_data=(x_test, y_test),
    epochs=20,
    batch_size=128,
    shuffle=True
)
```

Although a very simple example, the model achieves interesting performance, with accuracy on the validation set of around 90%.

In the notebook, we also provide a more complex network, using convolutional and max-pooling layers, which will be explained more in the following section. Feel free to try using and training more complex networks.

PyTorch

PyTorch is a library that is based on the Torch framework. Historically, Torch was initially developed at EPFL and it included a core implemented in C, wrapped by a framework in Lua. However, in 2016, Meta AI started to develop PyTorch, written in Python and offering an API more accessible to data scientists to train neural networks, still leveraging the same low-level core computational engine written in C. Since then, the framework has gained a lot of momentum. It is now part of the Linux Foundation and also implements APIs in other languages, such as C++.

Similar to TensorFlow, PyTorch revolves around the use and transformation of tensors. However, in PyTorch, tensors not only represent the inputs but also store the computational graph that is required to compute them. Notably, one of the key differences of PyTorch with respect to TensorFlow is that there is no static graph. In fact, in PyTorch, the computational graph is recreated from scratch at every iteration.

During the forward pass, PyTorch *both* performs the required computations *and* builds up the graph representing the operations that allow the gradient computation for any intermediate result/ parameter. Every tensor has an attribute (called grad_fn) that represents the entry point to this computational graph, and that is evaluated in the backward pass to compute the actual gradients.

Because of such a flexible design, PyTorch is naturally designed to deal with dynamic neural networks whose topology may change from one iteration to another. Moreover, such a design also generally feels more natural for beginners and new users since there is no strict separation between building the computational graph and executing/feeding data to it, as in TensorFlow v1. Because of its flexible design and ease of use, PyTorch has historically been very popular among researchers and people building prototypes.

Also, PyTorch is fairly lightweight, with several independent modules bundling datasets, transformations, and specific models for a given high-level topic, e.g., torchvision for image-processing tasks, torchtext for NLP tasks, torchaudio for audio and signal processing, torchrec for recommendation system use cases, and torchmetrics for metrics computation.

A simple classification example

We will now translate the previous example of classification with the Fashion-MNIST dataset to highlight the similarities and the key differences between the PyTorch API and syntax (owing to the different designs discussed above). As you will see shortly, since this is an image classification task, we will heavily rely on features of the torchvision library.

Similar to TensorFlow and Keras, PyTorch provides a simple API to load benchmark datasets. However, with PyTorch, we can also set a simple transformation (e.g., rescaling) during the input reading phase. To do so, we need to first define the transformation:

```
from torchvision import transforms
transformer=transforms.Compose([transforms.ToTensor()])
```

This allows us to convert the input images into a tensor with rescaled values. We can then use this transformation when reading the data (note that PyTorch is looking for the dataset in the ./ data folder and it will download the dataset if this folder does not exist):

```
train_dataset = datasets.FashionMNIST(
    './data', train=True, download=True, transform=transformer
)
test_dataset = datasets.FashionMNIST('./data', train=False,
transform=transformer)
```

Here, the dataset already comes with meta-information about the classes of the samples, which can be accessed through the `train_dataset.classes` and `train_dataset.class_to_idx` attributes.

Now that we have loaded the data, we can build the neural network architecture. In PyTorch, this is achieved by extending the `torch.nn.Module` class and implementing the forward method, with similar layers to the ones used previously in the TensorFlow example:

```python
import torch.nn as nn
import torch.nn.functional as F
class Model(nn.Module):
    def __init__(self):
        super().__init__()
        self.flatten = nn.Flatten()
        self.fc1 = nn.Linear(28*28, 128)
        self.dropout = nn.Dropout(0.2)
        self.fc2 = nn.Linear(128,10)
    def forward(self, x):
        x = self.flatten(x)
        x = F.relu(self.fc1(x))
        x = self.dropout(x)
        x = self.fc2(x)
        return F.log_softmax(x, dim=1)

# Instantiate the model
model=Model()
```

Once the model is defined, we need to specify the loss function as well as the optimization algorithm to fully specify our training process. Similar to TensorFlow, PyTorch provides some natively implemented functionalities for specifying these settings:

```python
import torch.optim as optim
criterion = nn.CrossEntropyLoss()
optimizer = optim.Adam(model.parameters())
```

In addition to using the cross-entropy loss, we will also gather some more insightful performance indicators during training and evaluation using the `torchmetrics` module:

```python
from torchmetrics.classification import MulticlassAccuracy
n_classes = len(train_dataset.classes)
accuracy = MulticlassAccuracy(num_classes = n_classes)
```

We now need to implement the training process. As pointed out earlier, PyTorch builds the computational graph at each iteration, and consequently, we will need to implement the gradient updates for each batch separately. PyTorch provides convenient utilities to split the dataset into batches, returning an iterable class:

```
trainloader = torch.utils.data.DataLoader(
    train_dataset, batch_size=128, shuffle=True
)
testloader = torch.utils.data.DataLoader(
    test_dataset, batch_size=test_dataset.data.shape[0]
)
```

Note that for tests, we have just created a single batch that includes the full test dataset. With this, we can decompose the training into its batch steps:

```
for epoch in range(n_epochs):  # Loop over the dataset multiple times
    for i, data in enumerate(trainloader, 0):
        # implement a backpropagation step
        ...


    # Evaluate accuracy at the end of epoch
    for inputs, labels in testloader:
        preds = model(inputs)
        print(f"Accuracy on validation set: {float(accuracy(preds,
labels))}")
```

PyTorch provides primitives that make it very easy and natural to implement the backpropagation step, explicitly defining the forward pass and the backward one:

```
inputs, labels = data
# Initialize parameters gradients
optimizer.zero_grad()
# Forward pass
outputs = model(inputs)
loss = criterion(outputs, labels)

# Backward pass
loss.backward() # Computes the gradient
optimizer.step() # Updates the parameters
```

That's it! The training of your first neural network in PyTorch is complete. You can find the full example in the notebook attached to this book. Feel free to run it and play around with some of the parameters (model architecture, optimizer specs, etc.).

Classification beyond fully connected layers

In addition to using fully connected layers, we can easily extend our model to incorporate convolutional layers, which are more effective for image data as they utilize the 2D structure of the input. CNNs apply filters across the input image to detect patterns such as edges, textures, or objects. Here's an example of how to build a simple CNN using Keras:

```
model = tf.keras.models.Sequential([
    tf.keras.layers.Conv2D(32, (3, 3), activation='relu', input_shape=(28,
28, 1)),
    tf.keras.layers.MaxPooling2D((2, 2)),
    tf.keras.layers.Conv2D(64, (3, 3), activation='relu'),
    tf.keras.layers.MaxPooling2D((2, 2)),
    tf.keras.layers.Conv2D(64, (3, 3), activation='relu'),
    tf.keras.layers.Flatten(),
    tf.keras.layers.Dense(64, activation='relu'),
    tf.keras.layers.Dense(10)
])
```

In this model, we first apply several convolutional layers followed by max pooling to reduce spatial dimensions, extract features, and pass them to a Dense layer for classification. You can compile and train this model in the same way as before, and it should yield improved performance due to the hierarchical feature extraction from the image.

In PyTorch, building a CNN is just as straightforward as using fully connected layers. CNNs are ideal for image classification tasks because they apply filters to detect patterns in the input images, such as edges or textures, hierarchically. Below is an example of a simple CNN architecture using `torch.nn.Conv2d` and `torch.nn.MaxPool2d` layers:

```
import torch.nn as nn
import torch.nn.functional as F
class CNNModel(nn.Module):
    def __init__(self):
        super(CNNModel, self).__init__()
        # 1 input channel (grayscale), 32 output filters
        self.conv1 = nn.Conv2d(1, 32, kernel_size=3)
```

```
        self.pool = nn.MaxPool2d(2, 2)  # Pooling with a 2x2 window
        self.conv2 = nn.Conv2d(32, 64, kernel_size=3)
        self.fc1 = nn.Linear(64 * 5 * 5, 64)
        # 64 filters output, 5x5 feature map

        self.fc2 = nn.Linear(64, 10)  # 10 output classes
    def forward(self, x):
        x = self.pool(F.relu(self.conv1(x)))
        x = self.pool(F.relu(self.conv2(x)))
        x = x.view(-1, 64 * 5 * 5)
        x = F.relu(self.fc1(x))
        x = self.fc2(x)
        return F.log_softmax(x, dim=1)
```

In this CNN model, we use two convolutional layers (Conv2d) to extract features from the images, followed by max-pooling layers to downsample the feature maps. The output is then flattened and passed through fully connected layers for classification. As before, you can define the loss function, optimizer, and training loop, but now you're using a more advanced architecture suited for image data.

You may be wondering why we are using images in a graph machine learning book. Just bear with us for now, and hopefully, by the end of this chapter, you'll realize that this example was not that off-path after all.

Introduction to GNNs

As you will see shortly, the concepts introduced in the previous section can be quite naturally extended to deal with graphs.

GNNs are deep learning methods that work on graph-structured data. This family of methods is sometimes also referred to as **geometric deep learning** and is gaining increasing popularity in a variety of applications, including social network analysis and computer graphics. The underlying idea of GNNs is a natural extension of **CNNs**, which we just used in the previous example to process images and have achieved impressive results when dealing with regular Euclidean spaces, such as text (one-dimensional), images (two-dimensional), and videos (three-dimensional). As shown in the previous example, CNN layers combine inputs from their neighborhood, applying a static kernel that is swept throughout the entire space – that is, every single pixel making up the image. Moreover, a classic CNN consists of a sequence of layers and each layer extracts multi-scale localized spatial features. Those features are exploited by deeper layers to construct more complex and highly expressive representations.

We can extend the same concept for graphs that are defined over a non-Euclidean space, for which neighboring information is not provided by the Euclidean distance between the points but provided by the connection information embedded in the graph, as shown in *Figure 3.3*. Indeed, the original formulation of GNNs (proposed by Scarselli et al. back in 2009) relies, similarly to CNNs, on the fact that each node can be described by its features and its neighborhood. Information coming from the neighborhood (which represents the concept of locality in the graph domain) can be aggregated and used to compute more complex and high-level features.

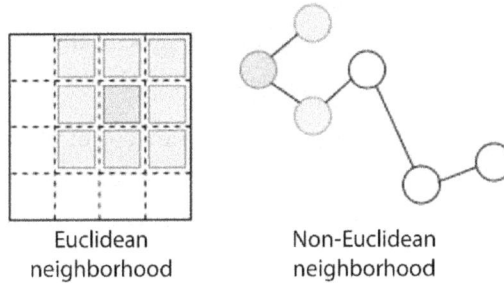

Euclidean
neighborhood

Non-Euclidean
neighborhood

Figure 3.3 – Visual difference between Euclidean and non-Euclidean neighborhoods

Each node v_i is therefore associated with a set of features or hidden state h_i^t, where t represents the t-th layer and i is the node's index. We will be ignoring node attributes for simplicity. At each layer, nodes accumulate input from their neighbors using a simple neural network layer:

$$h_i^t = \sum_{v_j \in N(v_i)} \sigma\left(w_j h_j^{t-1} + b\right)$$

Here, $W \in R^{d \times n}$ and $b \in R^d$ are trainable parameters (where n is the number of nodes and d the dimension of the features/hidden state). σ is a non-linear function. It's important to note that the summation only extends to the neighboring nodes. These equations can be applied recursively, where the previous state, computed at the previous layer, is used to calculate the hidden state of the next layer (similarly to what is generally done in recurrent neural networks).

As a result of this process, we can therefore imagine that we obtain a final hidden state or embedded representation, Z, which is a function of the input set of features X and the topology of the network, represented by the adjacency matrix A:

$$Z = \text{GNN}(X, A)$$

Here, Z can be seen as a compressed latent representation of the nodes. Note that Z is actually a new set of features for each node, which can be downstream to other (GNN) layers to create deeper representations. These features can be used to achieve a number of tasks, as we will see in the rest of the book, including classification and forecasting, among others.

Variants of GNNs

Several attempts have been made in recent years to address the problem of learning from graph data. Variants of the previously described GNN have been proposed, with the aim of improving its representation learning capability. Some of them are specifically designed to process specific types of graphs (direct, indirect, weighted, unweighted, static, dynamic, and so on).

Graph convolutional network (GCN)-based encoders are one of the most diffused variants of GNN for unsupervised learning. GCNs are GNN models inspired by many of the basic ideas behind CNNs. Filter parameters are typically shared over all locations in the graph and several layers are concatenated to form a deep network.

There are essentially two types of convolutional operations for graph data – namely, spectral approaches and non-spectral (spatial) approaches. The first, as the name suggests, defines convolution in the spectral domain (that is, decomposing graphs in a combination of simpler elements). Spatial convolution formulates the convolution as aggregating feature information from neighbors, as we previously described.

Also, several modifications have been proposed for the propagation step (convolution, gate mechanisms, attention mechanisms, and skip connections, among others), with the aim of improving representation at different levels. Also, different training methods have been proposed to improve learning.

Although these concepts could be implemented using the framework we presented earlier in the chapter (namely, *TensorFlow* and *PyTorch*) there also exist libraries built on top of them that make integrating GNNs (in all their flavors) extremely easy, seamless, and fast. In the next section, we will present the most relevant ones.

Frameworks for deep learning on graphs

Over the years, a few libraries have been developed to help data scientists integrate graph machine learning, and more specifically, GNNs, into their analytical pipelines. The existence of different libraries is due to different choices of the framework used in the backend (e.g., TensorFlow, PyTorch, etc.), the research group authoring the library (sometimes being an industrial group and sometimes an academic department), as well as some specific needs addressed by the library (e.g., large graphs, dynamic graphs, etc.).

In the following, we will consider three frameworks:

- **PyTorch Geometric (PyG)**:

    As the name suggests, PyG (`https://pytorch-geometric.readthedocs.io/en/latest/`) is a library that is built upon PyTorch, and it provides abstractions and functionalities for defining, training, and evaluating GNNs. More generally, it provides various methods for performing deep learning on graphs and irregular structures. It also constitutes a central repository for existing implementations of published papers and academic research.

- **StellarGraph**:

    StellarGraph was possibly one of the first libraries to appear in the landscape of graph machine learning and GNNs and was designed, developed, and supported by the Data61 team at CSIRO (`https://data61.csiro.au/`). StellarGraph was built upon TensorFlow 2.x and therefore uses Keras and scikit-learn APIs to provide a modular, extensible, and user-friendly user experience that can be easily integrated with common data analytics pipelines. Besides its analytical capabilities, StellarGraph also features a set of classes and utilities to represent and work with graph data, seamlessly parsing and importing **NetworkX** objects.

- **Deep Graph Library (DGL)**:

    DGL can work on top of multiple existing frameworks, currently supporting PyTorch, MXNet, and TensorFlow (although at the time of writing, the community is progressing more and more toward PyTorch). Moreover, it is powered by a group of committers spanning multiple organizations and universities, including AWS, NVIDIA, New York University, and Georgia Institute of Technology. More importantly, DGL also provides very structured and extensible functionalities to implement message passing and reduction steps, which, as you will see in the next chapters, are some fundamental building blocks for implementing more complex and scalable GNN layers and models.

Despite their unique features and design philosophies, these frameworks share a similar structure that facilitates a comparative analysis. In fact, all the frameworks provide primitives for the following:

- Graph representation
- Data loading
- Model definition
- Training loop

In the following subsections, we will provide an overview of the main similarities and differences that the various framework provides for implementing the operations above.

Graph representation

PyG, StellarGraph, and DGL have specific classes to model pairwise relations (edges) between objects (nodes). PyG represents graphs using PyTorch tensors and provides a versatile data structure for storing various types of graph-related information such as adjacency matrices, edge lists, and node features. StellarGraph uses optimized data structures to offer support for different graph formats including NetworkX graphs and pandas DataFrames. Moreover, it uses specialized structures such as the `StellarGraph` class and `StellarDiGraph` class for handling different types of graphs efficiently. DGL has its own graph object that can be created from diverse input formats such as adjacency matrices and edge lists. Its unified interface handles both static and dynamic graphs, and this helps ensure compatibility with deep learning models across different scenarios.

Nevertheless, despite the specific implementations of the classes, all three libraries offer methods and attributes to explore and access the graph structure (number of nodes, number of edges, node features, edge features, etc.).

Data loading

The way the graph structure is loaded into memory depends on several factors, including the specific training algorithm. For example, some models should be trained using samples of nodes, others need subgraphs as input or even the full graph. Moreover, especially when dealing with large-scale graphs that may not fit entirely into memory, designing proper data-loading strategies is crucial. Graph deep learning frameworks offer various strategies to tackle this challenge, enabling the processing of graphs of varying sizes while optimizing resource utilization.

For well-known graph models, the specific graph-splitting implementation (to properly generate training, validation, and test sets) may be already implemented for you. For example, for the well-known GraphSAGE, PyG implements the NeighborLoader and LinkNeighborLoader to choose from, thus making sure the input feeding the models is constructed correctly.

To summarize, the main data loading methods are:

- **Loading whole graphs:** The simplest approach is to load the entire graph into memory. This method is suitable for smaller graphs that can be accommodated within available memory constraints. All three frameworks support this method, providing utilities to load graph data directly into memory for processing.

- **Edge loading**: To handle larger graphs, batching the edges is a common strategy employed by these frameworks. In this approach, the graph is partitioned into smaller subgraphs based on edges. Each subgraph is then loaded into memory separately, allowing for efficient processing of large graphs by working on smaller chunks at a time.

- **Node loading**: Similar to edges, batching of nodes involves partitioning the graph based on nodes instead of edges. This method divides the graph into smaller subgraphs, each containing a subset of nodes and their associated edges. By loading these subgraphs individually, node splitting enables the processing of large graphs in a memory-efficient manner.

Moreover, in addition to traditional data-loading methods, these frameworks also support data augmentation techniques for generating synthetic graphs or augmenting existing graph data.

Note that the implementation of the data-loading process for each framework reflects the style of the framework itself (Keras for StellarGraph, PyTorch for PyG, and DGL). Therefore, you will find "generators" for Keras-inspired frameworks and "data loaders" for PyTorch-inspired ones. It is worth mentioning that, unlike traditional deep learning settings where splitting data into training and validation sets is straightforward, doing the same for graph data is not easy. When dealing with a graph-level task (your dataset consists of a set of graphs), splitting a single graph into training and validation sets involves sampling a set of edges or nodes in a proper way; otherwise, improper splitting can lead to data leakage. Some of these "splitting" (or sampling) strategies are already implemented in the various frameworks and are often embedded in the data-loading process. We will discuss this topic further and how to avoid data leakage in graph settings in *Chapter 6, Solving Common Graph-Based Machine Learning Problems*, when discussing the link prediction problem.

Model definition

Once you understand how to represent a graph in the specific framework and how to load it (or part of it), it is time to define the actual deep learning model. In the next chapters, you will learn about several variants of GNNs. For now, it is sufficient to know that each of these frameworks implements most of these variants and you can simply use them to solve your problem without dealing with their internal implementation (similarly to what you do when using standard CNN models and layers in Tensorflow and PyTorch). Some of these classes include graph convolutional layers, GraphSAGE, and graph attention layers, among others.

Training loop

Optimizing the model involves iterative batch processing, loss calculation, and parameter updates. These operations are supported by all three frameworks and are strictly related to the deep learning framework used as the backend (PyG and DGL use a PyTorch-like training approach, while StellarGraph exposes the "fit" method typical of Keras).

In the rest of the chapter, we will show you how easy it is to use each of these frameworks by implementing a simple "link prediction" task on top of the Cora dataset. Without going into much detail (you will be learning about several problems that can be solved with machine learning on graphs in the next chapters), the task of link prediction consists of predicting whether a link exists between two given nodes. The Cora dataset is a perfect example for this task, featuring a collaboration network of 2,700+ scientific publications, classified into 7 classes. The citation links consist of 5,400+ links and each node (representing a publication) is described by a 1,433-size vector indicating the presence/absence of a particular word from a restricted vocabulary of 1,433 words. This representation is a simplified version of the so-called bag-of-words representation. Therefore, the Cora dataset, thanks to the node features, represents a very suitable dataset on which a GNN and GCN can be implemented.

Interestingly, the link prediction task may also correspond to reconstructing the original graph from the compact representation of the nodes, Z. Indeed, we would be successful at our prediction task when the "predicted" edges correspond to the ones actually existing in the "target" graph. In view of this, the link prediction task is therefore closely connected to **graph auto-encoders (GAEs)**, which encode the nodes into a compact representation and then decode such information to reconstruct the original topology, minimizing the difference between the original and the reconstructed graph. We will digress more on auto-encoders in the next chapter, but to start, a simple way to reconstruct the adjacency matrix using the compressed representation Z is by means of the inner product:

$$S_{ij} = z_i \cdot z_j$$

Here, S_{ij} can be viewed as the probability that a link exists between node i and node j. The reconstructed matrix can be expressed as:

$$A = Z \cdot Z^T$$

In the next examples, you will see how this can easily be done with the frameworks.

PyG

PyG provides a simple API to load the Cora dataset. Similar to what we have done previously, we add some transformation in the data-loading phase to normalize the data and to split it (note that, this time, the function comes from PyG as it is specifically designed to work with graph datasets). Moreover, besides normalization, we generate a positive (existing) and negative (non-existing) list of edges, which we will use to make sure that the reconstructed graph correctly provides edges only where they exist. Because of the specific auto-encoder implementation in PyG, negative edges are not required for the training; therefore, we will only generate them for the test set:

```
import torch_geometric.transforms as T
from torch_geometric.datasets import Planetoid
transform = T.Compose([
    T.NormalizeFeatures(),
    T.RandomLinkSplit(
        num_val=0., num_test=0.1, is_undirected=True,
        split_labels=True, add_negative_train_samples=False
    ),
])
path = os.path.join(DATA_PATH, 'data')
dataset = Planetoid(path, "Cora", transform=transform)
train_data, val_data, test_data = dataset[0]
```

The datasets that have been generated (`train_data`, `val_data`, and `test_data`) bundle topology information, node features, and labels. For instance, the matrix with node features can be accessed via the following:

```
train_data.x # of shape (N_nodes, 1433)
```

The vector with label information can be accessed via:

```
train_data.y # of shape (N_nodes) with numerical values representing
             # the labels
```

The edges (topology) of the graph can be found with:

```
train_data.edge_index # of shape (2, N_edges)
```

Moreover, the `RandomLinkSplit` class also generates a list of positive and negative labels:

```
test_data.pos_edge_label_index # of shape (2, N_positive_edges)
test_data.neg_edge_label_index # of shape (2, N_negative_edges)
```

Note that since we specified add_negative_train_samples=False, negative edges are not generated for the training set.

To build the model, we will implement a GAE in PyG that will allow us to reconstruct the original graph. We first start by defining the structure for the encoder using a GNN based on two GCN layers. PyG makes this extremely easy. We first create the encoder layer:

```
from torch_geometric.nn import GCNConv
class GCNEncoder(torch.nn.Module):
    def __init__(self, num_node_features, num_embedding):
        super().__init__()
        self.conv1 = GCNConv(num_node_features, 2 * num_embedding)
        self.conv2 = GCNConv(2 * num_embedding, num_embedding)
    def forward(self, x, edge_index):
        x = self.conv1(x, edge_index).relu()
        return self.conv2(x, edge_index)
```

Note that the forward function of GCNs in PyG always takes two required arguments and one optional argument: the first required argument is the tensor of the node features, the second required argument is edge_indices (or the adjacency matrix), and the third, optional argument is the edge weights. Once the encoder is defined, the GAE can be created with one single command:

```
from torch_geometric.nn import GAE
model = GAE(GCNEncoder(1433, n_embeddings))
```

The GAE class already implements the decoder layer for us (using the inner product between the Z vector of embedding). It also provides some useful functions to compute the loss as well as evaluate the model during training, as we will see shortly.

As usual, to fully specify the training, we need to define the optimizer to be used (since we will be using the loss implemented in the GAE class):

```
optimizer = torch.optim.Adam(model.parameters(), lr=0.01)
```

Using this, we can implement the training procedure using the usual PyTorch functionalities:

```
for epoch in range(num_epochs):
    # zero the parameter gradients
    model.train()
    optimizer.zero_grad()
    z = model.encode(train_data.x, train_data.edge_index)
    loss = model.recon_loss(z, train_data.pos_edge_label_index)
    loss.backward()
    optimizer.step()
    # Test/Evaluate
    model.eval()
    z = model.encode(test_data.x, test_data.edge_index)
    auc, ap = model.test(
        z, test_data.pos_edge_label_index, test_data.neg_edge_label_index
    )
print(f"Performance on validation set => Area Under the ROC Curve: {auc}
Average Precision: {ap}")
```

During training, we can monitor prediction performances by using common metrics such as area under the ROC curve (AUC) or average precision. By maximizing one of these, we can build a robust model that may either prioritize precision or recall, or a combination of them, and that is able to best generalize on future cases.

The GAE is now implemented and trained, and its output can be used to generate some node embeddings, similar to what was previously done for images.

StellarGraph

To load the Cora dataset into a StellarGraph Graph object, you can simply run the following code:

```
from stellargraph import datasets
dataset = datasets.Cora()
G, _ = dataset.load()
```

G represents the graph structure with several useful methods readily available to the user, such as G.info() to display the main information of the dataset:

```
StellarGraph: Undirected multigraph
 Nodes: 2708, Edges: 5429

 Node types:
  paper: [2708]
    Features: float32 vector, length 1433
    Edge types: paper-cites->paper

 Edge types:
    paper-cites->paper: [5429]
        Weights: all 1 (default)
        Features: none
```

As you can see, StellarGraph supports both node and edge features (although, in this case, edges do not have any features) as well as the possibility of defining a taxonomy for the node and edge types.

Similar to what is provided in PyG, StellarGraph also provides functionality to split edges into positive and negative samples:

```
edge_splitter_test = EdgeSplitter(G)
G_test, edge_ids_test, edge_labels_test = edge_splitter_test.train_test_
split(
    p=0.1, method="global", keep_connected=True
)
edge_splitter_train = EdgeSplitter(G_test)
G_train, edge_ids_train, edge_labels_train = edge_splitter_train.train_
test_split(
    p=0.1, method="global", keep_connected=True
)
```

Note that, in this case, we also create the reduced graph where we remove the test edges (G_test) and the test and train edges (G_train). Since this is an auto-encoder, we will not use these graphs, but as you will see in the following code, in predictive tasks (e.g., node/edge classification), this will be very important to prevent data leakage from training to test sets.

StellarGraph also provides a simple API to create both the input tensors and the computational graph that will be fed into TensorFlow. In particular, `FullBatchLinkGenerator` allows us to supply both the node features as well as the graph data (i.e., adjacency matrix) to be used on different kinds of models. In the following code, since we will be using a GCN, we create the inputs as follows:

```
train_gen = FullBatchLinkGenerator(G, method="gcn")
train_flow = train_gen.flow(edge_ids_train, edge_labels_train)
```

`FullBatchLinkGenerator` specifically generates batches that include all the nodes in the graph in a single batch. This is in contrast to mini-batch training, where only a subset of the graph is used in each iteration. Full-batch training can be memory-intensive, but it's useful when you want to train a model on the whole graph at once. The flow method of the generator creates a data generator (a flow) that will feed the graph data to the model during training. This method prepares the actual data that will be passed to the GNN model, including the edges and their corresponding labels.

Creating a GCN model is extremely easy using the built-in functionalities:

```
gcn = GCN(
    layer_sizes=[16, 16], activations=["relu", "relu"], generator=train_
gen, dropout=0.3
)
```

The GCN class also allows us to extract the input and output tensors:

```
x_input, z = gcn.in_out_tensors()
```

If you inspect the x_input result, you will note that x_input is formed by four sub-inputs, corresponding to:

- The matrix of node features
- The edges we want to evaluate (whether they exist or not)
- The edge list of the graph
- The corresponding labels of the graph edges

These inputs also correspond to what is provided by the `train_flow.inputs` property, therefore making it seamless to create a full model of our auto-encoder. In fact, we only need to combine the embeddings using the inner product (as specified below by "**ip**") and reshape it to match the expected dimension:

```
prediction = LinkEmbedding(activation="relu", method="ip")(z)
prediction = keras.layers.Reshape((-1,))(prediction)
```

As with a normal TensorFlow model, we now just need to compile the model:

```
model = keras.Model(inputs=x_input, outputs=prediction)
model.compile(
    optimizer=keras.optimizers.Adam(lr=0.01),
    loss=keras.losses.binary_crossentropy,
    metrics=["binary_accuracy"]
)
```

Finally, the GNN can be simply trained using the standard API:

```
history = model.fit(train_flow, epochs=50)
```

Disclaimer

Owing to such a clean API and structured integration with TensorFlow (which is perhaps the standard de facto for production-ready applications based on neural networks), StellarGraph was the library chosen in the first edition of this book. Unfortunately, shortly after the publication of the first edition, this library stopped being actively maintained. Therefore, although the functionalities provided by the library are still extremely useful (especially for dynamic and temporal graphs), we advise you to consider this choice carefully if you need to provide long-term support for applications powered by your graph machine models. In this scenario, currently maintained frameworks such as PyG or DGL would be more appropriate, and whenever possible, examples using these frameworks will be provided alongside their StellarGraph implementations in the GitHub repository. However, it is important to note that StellarGraph still retains significant pedagogical value: its Keras-based coding style allows for concise and clear demonstrations of key concepts, making it an excellent tool for learning and experimentation. Therefore, the core ideas presented using StellarGraph can be easily translated to other frameworks, ensuring that the knowledge gained remains broadly applicable.

DGL

Similar to PyG and StellarGraph, DGL aims to provide a set of tools for an easy implementation of GNNs. To perform link prediction using DGL, first, you need to construct the graph object representing your dataset. DGL provides various methods to load and manipulate graph data efficiently. For example, to load a dataset like Cora into a DGL Graph object, you might use the following code:

```
import dgl
import dgl.data
dataset = dgl.data.CoraGraphDataset()
G = dataset[0]
```

Once you have your graph, let's split it into positive and negative examples and divide them into training and test sets, similar to other libraries like PyG. DGL provides some utilities to iterate over edges for link prediction tasks. One such way is to use a "sampler" to sample edges from the graph. With the following code, you will be creating a data loader that samples five negative examples per positive example:

```
sampler = dgl.dataloading.MultiLayerFullNeighborSampler(2)
sampler = dgl.dataloading.as_edge_prediction_sampler(sampler, negative_
sampler=dgl.dataloading.negative_sampler.Uniform(5))

train_seeds = torch.arange(G.num_nodes())
batch_size = 48
num_workers = 4

dataloader = dgl.dataloading.DataLoader(
G,
train_seeds,
sampler,
batch_size=batch_size,
shuffle=True,
drop_last=False,
num_workers=num_workers)
```

Next, you can define your model architecture. DGL supports various types of GNNs, including GCNs, **graph attention networks (GATs)**, and more (we will take a look at these types of models next, in *Chapter 4, Unsupervised Graph Learning*). For example, to create a GCN for link prediction, you can use the following code:

```python
import torch
import torch.nn as nn
import dgl.nn as dglnn
import torch.nn.functional as F


class StochasticTwoLayerGCN(nn.Module):
    def __init__(self, in_features, hidden_features, out_features):
        super().__init__()
        self.conv1 = dgl.nn.GraphConv(in_features, hidden_features)
        self.conv2 = dgl.nn.GraphConv(hidden_features, out_features)

def forward(self, blocks, x):
    x = F.relu(self.conv1(blocks[0], x))
    x = F.relu(self.conv2(blocks[1], x))
    return x


class ScorePredictor(nn.Module):
    def forward(self, edge_subgraph, x):
        with edge_subgraph.local_scope():
            edge_subgraph.ndata['x'] = x
            edge_subgraph.apply_edges(dgl.function.u_dot_v('x', 'x', 'score'))
            return edge_subgraph.edata['score']


class Model(nn.Module):
    def __init__(self, in_features, hidden_features, out_features):
        super().__init__()
        self.gcn = StochasticTwoLayerGCN(
            in_features, hidden_features, out_features)
        self.predictor = ScorePredictor()

    def forward(self, positive_graph, negative_graph, blocks, x):
        x = self.gcn(blocks, x)
        pos_score = self.predictor(positive_graph, x)
```

```
        neg_score = self.predictor(negative_graph, x)
        return pos_score, neg_score
```

After defining the model, you can train it using the standard PyTorch training loop, similar to other deep learning tasks:

```
def compute_loss(pos_score, neg_score):
    # an example hinge loss
    n = pos_score.shape[0]
    return (neg_score.view(n, -1) - pos_score.view(n, -1) +
1).clamp(min=0).mean()


in_features = G.ndata["feat"].shape[1]
model = Model(in_features, 256, 1)
opt = torch.optim.Adam(model.parameters())


for input_nodes, positive_graph, negative_graph, blocks in dataloader:
    blocks = [b for b in blocks]
    input_features = blocks[0].srcdata['feat']
    # inference and loss computation
    pos_score, neg_score = model(positive_graph, negative_graph, blocks,
input_features)
    loss = compute_loss(pos_score, neg_score)
    print(loss)
    opt.zero_grad()
    loss.backward()
    opt.step()
```

Notice that we skipped the validation part to improve readability. However, similar to the standard PyTorch training loop, you can implement it by defining a proper data loader and calling it in the training loop.

Be aware that, at the time of writing, DGL has just discontinued support on Darwin and Windows platforms, applying to versions higher than 2.2.1. In the following, we will often make use of version 2.4.0+. If you are running on one of the unsupported platforms, we recommend using the Docker images provided with the book.

Summary

In this chapter, you have seen an introduction to neural networks and how they can be used in practice using popular frameworks like TensorFlow and PyTorch. You have seen how to translate those concepts into the graph domain, and you have learned the basics of three modern frameworks for deep learning on graphs: PyG, StellarGraph, and DGL. It is worth noting that some concepts may seem unclear at this point. Don't worry! In the next chapters, all these concepts will be examined in more detail.

Get ready to embark on our journey into the GraphML landscape as we explore **Unsupervised Graph Learning** in the next chapter!

Part 2

Machine Learning on Graphs

In this part, you will be introduced to the main existing machine learning models for graph representation learning: their purpose, how they work, and how they can be implemented.

This part comprises the following chapters:

- *Chapter 4, Unsupervised Graph Learning*
- *Chapter 5, Supervised Graph Learning*
- *Chapter 6, Solving Common Graph-Based Machine Learning Problems*

4

Unsupervised Graph Learning

Unsupervised machine learning refers to the subset of machine learning algorithms that do not exploit any target information during training. Instead, they work on their own to find clusters, discover patterns, detect anomalies, and solve many other problems for which there is no teacher and no correct answer known *a priori*.

As per many other machine learning algorithms, unsupervised models have found large applications in the graph representation learning domain. Indeed, they represent an extremely useful tool for solving various downstream tasks, such as node classification and community detection, among others.

In this chapter, an overview of recent unsupervised graph embedding methods will be provided. When given a graph, the goal of these techniques is to automatically learn a latent representation of it, in which the key structural components are somehow preserved.

The following topics will be covered in this chapter:

- The unsupervised graph embedding roadmap
- Shallow embedding methods
- Autoencoders
- Graph neural networks

Technical requirements

All code files relevant to this chapter are available at https://github.com/PacktPublishing/ Graph-Machine-Learning/tree/main/Chapter04. Please refer to the *Practical exercises* section of *Chapter 1, Getting Started with Graphs,* for guidance on how to set up the environment to run the examples in this chapter, using either Poetry, `pip,` or docker.

For the more complex data visualization tasks provided in this chapter, Gephi (`https://gephi.org/`) may also be required. The installation manual is available here: `https://gephi.org/users/install/`.

The unsupervised graph embedding roadmap

Unsupervised machine learning involves algorithms that can be trained without the need for manually annotated data, making them especially valuable for identifying hidden structures and relationships in complex graph networks. Most of these models rely only on information in the adjacency matrix and the node features, without any knowledge of the downstream machine learning task.

How can this be done? One of the most common approaches is to learn embeddings that preserve the graph structure. The learned representation is usually optimized so that it can be used to reconstruct the pair-wise node similarity, for example, the **adjacency matrix**. These techniques bring an important feature: the learned representation can encode latent relationships among nodes or graphs, allowing us to discover hidden and complex novel patterns.

Many algorithms have been developed in relation to unsupervised graph machine learning techniques. However, as previously reported by different scientific papers (`https://arxiv.org/abs/2005.03675`), those algorithms can be grouped into three macro-groups: shallow embedding methods, autoencoders, and **Graph Neural Networks (GNNs)**, as described in the following chart:

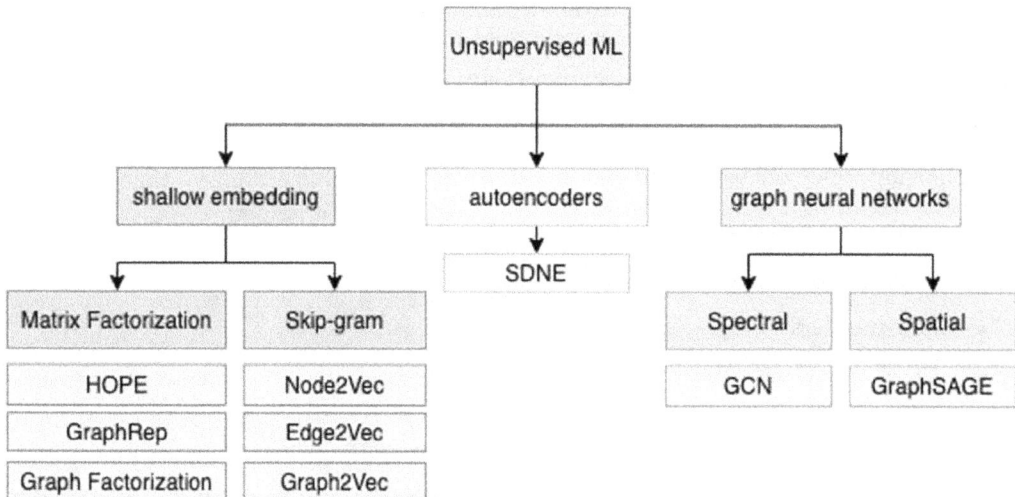

Figure 4.1: The hierarchical structure of the different unsupervised embedding algorithms
described in this book

In the following sections, you will learn about the main principles behind each group of algorithms. We will try to provide the idea behind the most well-known algorithms in the field as well as how they can be used to solve real problems.

Shallow embedding methods

As already introduced in *Chapter 2, Graph Machine Learning*, with shallow embedding methods, we identify a set of algorithms that are able to learn and return only the embedding values for the learned input data.

In this section, we will explore two main categories of these methods: matrix factorization-based approaches and skip-gram-based approaches. Matrix factorization methods decompose the adjacency matrix to capture latent patterns in the graph, while skip-gram methods, inspired by natural language processing, learn embeddings by predicting the likelihood of node co-occurrences. We will dive into these techniques in detail and provide Python examples for each, using libraries such as **Graph Embedding Methods (GEM)**, **Node to Vector (Node2Vec)**, and `karateclub`.

Matrix factorization

Matrix factorization is a general decomposition technique widely used in different domains. The technique has become very popular, especially in the context of recommendation engines (e.g., see Zhou et al., 2008), where the matrix to be decomposed is the so-called user-product matrix, having users, u_i, along the rows and products, p_j, along the columns, and the values in each cell, $s_{i,j}$, representing a score. Depending on the context, the score may have different forms: it may be a binary score representing whether the user bought a particular product, or it may also be a positive number representing how many times a user listened to a particular song. Interestingly, as we have seen in *Chapter 1, Getting Started with Graphs*, the user-product matrix could also be seen as the adjacency matrix of bipartite graphs, composed of two kinds of nodes: the users and the products/songs, and links (weighted) between them depending on whether the users bought a given product or listened to a given song. By re-mapping users and products into a Cartesian space (embeddings), matrix factorization techniques will provide a measure to quantify how similar users and/or products are, therefore providing the basis for a recommendation.

More generally, matrix factorization can be applied to graph structures to compute the various embeddings of a graph, and indeed a significant number of widely used graph embedding algorithms rely on this technique.

We will start by providing a general introduction to the matrix factorization problem. After the introduction of the basic principles, we will describe two algorithms, namely **graph factorization (GF)** and **High-Order Proximity Preserved Embedding (HOPE)**, which use matrix factorization to build the node embedding of a graph.

Let $W \in \mathbb{R}^{m \times n}$ be the input data. Matrix factorization decomposes $W \approx V \times H$, with $V \in \mathbb{R}^{m \times d}$ and $H \in \mathbb{R}^{d \times n}$ called the **source** matrix and the **abundance** matrix, respectively, and d is the number of dimensions of the generated embedding space. The matrix factorization algorithm learns the V and H matrices by minimizing a loss function that can change according to the specific problem we want to solve. In its general formulation, the loss function is defined by computing the reconstruction error using the Frobenius norm as $\|W - V \times H\|_F^2$.

Generally speaking, all the unsupervised embedding algorithms based on matrix factorization use the same principle. They all factorize an input graph expressed as a matrix in different components. The main difference between each method lies in the loss function used during the optimization process and the constraints/formulations posed for the V and H matrices. Indeed, different loss functions allow the creation of an embedding space that emphasizes specific properties of the input graph.

In the following sections, we will compare the different embedding algorithms using the same simple barbell network shown in *Figure 4.2*, to show the main similarities and differences. The graph can be generated with the following code snippet:

```
import networkx as nx
G = nx.barbell_graph(m1=10, m2=4)
```

When applied to this graph, all algorithms will output pairs of coordinates that will be plotted in an xy scatter plot. To help understand the results, we will color the points based on the color scheme depicted in the following figure:

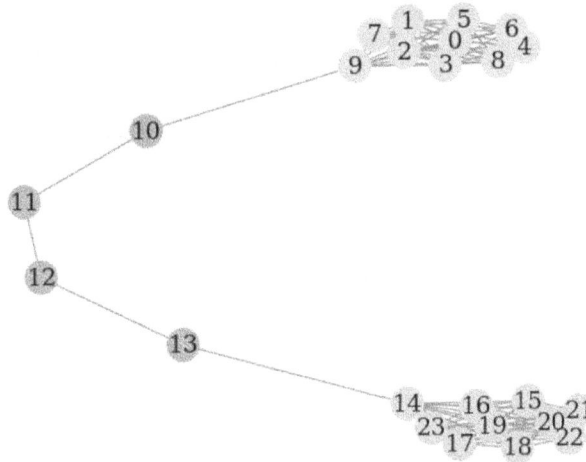

Figure 4.2: Color scheme used when comparing the different algorithms

Graph factorization

The GF algorithm was one of the first models to reach good computational performance in order to perform the node embedding of a given graph. By following the principle of matrix factorization that we previously described, the GF algorithm factorizes the adjacency matrix of a given graph.

Formally, let $G = (V, E)$ be the graph for which we want to compute the node embedding and let $A \in \mathbb{R}^{|V| \times |V|}$ be its adjacency matrix. The loss function (L) used in this matrix factorization problem is as follows:

$$L = \frac{1}{2} \sum_{(i,j) \in E} (A_{i,j} - Y_{i,:} Y_{j,:}^T)^2 + \frac{\lambda}{2} \sum_i \| Y_{i,:} \|^2$$

In the preceding equation, $(i, j) \in E$ represents one of the edges in G while $Y \in \mathbb{R}^{|V| \times d}$ is the matrix containing the d-dimensional embedding. Each row of the matrix represents the embedding of a given node. Moreover, a regularization term (λ) of the embedding matrix is used to ensure that the problem remains well posed even in the absence of sufficient data.

The loss function used in this method was mainly designed to improve GF performance and scalability. Indeed, the solution generated by this method could be noisy. Moreover, it should be noted, by looking at its matrix factorization formulation, that GF performs a strong symmetric factorization. This property is particularly suitable for undirected graphs, where the adjacency matrix is symmetric, but could be a potential limitation for directed graphs.

In the following code, we will show how to perform the node embedding for the barbell graph in *Figure 4.2* using Python and the GEM library:

```
from gem.embedding.gf import GraphFactorization
gf = GraphFactorization(d=2, data_set=None, max_iter=10000, eta=1*10**-4,
regu=1.0)
gf.learn_embedding(G)
embeddings = gf.get_embedding()
```

In the preceding example, the following have been done:

1. The GraphFactorization class is used to generate a **d=2**-dimensional embedding space.

2. The computation of the node embeddings of the input graph is performed using gf.learn_embedding(G).

3. The computed embeddings are extracted by calling the gf.get_embedding() method.

The results of the previous code are shown in the following graph:

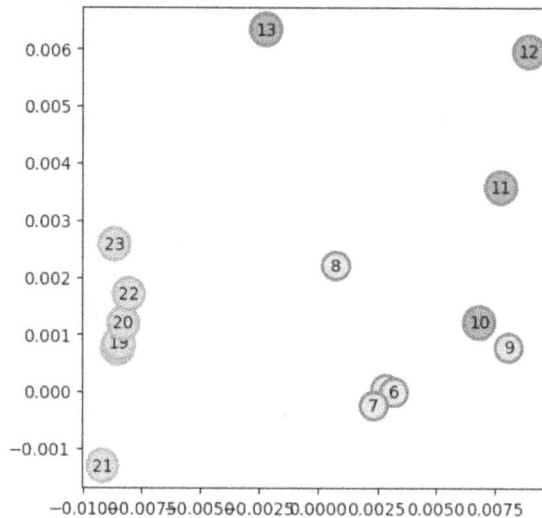

Figure 4.3: Application of the GF algorithm to generate the embedding vectors for the nodes
of the barbell graph shown in Figure 4.2. The color scheme depicted in Figure 4.2 has been
used. Axes represent the first two latent directions of the embedding algorithm.

From *Figure 4.3*, it is possible to see how nodes belonging to groups 1 and 3 are mapped together in the same region of space. Those points are separated by the nodes belonging to group 2. This mapping allows us to separate groups 1 and 3 from group 2 quite well. Unfortunately, there is no clear separation between groups 1 and 3.

Higher-order proximity preserved embedding

HOPE is another graph embedding technique based on the matrix factorization principle. This method allows for the preservation of higher-order proximity and does not force its embeddings to have any symmetric properties. Before starting to describe the method, let's understand what first-order proximity and high-order proximity mean:

- **First-order proximity**: Given a graph, $G = (V,E)$, where the edges have a weight, w_{ij}, for each vertex pair (v_i,v_j), we say they have a first-order proximity equal to w_{ij} if the edge $(v_i, v_j) \in E$. Otherwise, the first-order proximity between the two nodes is 0.
- **Second- and high-order proximity**: With the second-order proximity, we can capture the two-step relations between each pair of vertices. For each vertex pair (v_i,v_j), we can see the second-order proximity as a two-step transition from v_i to v_j. High-order proximity generalizes this concept and allows us to capture a more global structure. As a consequence, high-order proximity can be viewed as a k-step ($k \geq 3$) transition from v_i to v_j.

Given the definition of proximity, we can now describe the HOPE method. Formally, let $G = (V, E)$ be the graph we want to compute the embedding for. The loss function (L) used by this problem is as follows:

$$L = \left\| S - Y_S \times Y_t^T \right\|_F^2$$

In the preceding equation, $S \in \mathbb{R}^{|V| \times |V|}$ is a similarity matrix generated from graph G and its adjacency matrix $A \in \mathbb{R}^{|V| \times |V|}$, and $Y_S \in \mathbb{R}^{|V| \times d}$ and $Y_t \in \mathbb{R}^{|V| \times d}$ are two embedding matrices representing a d-dimensional embedding space. In more detail, Y_s represents the source embedding and Y_t represents the target embedding.

HOPE uses those two matrices in order to capture asymmetric proximity in directed networks where the direction from a source node and a target node is present. The final embedding matrix, Y, is obtained by simply concatenating, column-wise, the Y_s and Y_t matrices. Due to this operation, the final embedding space generated by HOPE will have 2 x d dimensions.

As we already stated, the S matrix is a similarity matrix obtained from the original graph, G. The goal of S is to obtain high-order proximity information. Formally, it is computed as $S = M_g . M_l$, where M_g and M_l are both polynomials of matrices.

In its original formulation, the authors of HOPE suggested different ways of computing M_g and M_l. Here, we report a common and easy method of computing those matrices, **Adamic-Adar (AA)**. In this formulation, $M_g = I$ (the identity matrix) while $M_l = ADA$, where D is a diagonal matrix (inverted) computed as $D_{ij} = \frac{1}{(\sum_j(A_{ij}+A_{ji}))}$. Other formulations to compute M_g and M_l are the **Katz index**, **Rooted PageRank (RPR)**, and **Common Neighbors (CN)**.

In the following code, we will show you how to perform the node embedding for the barbell graph in *Figure 4.2* using Python and the GEM library:

```
from gem.embedding.hope import HOPE
gf = HOPE(d=4, beta=0.01)
gf.learn_embedding(G)
embeddings = gf.get_embedding()
```

The preceding code is similar to the one used for GF. The only difference is in the class initialization since here we use **HOPE**. According to the implementation provided by GEM, the d parameter, representing the dimension of the embedding space, will define the number of columns of the final embedding matrix, Y, obtained after the column-wise concatenation of Y_s and Y_t.

As a consequence, the number of columns of Y_s and Y_t is defined by the floor division (the // operator in Python) of the value assigned to d. The results of the code are shown in the following graph:

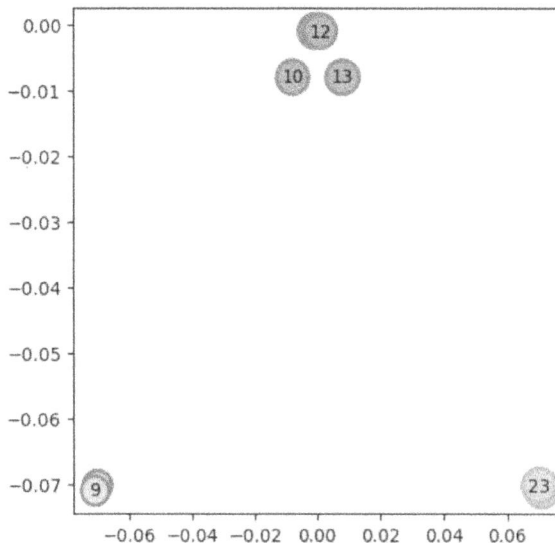

Figure 4.4: Application of the HOPE algorithm to generate the embedding vector for the nodes of the barbell graph shown in Figure 4.2. The color scheme depicted in Figure 4.2 has been used. Axes represent the first two latent directions of the embedding algorithm.

In this case, the graph is undirected and thus there is no difference between the source and target nodes. *Figure 4.4* shows the first two dimensions of the **embeddings** matrix representing Y_s. It is possible to see how the embedding space generated by HOPE provides, in this case, a better separation of the different nodes.

Graph representation with global structure information

Graph representation with global structure information (GraphRep), such as HOPE, allows us to preserve higher-order proximity without forcing its embeddings to have symmetric properties. Formally, let $G = (V, E)$ be the graph for which we want to compute the node embeddings and $A \in \mathbb{R}^{|V| \times |V|}$ be its adjacency matrix. The loss function (L) used by this problem is as follows:

$$L_k = \left\| X^k - Y_s{}^k \times Y_t{}^{k^T} \right\|_F^2 \quad 1 \leq k \leq K$$

In the preceding equation, $X^k \in \mathbb{R}^{|V| \times |V|}$ is a matrix generated from graph G in order to get the kth order of proximity between nodes.

$Y_s{}^k \in \mathbb{R}^{|V| \times d}$ and $Y_t{}^k \in \mathbb{R}^{|V| \times d}$ are two embedding matrices representing a d-dimensional embedding space of the kth order of proximity for the source and target nodes, respectively.

The X^k matrix is computed according to the following equation: $X^k = \prod_k (D^{-1}A)$. Here, D is a diagonal matrix known as the **degree matrix** computed using the following equation:

$$D_{ij} = \begin{cases} \sum_p A_{ip}, & i = j \\ 0, & i \neq j \end{cases}$$

$X^1 = D^{-1}A$ represents the (one-step) probability transition matrix, where X_{ij}^1 is the probability of a transition from v_i to vertex v_j within one step. In general, for a generic value of k, X_{ij}^k represents the probability of a transition from v_i to vertex v_j within k steps.

For each order of proximity, k, an independent optimization problem is fitted. All the k embedding matrices generated are then column-wise concatenated to get the final source embedding matrices.

In the following code, we will show how to perform the node embedding for the barbell graph in *Figure 4.2* using Python and the `karateclub` library:

```
from karateclub.node_embedding.neighbourhood.grarep import GraRep
gr = GraRep(dimensions=2, order=3)
gr.fit(G)
embeddings = gr.get_embedding()
```

We initialize the GraRep class from the karateclub library. In this implementation, the dimensions parameter represents the dimension of the embedding space, while the order parameter defines the maximum number of orders of proximity between nodes. The number of columns of the final embedding matrix (stored, in the example, in the embeddings variable) is dimension*order, since, as we said, for each proximity order an embedding is computed and concatenated in the final embedding matrix.

Since two dimensions are computed in the example, embeddings[:,:2] represents the embedding obtained for $k=1$, embeddings[:,2:4] for $k=2$, and embeddings[:,4:] for $k=3$. The results of the code are shown in the following graphs:

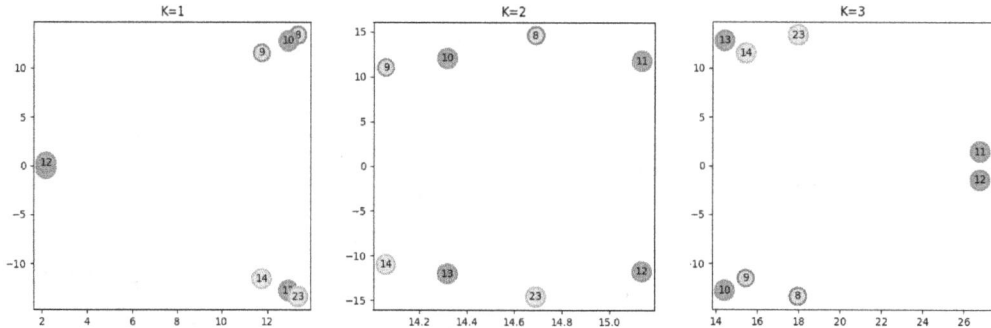

Figure 4.5: Application of the GraphRep algorithm to generate the embedding vectors for different values of k for the nodes of the barbell graph shown in Figure 4.2. The color scheme depicted in Figure 4.2 has been used. Axes represent the first two latent directions of the embedding algorithm.

From the preceding graphs, it is easy to see how different orders of proximity allow us to get different embeddings. Since the input graph is quite simple, in this case, already with $k=1$, a well-separated embedding space is obtained. The nodes belonging to groups 1 and 3 in all the proximity orders have the same embedding values (they are overlapping in the scatter plot).

Notice that the superior performance of $k=1$ can be attributed to the differences in how proximity is captured. While GF uses a regularization term to control the complexity of the model, this may oversimplify the representation in some cases, leading to less distinct embeddings. In contrast, the $k=1$ embedding focuses on local relationships, which might be sufficient to capture the necessary structure in this simple graph.

In this section, we have described some matrix factorization methods for unsupervised graph embedding. In the next section, we will introduce a different way of performing unsupervised graph embedding using skip-gram models.

Skip-gram

In this section, we will provide a quick description of the skip-gram model. Since it is widely used by different embedding algorithms, a high-level description is needed to better understand the different methods. Before going deep into a detailed description, let's look at a brief overview.

The skip-gram model is a simple neural network with one hidden layer trained in order to predict the probability of a given word being present when one or more input words are present. The neural network is trained by building the training data using a text corpus as a reference. This process is described in the following chart:

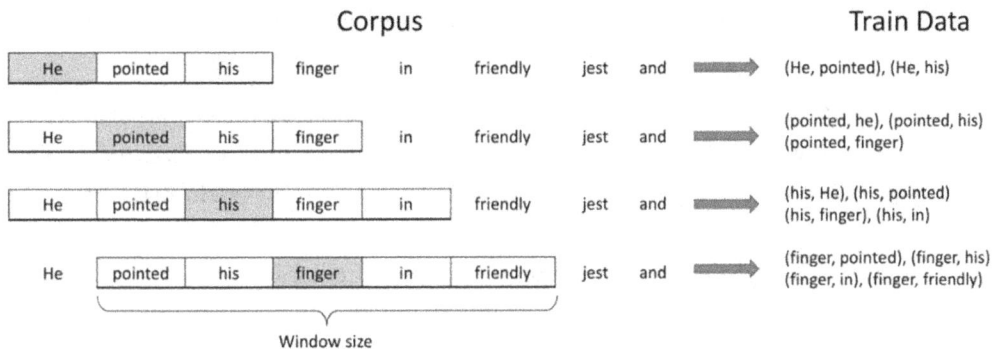

Figure 4.6: Example of the generation of training data from a given corpus. In the colored boxes are the target words. In the white boxes are the context words identified by a window size of length 2

The example described in *Figure 4.6* shows how the algorithm to generate the training data works. A *target* word is selected and a rolling window of fixed size *w* is built around that word. The words inside the rolling windows are known as *context* words. Multiple pairs of (*target word, context word*) are then built according to the words inside the rolling window.

Once the training data is generated from the whole corpus, the skip-gram model is trained to predict the probability of a word being a context word for the given target. During its training, the neural network learns a compact representation of the input words. This is why the skip-gram model is also known as **Word to Vector (Word2Vec)**.

The structure of the neural network representing the skip-gram model is described in the following chart:

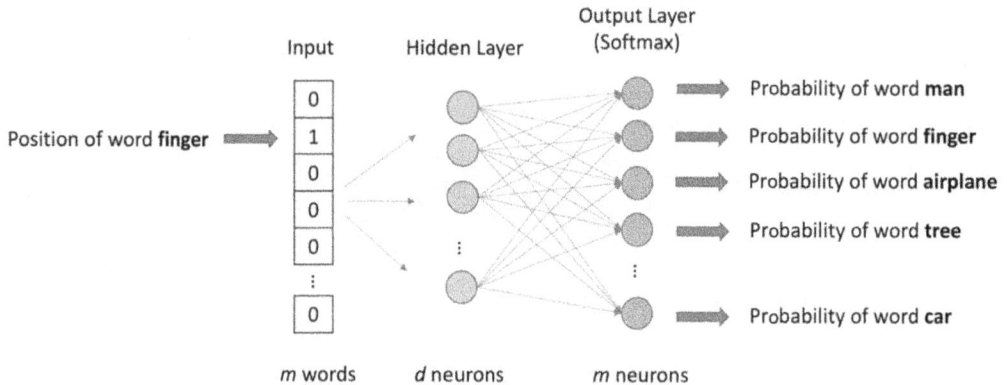

Figure 4.7: Structure of the neural network of the skip-gram model. The number of d neurons in the hidden layer represents the final size of the embedding space

The input of the neural network is a binary vector of size m. Each element of the vector represents a word in the dictionary of the language we want to embed the words in. When, during the training process, a (*target word, context word*) pair is given, the input array will have 0 in all its entries with the exception of the entry representing the target word, which will be equal to 1. The hidden layer has d neurons. The hidden layer will learn the embedding representation of each word, creating a d-dimensional embedding space.

Finally, the output layer of the neural network is a dense layer of m neurons (the same size as the input vector) with a *softmax* activation function. Each neuron represents a word in the dictionary. The value assigned by the neuron corresponds to the probability of that word being "related" to the input word. Since softmax can be hard to compute when the size of m increases, a *hierarchical softmax* approach is always used.

The final goal of the skip-gram model is not to actually learn the task we previously described but to build a compact d-dimensional representation of the input words. Thanks to this representation, it is possible to easily extract an embedding space for the words using the weight of the hidden layer. Another common approach to creating a skip-gram model, which will not be described here, is *context-based*: **Continuous Bag-of-Words (CBOW)**.

Since the basic concepts behind the skip-gram model have been introduced, we can start to describe a series of unsupervised graph embedding algorithms built upon this model. Generally speaking, all the unsupervised embedding algorithms based on the skip-gram model use the same principle.

Starting from an input graph, they extract from it a set of walks. Those walks can be seen as a text corpus where each node represents a word. Two words (representing nodes) are near each other in the text if they are connected by an edge in a walk. The main difference between each method lies in the way those walks are computed. Indeed, as we will see, different walk generation algorithms can emphasize particular local or global structures of the graph.

DeepWalk

The DeepWalk algorithm generates the node embedding of a given graph using the skip-gram model. In order to provide a better explanation of this model, we need to introduce the concept of **random walks.**

Let G be a graph and v_i be a vertex selected as the starting point. We select a neighbor of v_i at random and move toward it. From this point, we randomly select another point to move toward. This process is repeated t times. The random sequence of t vertices selected in this way is a random walk of length t. It is worth mentioning that the algorithm used to generate the random walks does not impose any constraint on how they are built. As a consequence, there is no guarantee that the local neighborhood of the node is well preserved.

Using the notion of random walks, the DeepWalk algorithm generates a random walk of a size of at most t for each node. Those random walks will be given as input to the skip-gram model. The embedding generated using skip-gram will be used as the final node embedding. In the following figure (*Figure 4.8*), we can see a step-by-step graphical representation of the algorithm:

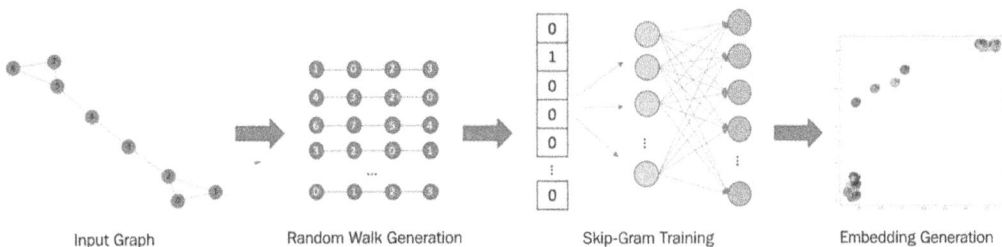

Figure 4.8: All the steps used by the DeepWalk algorithm to generate the node embedding of a given graph

Here is a step-by-step explanation of the algorithm graphically described in the preceding chart:

1. **Random walk generation**: For each node of input graph G, a set of γ random walks with a fixed maximum length (t) is computed. It should be noted that the length t is an upper bound. There are no constraints forcing all the paths to have the same length, especially in direct graphs where the random walks may end up on dead ends, i.e., nodes with no out-going edges.

2. **Skip-gram training**: Using all the random walks generated in the previous step, a skip-gram model is trained. As we described earlier, the skip-gram model works on words and sentences. When a graph is given as input to the skip-gram model, as visible in *Figure 4.8*, a graph can be seen as an input text corpus, while a single node of the graph can be seen as a word of the corpus.

3. A random walk can be seen as a sequence of words (a sentence). The skip-gram is then trained using the "fake" sentences generated by the nodes in the random walk. The parameters for the skip-gram model previously described (window size, w, and embed size, d) are used in this step.

4. **Embedding generation**: The information contained in the hidden layers of the trained skip-gram model is used in order to extract the embedding of each node.

In the following code, we will show how to perform the node embedding for the barbell graph in *Figure 4.2* using Python and the `karateclub` library:

```python
from karateclub.node_embedding.neighbourhood.deepwalk import DeepWalk
dw = DeepWalk(dimensions=2)
dw.fit(G)
embeddings = dw.get_embedding()
```

The code is quite simple. We initialize the `DeepWalk` class from the `karateclub` library. In this implementation, the `dimensions` parameter represents the dimension of the embedding space. Other parameters worth mentioning that the `DeepWalk` class accepts are as follows:

*   `walk_number`: The number of random walks to generate for each node
*   `walk_length`: The length of the generated random walks
*   `window_size`: The window size parameter of the skip-gram model

Finally, the model is fitted on graph G using `dw.fit(G)` and the embeddings are extracted using `dw.get_embedding()`.

The results of the code are shown in the following figure:

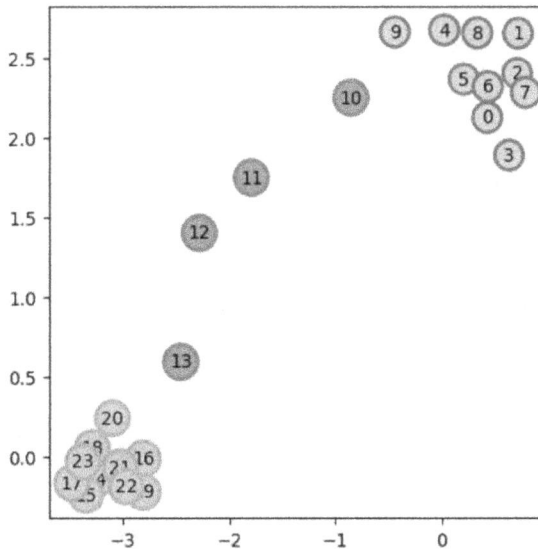

Figure 4.9: Application of the DeepWalk algorithm to generate the embedding vectors for the nodes of the barbell graph shown in Figure 4.2. The color scheme depicted in Figure 4.2 has been used. Axes represent the first two latent directions of the embedding algorithm.

From the previous graph, we can see how DeepWalk is able to separate region 1 from region 3. Those two groups are contaminated by the nodes belonging to region 2. Indeed, for those nodes, a clear distinction is not visible in the embedding space.

Node2Vec

The Node2Vec algorithm can be seen as an extension of DeepWalk. Indeed, as with DeepWalk, Node2Vec also generates a set of random walks used as input to a skip-gram model. Once trained, the hidden layers of the skip-gram model are used to generate the embedding of the node in the graph. The main difference between the two algorithms lies in the way the random walks are generated.

Indeed, if DeepWalk generates random walks without using any bias, in Node2Vec a new technique to generate biased random walks on the graph is introduced. The algorithm to generate the random walks combines graph exploration by merging **Breadth-First Search (BFS)** and **Depth-First Search (DFS)**. The way those two algorithms are combined in the random walk's generation is regularized by two parameters, p and q. p defines the probability of a random walk getting back to the previous node, while q defines the probability that a random walk can pass through a previously unseen part of the graph.

Due to this combination, Node2Vec can preserve high-order proximities by preserving local structures in the graph as well as global community structures. This new method of random walk generation allows solving the limitation of DeepWalk preserving the local neighborhood properties of the node.

In the following code, we will show how to perform the node embedding for the barbell graph in *Figure 4.2* using Python and the node2vec library:

```
from node2vec import Node2Vec
draw_graph(G)
node2vec = Node2Vec(G, dimensions=2)
model = node2vec.fit(window=10)
embeddings = model.wv
```

For Node2Vec, the code is straightforward. We initialize the Node2Vec class from the node2vec library. In this implementation, the dimensions parameter represents the dimension of the embedding space. The model is then fitted using node2vec.fit(window=10). Finally, the embeddings are obtained using model.wv.

It should be noted that model.wv is an object of the Word2VecKeyedVectors class. In order to get the embedding vector of a specific node with nodeid as the ID, we can use the trained model, as follows: model.wv[str(nodeId)]. Other parameters worth mentioning that the Node2Vec class accepts are as follows:

- num_walks: The number of random walks to generate for each node
- walk_length: The length of the generated random walks
- p, q: The *p* and *q* parameters of the random walk's generation algorithm

The results of the code are shown in *Figure 4.10*:

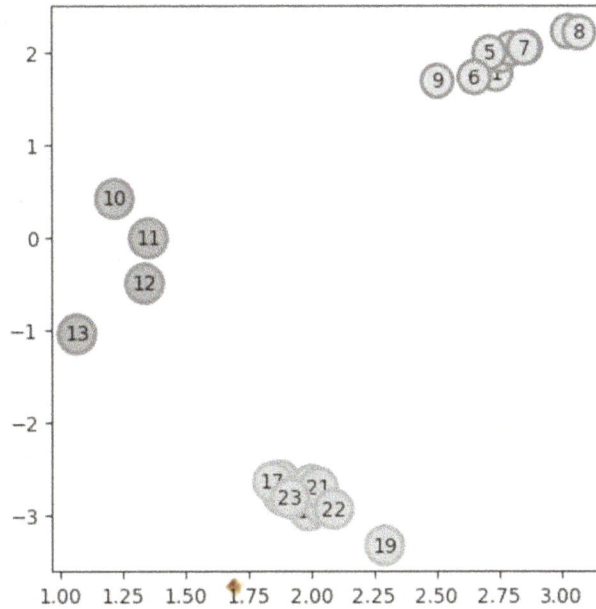

Figure 4.10: Application of the Node2Vec algorithm to generate the embedding vectors for the nodes of the barbell graph shown in Figure 4.2. The color scheme depicted in Figure 4.2 has been used. Axes represent the first two latent directions of the embedding algorithm.

As is visible from *Figure 4.10*, Node2Vec allows us to obtain a better separation between nodes in the embedding space compared to DeepWalk. Regions 1 and 3 are well clustered in two regions of space. Region 2 instead is well placed in the middle of the two groups without any overlap.

Edge2Vec

Contrary to the other embedding function, the **Edge to Vector (Edge2Vec)** algorithm generates the embedding space on edges, instead of nodes. This algorithm is a simple side effect of the embedding generated by using Node2Vec. The main idea is to use the node embedding of two adjacent nodes to perform some basic mathematical operations in order to extract the embedding of the edge connecting them.

Formally, let v_i and v_j be two adjacent nodes and let $f(v_i)$ and $f(v_j)$ be their embeddings computed with Node2Vec. The operators described in *Table 4.1* can be used in order to compute the embedding of their edge:

Operator	Equation	Class Name
Average	$\dfrac{f(v_i) + f(v_j)}{2}$	`AverageEmbedder`
Hadamard	$f(v_i) * f(v_j)$	`HadamardEmbedder`
Weighted-L1	$\lvert f(v_i) - f(v_j) \rvert$	`WeightedL1Embedder`
Weighted-L2	$\lvert f(v_i) - f(v_j) \rvert^2$	`WeightedL2Embedder`

Table 4.1: Edge embedding operators with their equation and class name in the Node2Vec library

In the following code, we will show how to perform the edge embedding using the Node2Vec embedding of the barbell graph shown in *Figure 4.2* using Python:

```
from node2vec.edges import HadamardEmbedder
embedding = HadamardEmbedder(keyed_vectors=model.wv)
```

The code is quite simple. The `HadamardEmbedder` class is instantiated with only the `keyed_vectors` parameter. The value of this parameter is the embedding model generated by Node2Vec. In order to use other techniques to generate the edge embedding, we just need to change the class and select one from those listed in *Table 4.1*. An example of the application of this algorithm is shown in the following figure, where each label represents the corresponding source and target IDs, and the fill and stroke colors are inherited from the source and target nodes, respectively.

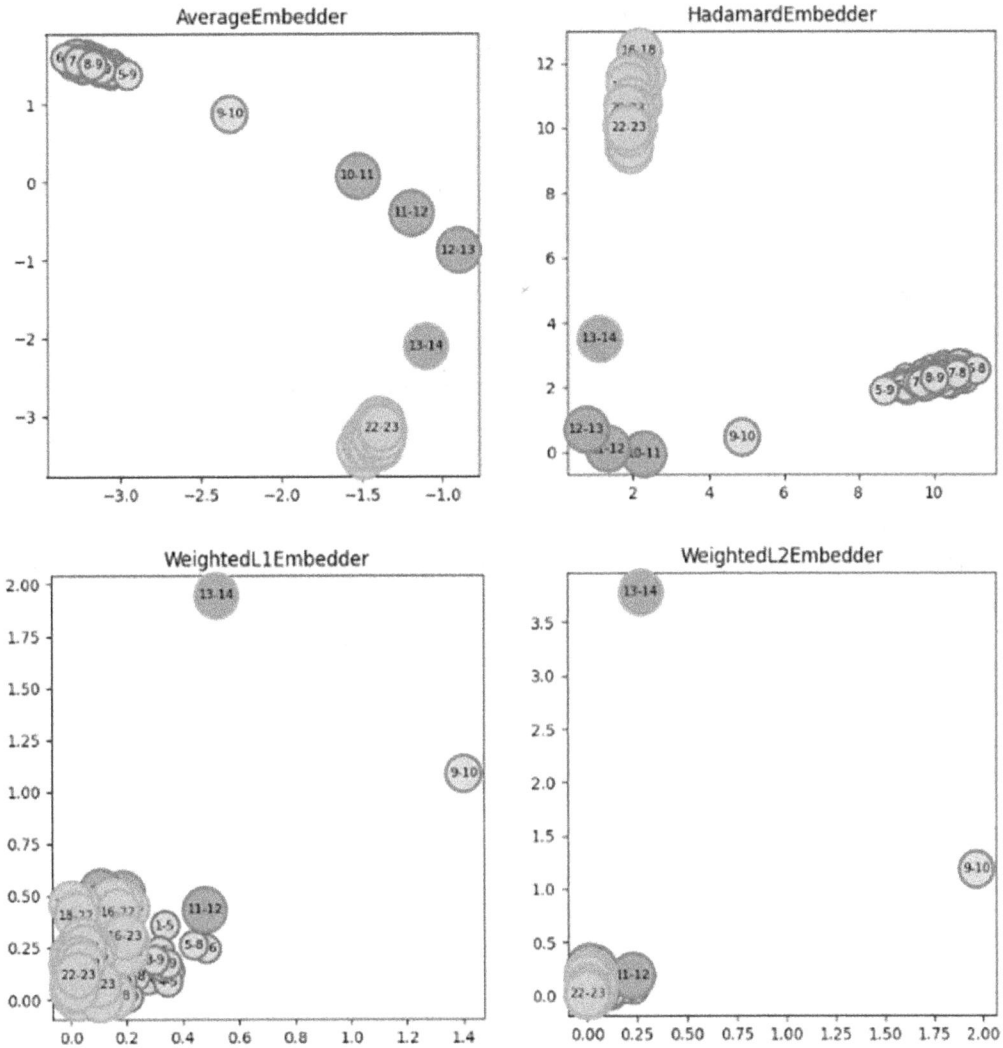

Figure 4.11: Application of the Edge2Vec algorithm to the Node2Vec embeddings of the barbell graph shown in Figure 4.2. Each node label presents the corresponding source and target IDs in the source-target format, and the color coding refers to the one shown in Figure 4.2, where the fill and stroke colors are based on the source and target, respectively. Axes represent the first two latent directions of the embedding algorithm.

From *Figure 4.11*, we can see how different embedding methods generate completely different embedding spaces. AverageEmbedder and HadamardEmbedder, in this example, generate well-separated embeddings for regions 1, 2, and 3.

For WeightedL1Embedder and WeightedL2Embedder, however, the embedding space is less well separated: the embedding is only able to differentiate the terminal edges attached to the cliques, whereas all the other nodes fall approximately on the same region.

Graph2Vec

The methods we previously described generated the embedding space for each node or edge on a given graph. **Graph to Vector (Graph2Vec)** generalizes this concept and generates embeddings for the whole graph.

Given a set of graphs, the Graph2Vec algorithms generate an embedding space where each point represents a graph. This algorithm generates its embedding using an evolution of the Word2Vec skip-gram model known as **Document to Vector (Doc2Vec)**. We can graphically see a simplification of this model in *Figure 4.12*:

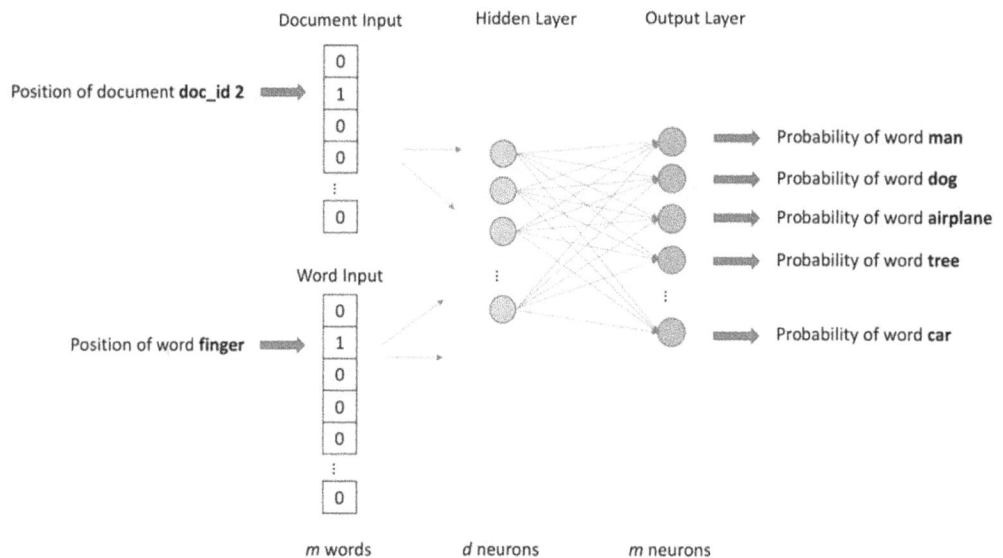

Figure 4.12: Simplified graphical representation of the Doc2Vec skip-gram model. The number of d neurons in the hidden layer represents the final size of the embedding space

Compared to the simple Word2Vec, Doc2Vec also accepts another binary array representing the document containing the input word. Given a "target" document and a "target" word, the model then tries to predict the most probable "context" word with respect to the input "target" word and document.

With the introduction of the Doc2Vec model, we can now describe the Graph2Vec algorithm. The main idea behind this method is to view an entire graph as a document and each of its subgraphs, generated as an ego graph (see *Chapter 1, Getting Started with Graphs*) of each node, as words that comprise the document.

In other words, a graph is composed of subgraphs as a document is composed of sentences. According to this description, the algorithm can be summarized into the following steps:

1. **Subgraph generation**: A set of rooted subgraphs is generated around every node.

2. **Doc2Vec training**: The Doc2Vec skip-gram is trained using the subgraphs generated by the previous step.

3. **Embedding generation**: The information contained in the hidden layers of the trained Doc2Vec model is used in order to extract the embedding of each node and graph.

In the following code, as we already did in *Chapter 2, Graph Machine Learning*, we will show how to perform the node embedding of a set of networkx graphs using Python and the karateclub library:

```
import matplotlib.pyplot as plt
from karateclub import Graph2Vec
n_graphs = 20
def generate_random():
    n = random.randint(6, 20)
    k = random.randint(5, n)
    p = random.uniform(0, 1)
    return nx.watts_strogatz_graph(n,k,p)
Gs = [generate_random() for x in range(n_graphs)]
model = Graph2Vec(dimensions=2)
model.fit(Gs)
embeddings = model.get_embedding()
```

In this example, the following have been done:

1. 20 Watts-Strogatz graphs have been generated with random parameters.

2. We then initialize the Graph2Vec class from the karateclub library with two dimensions. In this implementation, the dimensions parameter represents the dimension of the embedding space.

3. The model is then fitted on the input data using `model.fit(Gs)`.

4. The vector containing the embeddings is extracted using `model.get_embedding()`.

The results of the code are shown in the following figure:

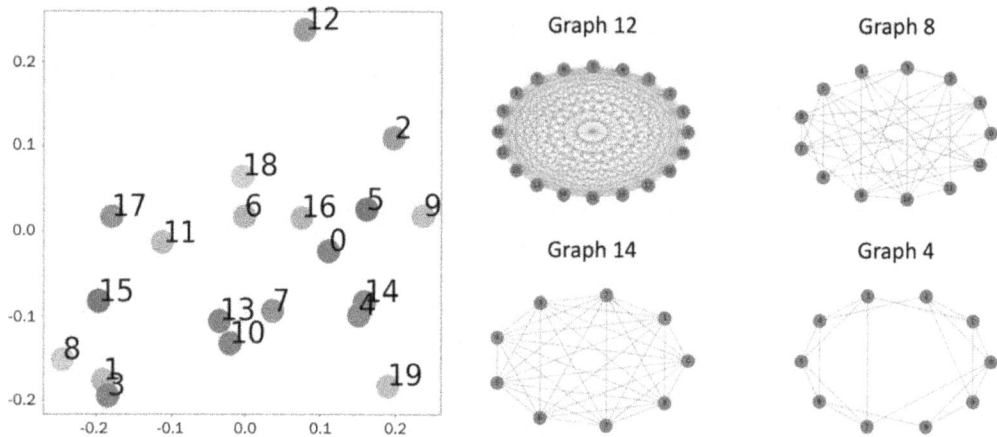

Figure 4.13: Application of the Graph2Vec algorithm to a graph (left) to generate the embedding vector of its nodes (right) using different methods

From *Figure 4.13*, it is possible to see the embedding space generated for the different graphs.

In this section, we described different shallow embedding methods based on matrix factorization and the skip-gram model. However, in the scientific literature, a lot of unsupervised embedding algorithms exist, such as Laplacian methods. We refer those of you who are interested in exploring those methods to look at the paper *Machine Learning on Graphs: A Model and Comprehensive Taxonomy*, available at https://arxiv.org/pdf/2005.03675.pdf.

We will continue our description of the unsupervised graph embedding method in the next sections. We will describe more complex graph embedding algorithms based on autoencoders.

Autoencoders

Autoencoders are an extremely powerful tool that can effectively help data scientists to deal with high-dimensional datasets. Although first presented around 30 years ago, in recent years, autoencoders have become more and more widespread in conjunction with the general rise of neural network-based algorithms. Besides allowing us to compact sparse representations, they can also be at the base of generative models, representing the first inception of the famous **Generative Adversarial Network (GAN)**, which is, using the words of Geoffrey Hinton:

"The most interesting idea in the last 10 years in machine learning"

Indeed, Generative AI has become very popular in recent years, thanks to the astonishing results provided by generative models, especially in image generation (such as DALL-E) and text generation (such as ChatGPT).

An autoencoder is a neural network where the inputs and outputs are basically the same, but that is characterized by a small number of units in the hidden layer. Loosely described, it is a neural network that is trained to reconstruct its inputs using a significantly lower number of variables and/or degree of freedom.

Since an autoencoder does not need a labeled dataset, it can be seen as an example of unsupervised learning and a dimensionality-reduction technique. However, different from other techniques such as **Principal Component Analysis (PCA)** and matrix factorization, which we discussed in the previous section, autoencoders can learn non-linear transformation thanks to the non-linear activation functions of their neurons:

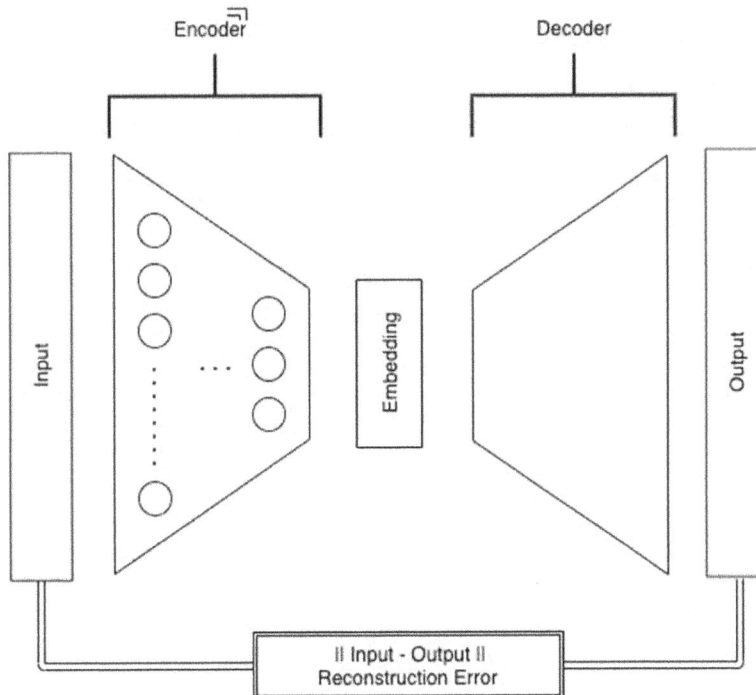

Figure 4.14: Diagram of the autoencoder structure

Figure 4.14 shows a simple example of an autoencoder. You can see how the autoencoder can generally be seen as composed of two parts:

- An encoder network that processes the input through one or more units and maps it into an encoded representation that reduces the dimension of the inputs (under-complete autoencoders) and/or constrains its sparsity (over-complete regularized autoencoders)
- A decoder network that reconstructs the input signal from the encoded representation of the middle layer

The encoder-decoder structure is then trained to minimize the ability of the full network to reconstruct the input. In order to completely specify an autoencoder, we need a loss function. The error between the inputs and the outputs can be computed using different metrics, and indeed the choice of the correct form for the "reconstruction" error is a critical point when building an autoencoder. Some common choices for the loss functions that measure the reconstruction error are **mean square error, mean absolute error, cross-entropy**, and **KL divergence**.

In the following sections, we will show you how to build an autoencoder, starting with some basic concepts, and then apply those concepts to graph structures.

Our first autoencoder

Let's start by implementing an autoencoder in its simplest form, that is, a simple feed-forward network trained to reconstruct its input (notice in autoencoders the image label is not used; we only aim to compress and decompress the input). We'll apply this to the Fashion-MNIST dataset already presented in *Chapter 3, Neural Networks and Graphs*.

First, let's load the dataset. We will be using the Keras library for our example:

```
from tensorflow.keras.datasets import fashion_mnist
(x_train, y_train), (x_test, y_test) = fashion_mnist.load_data()
```

And let's rescale the inputs:

```
x_train = x_train.astype('float32') / 255.
x_test = x_test.astype('float32') / 255.
```

Now that we have imported the inputs, we can build our autoencoder network by creating the encoder and the decoder. We will be doing this using the Keras functional API, which provides more generality and flexibility compared to the so-called Sequential API. We start by defining the encoder network:

```
from tensorflow.keras.layers import Conv2D, Dropout, MaxPooling2D,
UpSampling2D, Input
input_img = Input(shape=(28, 28, 1))
x = Conv2D(16, (3, 3), activation='relu', padding='same')(input_img)
x = MaxPooling2D((2, 2), padding='same')(x)
x = Conv2D(8, (3, 3), activation='relu', padding='same')(x)
x = MaxPooling2D((2, 2), padding='same')(x)
x = Conv2D(8, (3, 3), activation='relu', padding='same')(x)
encoded = MaxPooling2D((2, 2), padding='same')(x)
```

Our network is composed of a stack of three levels of the same pattern composed of the same two-layer building block:

- **Conv2D**, a two-dimensional convolutional kernel that is applied to the input and effectively corresponds to having weights shared across all the input neurons. After applying the convolutional kernel, the output is transformed using the ReLU activation function. This structure is replicated for n hidden planes, with n being 16 in the first stacked layer and 8 in the second and third stacked layers.

- **MaxPooling2D**, which down-samples the inputs by taking the maximum value over the specified window (2x2 in this case).

Using the Keras API, we can also have an overview of how the layers transformed the inputs using the Model class, which converts the tensors into a user-friendly model ready to be used and explored:

```
Model(input_img, encoded).summary()
```

This provides a summary of the encoder network:

```
Model: "model_1"

Layer (type)                  Output Shape              Param #
=================================================================
input_1 (InputLayer)          [(None, 28, 28, 1)]       0

conv2d (Conv2D)               (None, 28, 28, 16)        160

max_pooling2d (MaxPooling2D)  (None, 14, 14, 16)        0

conv2d_1 (Conv2D)             (None, 14, 14, 8)         1160
```

```
max_pooling2d_1 (MaxPooling2 (None, 7, 7, 8)            0

conv2d_2 (Conv2D)            (None, 7, 7, 8)            584

max_pooling2d_2 (MaxPooling2 (None, 4, 4, 8)            0
=================================================================
Total params: 1,904
Trainable params: 1,904
Non-trainable params: 0
```

As can be seen, at the end of the encoding phase, we have a (4, 4, 8) tensor, which is more than six times smaller than our original initial inputs (28x28). We can now build the decoder network. Note that the encoder and decoder do not need to have the same structure and/or shared weights:

```
x = Conv2D(8, (3, 3), activation='relu', padding='same')(encoded)
x = UpSampling2D((2, 2))(x)
x = Conv2D(8, (3, 3), activation='relu', padding='same')(x)
x = UpSampling2D((2, 2))(x)
x = Conv2D(16, (3, 3), activation='relu')(x)
x = UpSampling2D((2, 2))(x)
decoded = Conv2D(1, (3, 3), activation='sigmoid', padding='same')(x)
```

In this case, the decoder network resembles the encoder structure where the down-sampling of the input achieved using the **MaxPooling2D** layer has been replaced by the **UpSampling2D** layer, which basically repeats the input over a specified window (2x2 in this case, effectively doubling the tensor in each direction).

We have now fully defined the network structure with the encoder and decoder layers. In order to completely specify our autoencoder, we also need to specify a loss function. Moreover, to build the computational graph, Keras also needs to know which algorithms should be used in order to optimize the network weights. Both are provided to Keras when *compiling* the model:

```
autoencoder = Model(input_img, decoded)
autoencoder.compile(optimizer='adam', loss='binary_crossentropy')
```

We can now finally train our autoencoder. Keras Model classes provide APIs that are similar to scikit-learn, with a fit method to be used to train the neural network. Note that, owing to the nature of the autoencoder, we are using the same information as the input and output of our network:

```
autoencoder.fit(x_train, x_train,
                epochs=50,
                batch_size=128,
                shuffle=True,
                validation_data=(x_test, x_test))
```

Once the training is finished, we can examine the ability of the network to reconstruct the inputs by comparing input images with their reconstructed version, which can be easily computed using the `predict` method of the Keras `Model` class as follows:

```
decoded_imgs = autoencoder.predict(x_test)
```

In *Figure 4.15*, we show the reconstructed images. As you can see, the network is quite good at reconstructing unseen images, especially when considering the large-scale features. Details might have been lost in the compression (see, for instance, the logos on the T-shirts), but the overall relevant information has indeed been captured by our network:

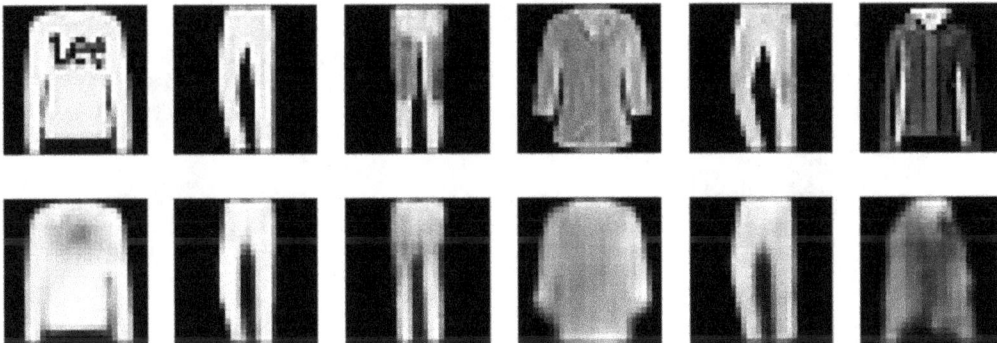

Figure 4.15: Examples of the reconstruction done on the test set by the trained autoencoder

It can also be very interesting to represent the encoded version of the images in a two-dimensional plane using T-SNE:

```
from tensorflow.keras.layers import Flatten
embed_layer = Flatten()(encoded)
embeddings = Model(input_img, embed_layer).predict(x_test)
tsne = TSNE(n_components=2)
emb2d = tsne.fit_transform(embeddings)
x, y = np.squeeze(emb2d[:, 0]), np.squeeze(emb2d[:, 1])
```

The coordinates provided by T-SNE are shown in *Figure 4.16*, colored by the class the sample belongs to. The clustering of the different clothing can clearly be seen, particularly for some classes that are very well separated from the rest:

Figure 4.16: T-SNE transformation of the embeddings extracted from the test set, colored by the class that the sample belongs to

Autoencoders are, however, rather prone to overfitting, as they tend to re-create exactly the images of the training and not generalize well. In the following subsection, we will see how overfitting can be prevented in order to build more robust and reliable dense representations.

Denoising autoencoders

Besides allowing us to compress a sparse representation into a denser vector, autoencoders are also widely used to process a signal in order to filter out noise and extract only a relevant (characteristic) signal. This can be very useful in many applications, especially when identifying anomalies and outliers.

Denoising autoencoders are a small variation of what has already been implemented. As described in the previous section, basic autoencoders are trained using the same image as input and output. Denoising autoencoders corrupt the input using some noise of various intensity, while keeping the same noise-free target. This could be achieved by simply adding some Gaussian noise to the inputs:

```
noise_factor = 0.1
x_train_noisy = x_train + noise_factor * np.random.normal(loc=0.0,
scale=1.0, size=x_train.shape)
x_test_noisy = x_test + noise_factor * np.random.normal(loc=0.0,
scale=1.0, size=x_test.shape)
x_train_noisy = np.clip(x_train_noisy, 0., 1.)
x_test_noisy = np.clip(x_test_noisy, 0., 1.)
```

The network can then be trained using the corrupted input, while for the output, the noise-free image is used:

```
noisy_autoencoder.fit(x_train_noisy, x_train,
                epochs=50,
                batch_size=128,
                shuffle=True,
                validation_data=(x_test_noisy, x_test))
```

Such an approach is generally valid when datasets are large and when the risk of overfitting the noise is rather limited. When datasets are smaller, an alternative to avoid the network "learning" the noise as well (thus learning the mapping between a static noisy image to its noise-free version) is to add training stochastic noise using a GaussianNoise layer.

Note that in this way, the noise may change between epochs and prevent the network from learning a static corruption superimposed onto our training set. In order to do so, we change the first layers of our network in the following way:

```
input_img = Input(shape=(28, 28, 1))
noisy_input = GaussianNoise(0.1)(input_img)
x = Conv2D(16, (3, 3), activation='relu', padding='same')(noisy_input)
```

The difference is that instead of having statically corrupted samples (that do not change in time), the noisy inputs now keep changing between epochs, thus avoiding the network learning the noise as well.

The GaussianNoise layer is an example of a regularization layer, that is, a layer that helps reduce the overfitting of a neural network by inserting a random part in the network. GaussianNoise layers make models more robust and able to generalize better, avoiding autoencoders learning the identity function.

Another common example of a regularization layer is the dropout layers that effectively set to 0 some of the inputs (at random with a probability, ρ_0) and rescale the other inputs by a $\frac{1}{1-\rho_0}$ factor in order to (statistically) keep the sum over all the units constant, with and without dropout.

Dropout corresponds to randomly killing some of the connections between layers in order to reduce output dependency to specific neurons. Always keep in mind that regularization layers are only active during training, while at test time they simply correspond to identity layers.

In *Figure 4.17*, we compare the network reconstruction of a noisy input (input) for the previous unregularized trained network and the network with a GaussianNoise layer. As can be seen (compare, for instance, the images of trousers), the model with regularization tends to develop stronger robustness and reconstructs the noise-free outputs:

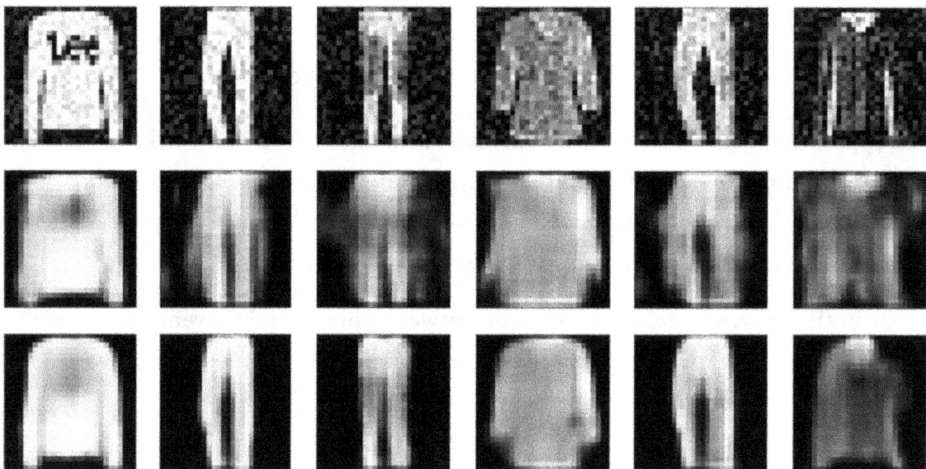

Figure 4.17: Comparison with reconstruction for noisy samples. Top row: noisy input; middle row: reconstructed output using a vanilla autoencoder; bottom row: reconstructed output using a denoising autoencoder

Regularization layers are often used when dealing with deep neural networks that tend to overfit and are able to learn identity functions for autoencoders. Often, dropout or GaussianNoise layers are introduced, repeating a similar pattern composed of regularization and learnable layers that we usually refer to as **stacked denoising layers**.

Graph autoencoders

Once the basic concepts of autoencoders are understood, we can turn to apply this framework to graph structures. If, on one hand, the network structure, decomposed into an encoder-decoder structure with a low-dimensional representation in between, still applies, the definition of the loss function to be optimized needs a bit of caution when dealing with networks. First, we need to adapt the reconstruction error to a meaningful formulation that can adapt to the peculiarities of graph structures.

When applying autoencoders to graph structures, the input and output of the network should be a graph representation, for instance, the adjacency matrix. The reconstruction loss could then be defined as the Frobenius norm of the difference between the input and output matrices. However, when applying autoencoders to such graph structures and adjacency matrices, two critical issues arise:

- Whereas the presence of links indicates a relation or similarity between two vertices, their absence does not generally indicate a dissimilarity between vertices.
- The adjacency matrix is extremely sparse and therefore the model will naturally tend to predict a 0 rather than a positive value.

To address such peculiarities of graph structures, when defining the reconstruction loss, we need to penalize more errors for the non-zero elements than for zero elements. This can be done using the following loss function:

$$\mathcal{L}_{2nd} = \sum_{i=1}^{n} \left\| (\tilde{A}_l - A_i) \odot b_i \right\|$$

Here, $\tilde{A}_l$ represents the reconstructed adjacency matrix produced by the autoencoder, and A_i represents the original adjacency matrix for node i (adjacency vector). $\odot$ is the Hadamard element-wise product, where $b_{ij} = \beta > 1$ if there is an edge between nodes i and j, and 0 otherwise. The preceding loss guarantees that vertices that share a neighborhood (that is, their adjacency vectors are similar) will also be close in the embedding space. Thus, the preceding formulation will naturally preserve second-order proximity for the reconstructed graph.

On the other hand, you can also promote first-order proximity in the reconstructed graph, thus enforcing connected nodes to be close in the embedding space. This condition can be enforced by using the following loss:

$$\mathcal{L}_{1st} = \sum_{i,j=1}^{n} S_{ij} \left\| y_j - y_i \right\|_2^2$$

Here, y_i and y_j are the two representations of nodes i and j in the embedding space. The term S_{ij} represents the strength of the connection between nodes i and j. Typically, S_{ij} is derived from the adjacency matrix or other similarity measures, where a larger S_{ij} indicates a stronger or more important connection between the two nodes. This loss function forces neighboring nodes to be close in the embedding space. In fact, if two nodes are tightly connected, S_{ij} will be large. As a consequence, their difference in the embedding space, $\left\| y_j - y_i \right\|_2^2$, should be limited (indicating the two nodes are close in the embedding space) to keep the loss function small. The two losses can also be combined into a single loss function, where, in order to prevent overfitting, a regularization loss can be added that is proportional to the norm of the weight coefficients:

$$\mathcal{L}_{tot} = \mathcal{L}_{2nd} + \alpha\mathcal{L}_{1st} + v\mathcal{L}_{reg} = \mathcal{L}_{2nd} + \alpha\mathcal{L}_{1st} + v \parallel W \parallel_F^2$$

In the preceding equation, W represents all the weights used across the network. The preceding formulation was proposed in 2016 by Wang et al., and it is now known as **Structural Deep Network Embedding (SDNE)**.

Although the preceding loss could also be directly implemented with TensorFlow and Keras, you can already find this network integrated in the GEM package we referred to previously. As before, extracting the node embedding can be done similarly in a few lines of code, as follows:

```
G=nx.karate_club_graph()
sdne=SDNE(d=2, beta=5, alpha=1e-5, nu1=1e-6, nu2=1e-6,
        K=3,n_units=[50, 15,], rho=0.3, n_iter=10,
        xeta=0.01,n_batch=100,
        modelfile=['enc_model.json','dec_model.json'],
        weightfile=['enc_weights.hdf5','dec_weights.hdf5'])
sdne.learn_embedding(G)
embeddings = m1.get_embedding()
```

Although very powerful, these graph autoencoders encounter some issues when dealing with large graphs. For these cases, the input of our autoencoder is one row of the adjacency matrix that has as many elements as the nodes in the network. In large networks, this size can easily be of the order of millions or tens of millions.

In the next section, we describe a different strategy for encoding the network information that in some cases may iteratively aggregate embeddings only over local neighborhoods, making it scalable to large graphs.

Graph neural networks

We introduced in *Chapter 3, Neural Networks and Graphs*, the concept of GNNs, a deep learning method that works on graph-structured data. It is now time to go into the details of how GNNs can be used for unsupervised learning. Similar to other techniques seen in this chapter (e.g., matrix factorization), GNNs will also allow us to obtain embeddings of our graph elements (nodes, edges, and graphs). However, it is worth noting that different from shallow embeddings, some of the GNN techniques are inductive, and therefore allow the embeddings to also be applied to unseen data.

As shown in the previous chapter, **Convolutional Neural Networks (CNNs)** are widely used in images to extract multi-scale localized spatial features that are exploited by deeper layers to construct more complex and highly expressive representations.

In recent years, it has been observed that concepts such as multi-layer and locality are also useful for processing graph-structured data. However, graphs are defined over a *non-Euclidean space*.

The original formulation of GNNs was proposed by Scarselli et al., back in 2009. It relies on the fact that each node can be described by its features and its neighborhood. Information coming from the neighborhood (which represents the concept of locality in the graph domain) can be aggregated and used to compute more complex and high-level features.

At the beginning, each node, v_i, is associated with a state. Let's start with a random embedding, h_i^t (ignoring node attributes for simplicity). At each iteration of the algorithm, nodes accumulate input from their neighbors using a simple neural network layer:

$$h_i^t = \sum_{vj \in N(v_i)} \sigma\left(W h_j^{t-1} + b\right)$$

Here, $W \in \mathbb{R}^{d \times d}$ and $b \in \mathbb{R}^d$ are trainable parameters (where d is the dimension of the embedding), σ is a non-linear function, and t represents the tth iteration of the algorithm. The equation is applied recursively until a particular objective is reached. Note that, at each iteration, the *previous state* (computed at the previous iteration) of neighbors is exploited in order to compute the new state, similar to *recurrent neural networks*.

Starting from this first idea, several attempts have been made in recent years to re-address the problem of learning from graph data. In particular, variants of the previously described GNN have been proposed, with the aim of improving its representation learning capability. Some of them are specifically designed to process specific types of graphs, such as direct, indirect, weighted, unweighted, static, and dynamic.

Several modifications have been proposed for the propagation step (convolution, gate mechanisms, attention mechanisms, and skip connections, among others), with the aim of improving the representation at different levels. Also, different training methods have been proposed to improve learning.

There are essentially two types of convolutional operations for graph data, namely **spectral approaches** and **non-spectral (spatial)** approaches. The first, as the name suggests, defines convolution in the spectral domain (that is, decomposing graphs in a combination of simpler elements). Spatial convolution, on the other hand, formulates the convolution as aggregating feature information from neighbors.

Spectral graph convolution

Spectral approaches are related to spectral graph theory, the study of the characteristics of a graph in relation to the characteristic polynomial, eigenvalues, and eigenvectors of the matrices associated with the graph. The convolution operation is defined as the multiplication of a signal (node features) by a kernel. In more detail, it is defined in the Fourier domain by determining the *eigendecomposition of the graph Laplacian* (think about the graph Laplacian as an adjacency matrix normalized in a special way).

While this definition of spectral convolution has a strong mathematical foundation, the operation is computationally expensive. For this reason, several works have been done to approximate it in an efficient way. *ChebNet* by *Defferrard* et al., for instance, is one of the first seminal works on spectral graph convolution. Here, the operation is approximated by using the concept of the Chebyshev polynomial of order K (a special kind of polynomial used to efficiently approximate functions).

Here, K is a very useful parameter because it determines the locality of the filter. Intuitively, for $K = 1$, only the node features are fed into the network. With $K = 2$, we average over two-hop neighbors (neighbors of neighbors) and so on.

Let $X \in \mathbb{R}^{N \times d}$ be the matrix of node features, where N is the number of nodes. In classical neural network processing, this signal would be composed of layers of the following form:

$$H^1 = \sigma(XW)$$

Here, $W \in \mathbb{R}^{N \times N}$ is the layer weights and σ represents some non-linear activation function. The drawback of this operation is that it processes each node signal independently without taking into account connections between nodes. To overcome this limitation, a simple (yet effective) modification can be done, as follows:

$$H^1 = \sigma(AXW)$$

By introducing the adjacency matrix, $A \in \mathbb{R}^{N \times N}$, a new linear combination between each node and its corresponding neighbors is added. This way, the information depends only on the neighborhood and parameters are applied to all the nodes, simultaneously.

It is worth noting that this operation can be repeated in sequence several times, thus creating a deep network. At each layer, the node descriptors, X, will be replaced with the output of the previous layer, H^{l-1}, feeding the computation of H^l.

The preceding presented equation, however, has some limitations and cannot be applied as it stands. The first limitation is that by multiplying by A, we consider all the neighbors of the node but not the node itself. This problem can be easily overcome by adding self-loops in the graph, that is, adding the $\hat{A} = A + I$ identity matrix.

The second limitation is related to the adjacency matrix itself. Since it is typically not normalized, we will observe large values in the feature representation of high-degree nodes and small values in the feature representation of low-degree nodes. This will lead to several problems during training since optimization algorithms are often sensitive to feature scale. Several methods have been proposed for normalizing A.

In Kipf and Welling, 2017 (one of the well-known GCN models), for example, the normalization is performed by multiplying A by the *diagonal node degree matrix D*, such that all the rows sum to 1: $D^{-1}A$. More specifically, they used symmetric normalization ($D^{-1/2}AD^{-1/2}$), such that the proposed propagation rule becomes as follows:

$$H^l = \sigma(\widehat{D}^{-\frac{1}{2}}\hat{A}\widehat{D}^{-\frac{1}{2}}XW)$$

Here, $\widehat{D}$ is the diagonal node degree matrix of $\hat{A}$.

In the following example, we will create a GCN as defined in Kipf and Welling and apply this propagation rule for embedding a simple graph:

1. To begin, it is necessary to import all the Python modules. We will use `networkx` to load the *barbell graph*:

    ```
    import networkx as nx
    import numpy as np
    G = nx.barbell_graph(m1=10,m2=4)
    ```

2. To implement the GC propagation rule, we need an adjacency matrix representing G. Since this network does not have node features, we will use the $I \in \mathbb{R}^{N \times N}$ identity matrix as the node descriptor:

```
A = nx.to_numpy_matrix(G)
I = np.eye(G.number_of_nodes())
```

3. We now add the self-loop and prepare the diagonal node degree matrix:

```
from scipy.linalg import sqrtm
A_hat = A + I
D_hat = np.array(np.sum(A_hat, axis=0))[0]
D_hat = np.array(np.diag(D_hat))
D_hat = np.linalg.inv(sqrtm(D_hat))
A_norm = D_hat @ A_hat @ D_hat
```

4. Our GCN will be composed of two layers. Let's define the layers' weights and the propagation rule. Layer weights, *W*, will be initialized using *Glorot uniform initialization* (even if other initialization methods can be also used, for example, by sampling from a Gaussian or uniform distribution):

```
def glorot_init(nin, nout):
    sd = np.sqrt(6.0 / (nin + nout))
    return np.random.uniform(-sd, sd, size=(nin, nout))
class GCNLayer():
  def __init__(self, n_inputs, n_outputs):
      self.n_inputs = n_inputs
      self.n_outputs = n_outputs
      self.W = glorot_init(self.n_outputs, self.n_inputs)
      self.activation = np.tanh
  def forward(self, A, X):
      self._X = (A @ X).T
      H = self.W @ self._X
      H = self.activation(H)
      return H.T # (n_outputs, N)
```

5. Finally, let's create our network and compute the forward pass, that is, propagate the signal through the network:

```
gcn1 = GCNLayer(G.number_of_nodes(), 8)
gcn2 = GCNLayer(8, 4)
```

```
gcn3 = GCNLayer(4, 2)
H1 = gcn1.forward(A_norm, I)
H2 = gcn2.forward(A_norm, H1)
H3 = gcn3.forward(A_norm, H2)
```

H3 now contains the embedding computed using the GCN propagation rule. Note that we chose 2 as the number of outputs, meaning that the embedding is bi-dimensional and can be easily visualized. In *Figure 4.18*, you can see the output:

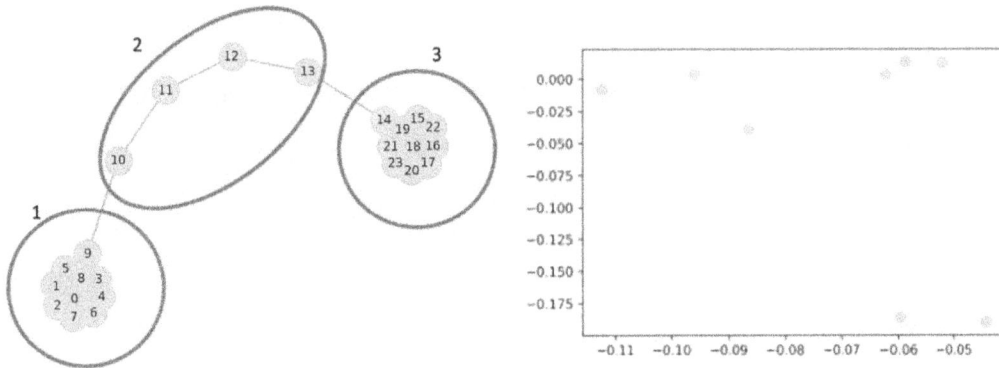

Figure 4.18: Application of the graph convolutional layer to a graph (left) to generate the embedding vector of its nodes (right)

You can observe the presence of two quite well-separated communities. This is a nice result, considering that we have not trained the network yet!

Spectral graph convolution methods have achieved noteworthy results in many domains. However, they present some drawbacks. Consider, for example, a very big graph with billions of nodes: a spectral approach requires the graph to be processed simultaneously, which can be impractical from a computational point of view.

Furthermore, spectral convolution often assumes a fixed graph, leading to poor generalization capabilities on new, unseen samples. To overcome these issues, spatial graph convolution represents an interesting alternative.

Spatial graph convolution

Spatial graph convolutional networks perform the operations directly on the graph by aggregating information from spatially close neighbors. Spatial convolution has many advantages: weights can be easily shared across different locations of the graph, leading to a good generalization capability on different graphs. Furthermore, the computation can be done by considering subsets of nodes instead of the entire graph, potentially improving computational efficiency.

GraphSAGE is one of the algorithms that implement spatial convolution. One of its main characteristics is its ability to scale over various types of networks. We can think of GraphSAGE as composed of three steps:

1. **Neighborhood sampling**: For each node in a graph, the first step is to find its k-neighborhood, where k is defined by the user for determining how many hops to consider (neighbors of neighbors).

2. **Aggregation**: The second step is to aggregate, for each node, the node features describing the respective neighborhood. Various types of aggregation can be performed, including average, pooling (for example, taking the best feature according to certain criteria), or an even more complicated operation, such as using recurrent units (such as LSTM).

3. **Prediction**: Each node is equipped with a simple neural network that learns how to perform predictions based on the aggregated features from the neighbors.

GraphSAGE is often used in supervised learning tasks, as explored in *Chapter 5, Supervised Graph Learning*. However, it is also suitable for unsupervised learning. In unsupervised settings, GraphSAGE employs a loss function based on node similarity, which is derived from random walks. Specifically, as outlined in the original paper by Hamilton, Ying, and Leskovec (2018), fixed-length random walks are used to identify pairs of nodes that are likely to share similar representations. The learned embeddings are optimized to ensure that nodes close in these random-walk neighborhoods have similar embeddings, using a loss function that measures similarity (a dot product, for example) between these embeddings. This approach enables GraphSAGE to learn effective node embeddings even without explicit labels.

Graph convolution in practice

In practice, GNNs have been implemented in many machine learning and deep learning frameworks, such as StellarGraph, PyG, and DGL. For the next example, we will be using StellarGraph. A similar example using other frameworks can be found in the GitHub repository. We will learn about embedding vectors in an unsupervised manner, without a target variable. The method is inspired by Bai et al., 2019 and is based on the simultaneous embedding of pairs of graphs. This embedding should match a ground-truth distance between graphs:

1. First, let's load the required Python modules:

```
import numpy as np
import stellargraph as sg
from stellargraph.mapper import FullBatchNodeGenerator
from stellargraph.layer import GCN
```

```
import tensorflow as tf
from tensorflow.keras import layers, optimizers, losses, metrics,
Model
```

2. We will be using the PROTEINS dataset for this example, which is available in StellarGraph and consists of 1,114 graphs with 39 nodes and 73 edges on average for each graph. Each node is described by four attributes and belongs to one of two classes:

```
dataset = sg.datasets.PROTEINS()
graphs, graph_labels = dataset.load()
```

3. The next step is to create the model. It will be composed of two GC layers with 64 and 32 output dimensions followed by ReLU activation, respectively. The output will be computed as the Euclidean distance of the two embeddings:

```
generator = sg.mapper.PaddedGraphGenerator(graphs)
# define a GCN model containing 2 layers of size 64 and 32,
# respectively.ReLU activation function is used to add
# non-linearity between layers
gc_model = sg.layer.GCNSupervisedGraphClassification(
[64, 32], ["relu", "relu"], generator, pool_all_layers=True)
# retrieve the input and the output tensor of the GC layer
# such that they can be connected to the next layer
inp1, out1 = gc_model.in_out_tensors()
inp2, out2 = gc_model.in_out_tensors()
vec_distance = tf.norm(out1 - out2, axis=1)
# create the model. It is also useful to create a specular model in
# order to easily retrieve the embeddings
pair_model = Model(inp1 + inp2, vec_distance)
embedding_model = Model(inp1, out1)
```

4. It is now time to prepare the dataset for training. To each pair of input graphs, we will assign a similarity score. Notice that any notion of graph similarity can be used in this case, including graph edit distances. For simplicity, we will be using the distance between the spectrum of the Laplacian of the graphs:

```
def graph_distance(graph1, graph2):
    spec1 = nx.laplacian_spectrum(graph1.to_networkx(feature_
attr=None))
    spec2 = nx.laplacian_spectrum(graph2.to_networkx(feature_
```

```
attr=None))
    k = min(len(spec1), len(spec2))
    return np.linalg.norm(spec1[:k] - spec2[:k])
graph_idx = np.random.RandomState(0).randint(len(graphs), size=(100,
2))
targets = [graph_distance(graphs[left], graphs[right]) for left,
right in graph_idx]
train_gen = generator.flow(graph_idx, batch_size=10,
targets=targets)                                          .
```

5. Finally, let's compile and train the model. We will be using the Adaptive Moment Estimation (Adam) optimizer with the learning rate parameter set to 1e-2. The loss function we will be using is defined as the minimum squared error between the prediction and the ground-truth distance computed as previously. The model will be trained for 500 epochs:

```
pair_model.compile(optimizers.Adam(1e-2), loss="mse")
pair_model.fit(train_gen, epochs=500, verbose=0)
```

After training, we are now ready to inspect and visualize the learned representation. Since the output is 32-dimensional, we need a way to qualitatively evaluate the embeddings, for example, by plotting them in a bi-dimensional space. We will use T-SNE for this purpose:

```
# retrieve the embeddings
embeddings = embedding_model.predict(generator.flow(graphs))
# TSNE is used for dimensionality reduction
from sklearn.manifold import TSNE
tsne = TSNE(2)
two_d = tsne.fit_transform(embeddings)
```

Let's plot the embeddings. In the plot, each point (embedded graph) is colored according to the corresponding label (blue=0, red=1). The results are visible in *Figure 4.19*:

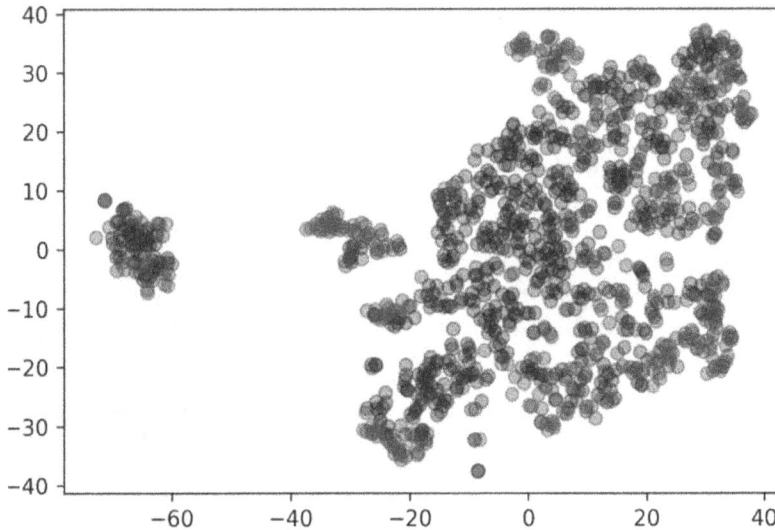

Figure 4.19: The PROTEINS dataset embedding using GCNs. Axes represent the first two latent directions of the embedding algorithm.

As you can see, the obtained embeddings seem not to be related to the actual label distribution. Since this is an unsupervised method, this may happen because the network is learning patterns it thinks are useful. However, this is not necessarily a bad thing! By studying these "new" clusters, we may discover something new or unexpected in our data.

Moreover, this is just one of the possible methods for learning embeddings for graphs. More advanced solutions can be experimented with to better fit the problem of interest.

Summary

In this chapter, we have learned how unsupervised machine learning can be effectively applied to graphs to solve real problems, such as node and graph representation learning. In particular, we first analyzed shallow embedding methods, a set of algorithms that are able to learn and return only the embedding values for the learned input data.

We then learned how autoencoder algorithms can be used to encode the input by preserving important information in a lower-dimensional space. We have also seen how this idea can be adapted to graphs, by learning about embeddings that allow us to reconstruct the pair-wise node/graph similarity. Finally, we introduced the main concepts behind GNNs. We have seen how well-known concepts, such as convolution, can be applied to graphs.

In the next chapter, we will revise these concepts in a supervised setting. There, a target label is provided and the objective is to learn a mapping between the input and the output.

5

Supervised Graph Learning

Supervised learning likely represents the majority of practical **machine learning** (**ML**) tasks. Thanks to more and more active and effective data collection activities, it is very common nowadays to deal with labeled datasets.

This is also true for graph data, where labels can be assigned to nodes, communities, or even an entire structure. The task, then, is to learn a mapping function between the input and the label (also known as a target or an annotation).

For example, given a graph representing a social network, we might be asked to guess which user (node) will close their account. Or, similarly, we might be asked to predict the number of posts a user will make over the next month. We can learn these predictive functions by training graph ML on **retrospective data**, where each user is labeled as "faithful" or "quitter" based on whether they closed their account after a few months. Similarly, users could be associated with the number of posts they published in the previous month.

In this chapter, we will explore the concept of **supervised graph learning**. Therefore, we will also be providing an overview of the main supervised graph embedding methods. The following topics will be covered:

- The supervised graph embedding roadmap
- Feature-based methods
- Shallow embedding methods
- Graph regularization methods
- Graph **convolutional neural networks** (**CNNs**)

Technical requirements

All code files relevant to this chapter are available at https://github.com/PacktPublishing/ Graph-Machine-Learning/tree/main/Chapter05. Please refer to the *Practical exercises* section of *Chapter 1, Getting Started with Graphs,* for guidance on how to set up the environment to run the examples in this chapter, using either Poetry, pip or docker.

The supervised graph embedding roadmap

In supervised learning, a training set consists of a sequence of ordered pairs (x, y), where x is a set of input features (often signals defined on graphs) and y is the output label assigned to it. The goal of ML models, then, is to learn the function mapping each x value to each y value. Here, y can be either a categorical or a continuous variable, depending on whether we are addressing a classification or a regression problem. Common supervised tasks include predicting user properties in a large social network or predicting molecules' attributes, where each molecule is a graph. Sometimes, however, not all instances can be provided with a label. In this scenario, a typical dataset consists of a small set of labeled instances and a larger set of unlabeled instances. For such situations, **semi-supervised learning** is proposed, whereby algorithms aim to exploit label dependency information reflected by available label information in order to learn the predicting function for the unlabeled samples.

With regard to supervised graph ML techniques, many algorithms have been developed, aimed at learning an embedding function for graphs while solving a certain supervised task. However as previously reported by different scientific papers (https://arxiv.org/abs/2005.03675), they can be grouped into macro-groups, depending on how each method encodes the graph information to produce the embeddings. These groups are:

- Feature-based methods
- Shallow embedding methods
- Regularization methods
- **Graph neural networks (GNNs)**

They are graphically depicted in the following diagram:

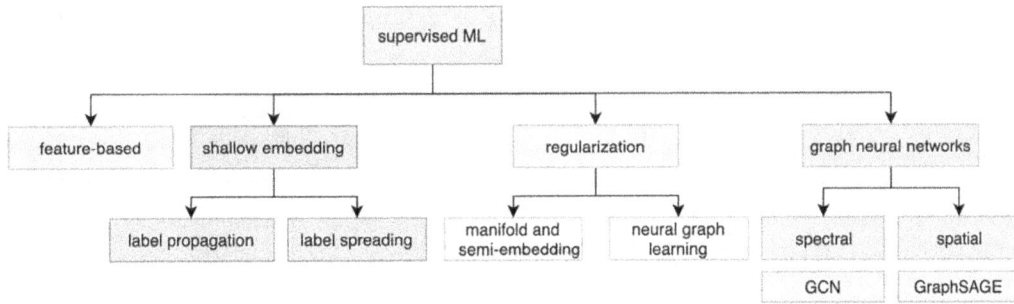

Figure 5.1: Hierarchical structure of the different supervised embedding algorithms described in this book

In the following sections, you will learn the main principles behind each group of algorithms. We will try to provide insight into the most well-known algorithms in the field as well, as these can be used to solve real-world problems.

Feature-based methods

One very simple (yet powerful) method for applying ML on graphs is to consider the encoding function as a simple embedding lookup. When dealing with supervised tasks, one simple way of doing this is to exploit graph properties. In *Chapter 1, Getting Started with Graphs*, we learned how graphs (or nodes in a graph) can be described by means of structural properties, each "encoding" important information from the graph itself.

Let's forget graph ML for a moment; in classical supervised ML, the task is to find a function that maps a set of (descriptive) features of an instance to a particular output. Such features should be carefully engineered so that they are sufficiently representative to learn that concept. Therefore, as the number of petals and the sepal length might be good descriptors for a flower, when describing a graph, we might rely on its average degree, its global efficiency, and its characteristic path length.

This naïve approach acts in two steps, outlined as follows:

1. Select a set of *good* descriptive graph properties.
2. Use such properties as input for a traditional ML algorithm.

Unfortunately, there is no general definition of *good* descriptive properties, and their choice strictly depends on the specific problem to solve. However, you can still compute a wide variety of graph properties and then perform *feature selection* to select the most informative ones. **Feature selection** is a widely studied topic in ML, but providing details about the various methods is outside the scope of this book. However, we refer you to the book *Machine Learning Algorithms – Second Edition* (`https://subscription.packtpub.com/book/big_data_and_business_intelligence/9781789347999`), published by Packt Publishing, for further reading on this subject.

Let's now see a practical example of how such a basic method can be applied. We will be performing a supervised graph classification task by using the `PROTEINS` dataset. The `PROTEINS` dataset contains several graphs representing protein structures. Each graph is labeled, defining whether the protein is an enzyme or not. We will follow these steps:

1. First, let's load the dataset through the `stellargraph` Python library, as follows:

   ```
   from stellargraph import datasets
   from IPython.display import display, HTML
   dataset = datasets.PROTEINS()
   graphs, graph_labels = dataset.load()
   ```

2. For computing graph properties, we will be using `networkx`, as described in *Chapter 1, Getting Started with Graphs*. To that end, we need to convert graphs from the `stellargraph` format to the `networkx` format. This can be done in two steps: first, convert the graphs from the `stellargraph` representation to *numpy* adjacency matrices. Then, use the adjacency matrices to retrieve the `networkx` representation. In addition (as we will see in the next steps), we also transform the labels (which are stored as a pandas Series) to a numpy array, which can be better exploited by the evaluation functions. The code is illustrated in the following snippet:

   ```
   # convert from StellarGraph format to numpy adj matrices
   adjs = [graph.to_adjacency_matrix().A for graph in graphs]
   # convert labels from Pandas.Series to numpy array
   labels = graph_labels.to_numpy(dtype=int)
   ```

3. Then, for each graph, we compute global metrics to describe it. For this example, we have chosen the number of edges, the average cluster coefficient, and the global efficiency. However, we suggest you compute several other properties you may find worth exploring. We can extract the graph metrics using `networkx`, as follows:

```
import numpy as np
import networkx as nx
metrics = []
for adj in adjs:
  # from numpy to networkx
  G = nx.from_numpy_matrix(adj)
  # basic properties
  num_edges = G.number_of_edges()
  # clustering measures
  cc = nx.average_clustering(G)
  # measure of efficiency
  eff = nx.global_efficiency(G)
  metrics.append([num_edges, cc, eff])
```

4. We can now exploit scikit-learn utilities to create train and test sets. In our experiments, we will be using 70% of the dataset as the training set and the remainder as the test set. We can do that by using the `train_test_split` function provided by scikit-learn, as follows:

```
from sklearn.model_selection import train_test_split
X_train, X_test, y_train, y_test = train_test_split(metrics, labels,
test_size=0.3, random_state=42)
```

5. It's now time to train a proper ML algorithm. We chose a **support vector machine (SVM)** for this task. More precisely, the SVM is trained to minimize the difference between the predicted labels and the actual labels (the ground truth). We can do this by using the SVC module of scikit-learn. In addition, we use the accuracy, precision, recall, and F1-score to evaluate how well the algorithm is performing on the test set:

```
from sklearn import svm
from sklearn.metrics import accuracy_score, precision_score, recall_
score, f1_score
clf = svm.SVC()
clf.fit(X_train, y_train)
y_pred = clf.predict(X_test)
print('Accuracy', accuracy_score(y_test,y_pred))
print('Precision', precision_score(y_test,y_pred))
print('Recall', recall_score(y_test,y_pred))
print('F1-score', f1_score(y_test,y_pred))
```

This should be the output of the previous snippet of code:

```
Accuracy 0.7455
Precision 0.7709
Recall 0.8413
F1-score 0.8045
```

We achieved about 80% for the F1 score, which is already quite good for such a naïve task. Moreover, we can observe from the accuracy, precision and recall that performance is quite balanced between classes!

Using graph properties as descriptive features for traditional supervised learning has several strengths. It is straightforward to implement, leverages well-established ML techniques, and allows for interpretability through human-understandable features like average degree or path length. However, it heavily relies on the careful selection and engineering of graph properties, which can be time-consuming and domain-specific. Additionally, this approach may struggle at capturing complex relationships inherent in graph data.

These limitations pave the way for shallow embedding methods, which automate the process of feature extraction by learning representations directly from the graph structure.

Shallow embedding methods

As we already described in *Chapter 4*, *Unsupervised Graph Learning*, shallow embedding methods are a subset of graph embedding methods that learn node, edge, or graph representations for only a finite set of input data. They are transductive methods, since they cannot be applied to other instances different from the ones used to train the model. Before starting our discussion, it is important to define how supervised and unsupervised shallow embedding algorithms differ.

The main difference between unsupervised and supervised embedding methods essentially lies in the task they attempt to solve. Indeed, if unsupervised shallow embedding algorithms try to learn a good graph, node, or edge representation in order to understand the underlying structure, the supervised algorithms try to find the best solution for a prediction task such as node classification, label prediction, or graph classification.

In this section, we will explain in detail some of those supervised shallow embedding algorithms. Moreover, we will enrich our description by providing several examples of how to use those algorithms in Python. For all the algorithms described in this section, we will present a custom implementation using the base classes available in the scikit-learn library.

Label propagation algorithm

The label propagation algorithm is a well-known semi-supervised algorithm widely applied in data science and used to solve the node classification task. More precisely, the algorithm *propagates* the label of a given node to its neighbors or to nodes with a high probability of being reached from that node.

The general idea behind this approach is quite simple: given a graph with a set of labeled and unlabeled nodes, the labeled nodes propagate their label to the nodes with the highest probability of being reached. In the following diagram, we can see an example of a graph with labeled and unlabeled nodes:

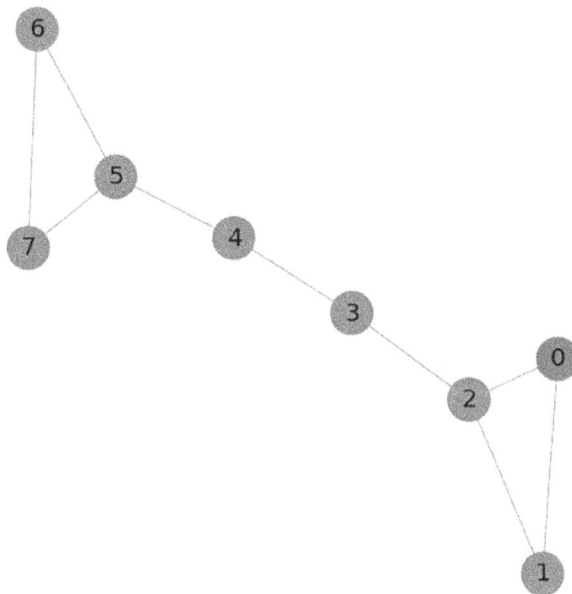

Figure 5.2: Example of a graph with two labeled nodes (class 0 in red and class 1 in green) and six unlabeled nodes

According to *Figure 5.2*, using the information of the labeled nodes (node **0** and **6**), the algorithm will calculate the probability of moving to another unlabeled node. The nodes with the highest probability from a labeled node will get the label of that node.

Formally, let $G = (V, E)$ be a graph and let $Y = \{y_1, \dots, y_p\}$ be a set of labels. Since the algorithm is semi-supervised, just a subset of nodes will have an assigned label. Moreover, let $A \in \mathbb{R}^{|V| \times |V|}$ be the adjacency matrix of the input graph G and $D \in \mathbb{R}^{|V| \times |V|}$ be the diagonal degree matrix where each element $d_{ij} \in D$ is defined as follows:

$$d_{ij} = \begin{cases} 0 \text{ if } i \neq j \\ \deg(v_i) \text{ if } i = j \end{cases}$$

Where $\deg(v_i)$ is the degree of the node v_i. In other words, the only nonzero elements of the degree matrix are the diagonal elements whose values are given by the degree of the node represented by the row. In the following figure, we can see the diagonal degree matrix of the graph represented in *Figure 5.2*:

$$D = \begin{bmatrix} 2 & 0 & 0 & 0 & 0 & 0 & 0 & 0 \\ 0 & 2 & 0 & 0 & 0 & 0 & 0 & 0 \\ 0 & 0 & 3 & 0 & 0 & 0 & 0 & 0 \\ 0 & 0 & 0 & 2 & 0 & 0 & 0 & 0 \\ 0 & 0 & 0 & 0 & 2 & 0 & 0 & 0 \\ 0 & 0 & 0 & 0 & 0 & 3 & 0 & 0 \\ 0 & 0 & 0 & 0 & 0 & 0 & 2 & 0 \\ 0 & 0 & 0 & 0 & 0 & 0 & 0 & 2 \end{bmatrix}$$

Figure 5.3: Diagonal degree matrix for the graph in Figure 5.2

From *Figure 5.3*, it is possible to see how only the diagonal elements of the matrix contain nonzero values, and those values represent the degree of the specific node. We also need to introduce the transition matrix $L = D^{-1} A$. This matrix defines the probability of a node being reached from another node. More precisely, $l_{ij} \in L$ is the probability of reaching node v_j from node v_i. The following figure shows the transition matrix L for the graph depicted in *Figure 5.2*:

$$L = \begin{bmatrix} 0 & 0.5 & 0.5 & 0 & 0 & 0 & 0 & 0 \\ 0.5 & 0 & 0.5 & 0 & 0 & 0 & 0 & 0 \\ 0.33 & 0.33 & 0 & 0.33 & 0 & 0 & 0 & 0 \\ 0 & 0 & 0.5 & 0 & 0.5 & 0 & 0 & 0 \\ 0 & 0 & 0 & 0.5 & 0 & 0.5 & 0 & 0 \\ 0 & 0 & 0 & 0 & 0.33 & 0 & 0.33 & 0.33 \\ 0 & 0 & 0 & 0 & 0 & 0.5 & 0 & 0.5 \\ 0 & 0 & 0 & 0 & 0 & 0.5 & 0.5 & 0 \end{bmatrix}$$

Figure 5.4: Transition matrix for the graph in Figure 5.2

In *Figure 5.4*, the matrix shows the probability of reaching an end node given a start node. For instance, from the first row of the matrix, we can see how from node 0 it is possible to reach, with equal probability of 0.5, only nodes 1 and 2. If we defined with Y^0 the initial label assignment, the probability of label assignment for each node obtained using the L matrix can be computed as $Y^1 = LY^0$. The Y^1 matrix computed for the graph in *Figure 5.2* is shown in the following figure:

$$Y^1 = \begin{bmatrix} 0 & 0.5 & 0.5 & 0 & 0 & 0 & 0 & 0 \\ 0.5 & 0 & 0.5 & 0 & 0 & 0 & 0 & 0 \\ 0.33 & 0.33 & 0 & 0.33 & 0 & 0 & 0 & 0 \\ 0 & 0 & 0.5 & 0 & 0.5 & 0 & 0 & 0 \\ 0 & 0 & 0 & 0.5 & 0 & 0.5 & 0 & 0 \\ 0 & 0 & 0 & 0 & 0.33 & 0 & 0.33 & 0.33 \\ 0 & 0 & 0 & 0 & 0 & 0.5 & 0 & 0.5 \\ 0 & 0 & 0 & 0 & 0 & 0.5 & 0.5 & 0 \end{bmatrix} * \begin{bmatrix} 1 & 0 \\ 0 & 0 \\ 0 & 0 \\ 0 & 0 \\ 0 & 0 \\ 0 & 0 \\ 0 & 1 \\ 0 & 0 \end{bmatrix} = \begin{bmatrix} 0 & 0 \\ 0.5 & 0 \\ 0.33 & 0 \\ 0 & 0 \\ 0 & 0 \\ 0 & 0.33 \\ 0 & 0 \\ 0 & 0.5 \end{bmatrix}$$

Figure 5.5: Solution obtained using the matrix for the graph in Figure 5.2

From *Figure 5.5*, we can see that using the transition matrix, node 1 and node 2 have a probability of being assigned to the [1 0] label of 0.5 and 0.33, respectively, while node 5 and node 6 have a probability of being assigned to the [0 1] label of 0.33 and 0.5, respectively.

Moreover, if we better analyze *Figure 5.5*, we can see two main problems, as follows:

- With this solution, it is possible to assign only to nodes [1 2] and [5 7] a probability associated with a label.
- The initial labels of nodes 0 and 6 are different from the one defined in Y^0.

In order to solve the first point, the algorithm will perform n different iterations; at each iteration t, the algorithm will compute the solution for that iteration, as follows:

$$Y^t = LY^{t-1}$$

Where L is the transition matrix and Y^{t-1} is the label assignment at time *t-1*. The algorithm stops its iteration when a certain condition is met, such as when the labels of all nodes remain unchanged between iterations, or a maximum number of iterations is reached. The second problem is solved by the label propagation algorithm by imposing, in the solution of a given iteration t, the labeled nodes to have the initial class values. For example, after computing the result visible in *Figure 5.5*, the algorithm will force the first line of the result matrix to be [1 0] and the seventh line of the matrix to be [0 1].

Here, we propose a modified version of the LabelPropagation class available in the scikit-learn library. The main reason behind this choice is given by the fact that the LabelPropagation class takes as input a matrix representing a dataset. Each row of the matrix represents a sample, and each column represents a feature.

Before performing a fit operation, the LabelPropagation class internally executes the _build_ graph function. This function builds a graph representation of the input dataset by applying a parametric kernel, such as **k-nearest neighbors (kNNs)** or radial basis functions.

The kernel can be specified using the _get_kernel function. As a result, the original dataset is transformed into a graph (in its adjacency matrix representation) where each node is a sample (a row of the input dataset) and each edge is an *interaction* between the samples.

In our specific case, the input dataset is already a graph, so we need to define a new class capable of dealing with a networkx graph and performing the computation operation on the original graph. The goal is achieved by creating a new class—namely, GraphLabelPropagation—by extending the ClassifierMixin, BaseEstimator, and ABCMeta base classes. The algorithm proposed here is mainly used in order to help you understand the concept behind the algorithm. The whole algorithm is provided in the 05_supervised_graph_machine_learning/02_Shallow_embeddings. ipynb notebook available in the GitHub repository of this book. In order to describe the algorithm, we will use only the fit(X,y) function as a reference. The code is illustrated in the following snippet:

```python
class GraphLabelPropagation(ClassifierMixin, BaseEstimator,
metaclass=ABCMeta):
    def fit(self, X, y):
        X, y = self._validate_data(X, y)
        self.X_ = X
        check_classification_targets(y)
        D = [X.degree(n) for n in X.nodes()]
        D = np.diag(D)
        # Label construction
        # construct a categorical distribution for classification only
        unlabeled_index = np.where(y==-1)[0]
        labeled_index = np.where(y!=-1)[0]
        unique_classes = np.unique(y[labeled_index])
        self.classes_ = unique_classes
        Y0 = np.array([self.build_label(y[x], len(unique_classes)) if x in
labeled_index else np.zeros(len(unique_classes)) for x in range(len(y))])
        A = inv(D)*nx.to_numpy_matrix(G)
        Y_prev = Y0
        it = 0
        c_tool = 10
```

```
    while it < self.max_iter & c_tool > self.tol:
        Y = A*Y_prev
        Y = Y / (Y.sum(axis=1) + 1-e7)   # Normalize

        # force labeled nodes
        Y[labeled_index] = Y0[labeled_index]
        it +=1
        c_tol = np.sum(np.abs(Y-Y_prev))
        Y_prev = Y
    self.label_distributions_ = Y
    return self
```

The fit(X,y) function takes as input a networkx graph X and an array Y representing the labels assigned to each node. Nodes without labels should have a representative value of -1. The while loop performs the real computation. More precisely, it computes the Y^t value at each iteration and forces the labeled nodes in the solution to be equal to their original input value. The algorithm performs the computation until the two stop conditions are satisfied. In this implementation, the following two criteria have been used:

- **Number of iterations:** The algorithm runs the computation until a given number of iterations has been performed.

- **Solution tolerance error:** The algorithm runs the computation until the absolute difference of the solution obtained in two consecutive iterations, Y^{t-1} and Y^t, is lower than a given threshold value.

The algorithm can be applied to the example graph depicted in *Figure 5.2* using the following code:

```
glp = GraphLabelPropagation()
y = np.array([-1 for x in range(len(G.nodes()))])
y[0] = 0
y[6] = 1
glp.fit(G,y)
glp.predict_proba(G)
```

The result obtained by the algorithm is shown in the following diagram:

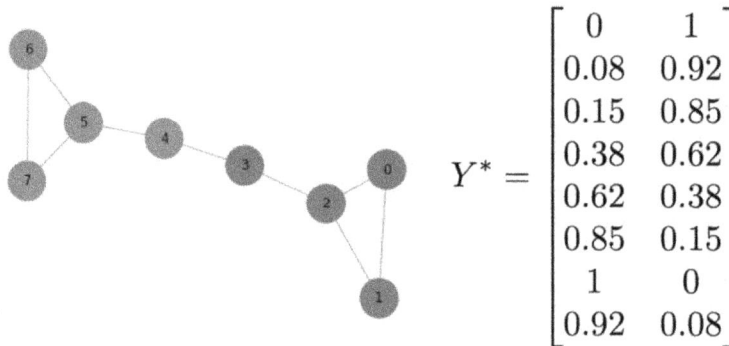

$$Y^* = \begin{bmatrix} 0 & 1 \\ 0.08 & 0.92 \\ 0.15 & 0.85 \\ 0.38 & 0.62 \\ 0.62 & 0.38 \\ 0.85 & 0.15 \\ 1 & 0 \\ 0.92 & 0.08 \end{bmatrix}$$

Figure 5.6: The final labeled graph (left) and the final probability assignment matrix (right)
generated after applying the label propagation algorithm on the graph shown in Figure 5.2

In *Figure 5.6*, we can see the results of the algorithm applied to the example shown in *Figure 5.2*. From the final probability assignment matrix, it is possible to see how the probability of the initial labeled nodes is 1 due to the constraints of the algorithm and how nodes that are "near" to labeled nodes get their label.

Label spreading algorithm

The label spreading algorithm is another semi-supervised shallow embedding algorithm. It was built in order to overcome one big limitation of the label propagation method: the **initial labeling**. Indeed, according to the label propagation algorithm, the initial labels cannot be modified in the training process and, in each iteration, they are forced to be equal to their original value. This constraint could generate incorrect results when the initial labeling is affected by errors or noise. As a consequence, the error will be propagated in all nodes of the input graph.

In order to solve this limitation, the label spreading algorithm tries to relax the constraint of the original labeled data, allowing the labeled input nodes to change their label during the training process.

Formally, let $G = (V, E)$ be a graph and let $Y = \{y_1, \dots, y_p\}$ be a set of labels (since the algorithm is semi-supervised, just a subset of nodes will have an assigned label), and let $A \in \mathbb{R}^{|V| \times |V|}$ and $D \in \mathbb{R}^{|V| \times |V|}$ be the adjacency matrix and diagonal degree matrix of graph G, respectively. Instead of computing the probability transition matrix, the label spreading algorithm uses the normalized graph **Laplacian matrix**, defined as follows:

$$\mathcal{L} = D^{-1/2} A D^{-1/2}$$

Where A is the adjacency matrix and D is the diagonal matrix. As with label propagation, this matrix can be seen as a sort of compact low-dimensional representation of the connections defined in the whole graph. This matrix can be easily computed using networkx with the following code:

```
from scipy.linalg import fractional_matrix_power
D_inv = fractional_matrix_power(D, -0.5)
L = D_inv*nx.to_numpy_matrix(G)*D_inv
```

As a result, we get the following:

$$\mathcal{L} = \begin{bmatrix} 0 & 0.5 & 0.40824829 & 0 & 0 & 0 & 0 & 0 \\ 0.5 & 0 & 0.40824829 & 0 & 0 & 0 & 0 & 0 \\ 0.40824829 & 0.40824829 & 0 & 0.40824829 & 0 & 0 & 0 & 0 \\ 0 & 0 & 0.40824829 & 0 & 0.5 & 0 & 0 & 0 \\ 0 & 0 & 0 & 0.5 & 0 & 0.40824829 & 0 & 0 \\ 0 & 0 & 0 & 0 & 0.40824829 & 0 & 0.40824829 & 0.40824829 \\ 0 & 0 & 0 & 0 & 0 & 0.40824829 & 0 & 0.5 \\ 0 & 0 & 0 & 0 & 0 & 0.40824829 & 0.5 & 0 \end{bmatrix}$$

Figure 5.7: The normalized graph Laplacian matrix

The most important difference between the label spreading and label propagation algorithms is related to the function used to extract the labels. If we define with Y^0 the initial label assignment, the probability of a label assignment for each node obtained using the $\mathcal{L}$ matrix can be computed as follows:

$$Y^1 = \alpha \mathcal{L} Y^0 + (1 - \alpha)Y^0$$

Where α $(0 \leq \alpha \leq 1)$ is the regularization parameter that controls the influence of the graph structure versus the initial label assignment. As with label propagation, label spreading has an iterative process to compute the end solution. The algorithm will perform n different iterations; in each iteration t, the algorithm will compute the solution for that iteration, as follows:

$$Y^t = \alpha \mathcal{L} Y^{t-1} + (1 - \alpha)Y^0$$

The algorithm stops its iteration when a certain condition is met. Common stopping conditions include convergence, where the labels no longer change between iterations, or reaching a maximum number of iterations specified by the user. It is important to underline the term $(1 - \alpha)Y^0$ of the equation. Indeed, as we said, label spreading does not force the labeled element of the solution to be equal to its original value. Instead, the algorithm uses a regularization parameter $\alpha \in [0,1)$ to weigh the influence of the original solution at each iteration. This allows us to explicitly impose the "quality" of the original solution and its influence on the end solution.

As with the label propagation algorithm, in the following code snippet, we propose a modified version of the LabelSpreading class available in the scikit-learn library due to the motivations we already mentioned in the previous section. We propose the GraphLabelSpreading class by extending our GraphLabelPropagation class, since the only difference will be in the fit() method of the class. The whole algorithm is provided in the 05_supervised_graph_machine_learning/02_Shallow_embeddings.ipynb notebook available in the GitHub repository of this book:

```
class GraphLabelSpreading(GraphLabelPropagation):
    def fit(self, X, y):
        X, y = self._validate_data(X, y)
        self.X_ = X
        check_classification_targets(y)
        D = [X.degree(n) for n in X.nodes()]
        D = np.diag(D)
        D_inv = np.matrix(fractional_matrix_power(D,-0.5))
        L = D_inv*nx.to_numpy_matrix(G)*D_inv
        # label construction
        # construct a categorical distribution for classification only
        labeled_index = np.where(y!=-1)[0]
        unique_classes = np.unique(y[labeled_index])
        self.classes_ = unique_classes
        Y0 = np.array([self.build_label(y[x], len(unique_classes)) if x in
labeled_index else np.zeros(len(unique_classes)) for x in range(len(y))])
        Y_prev = Y0
        it = 0
        c_tool = 10
        while it < self.max_iter & c_tool > self.tol:
            Y = (self.alpha*(L*Y_prev))+((1-self.alpha)*Y0)
            it +=1
            Y = Y / (Y.sum(axis=1) + 1e-7)  # Normalize
            c_tol = np.sum(np.abs(Y-Y_prev))
            Y_prev = Y
        self.label_distributions_ = Y
        return self
```

Also in this class, the `fit()` function is the focal point. The function takes as input a networkx graph *X* and an array *Y* representing the labels assigned to each node. Nodes without labels should have a representative value of -1. The `while` loop computes the Y^t value at each iteration, weighting the influence of the initial labeling via the parameter α. Also, for this algorithm, the number of iterations and the difference between two consecutive solutions are used as stop criteria.

The algorithm can be applied to the example graph depicted in *Figure 5.2* using the following code:

```
gls = GraphLabelSpreading()
y = np.array([-1 for x in range(len(G.nodes()))])
y[0] = 0
y[6] = 1
gls.fit(G,y)
gls.predict_proba(G)
```

In the following diagram, the result obtained by the algorithm is shown:

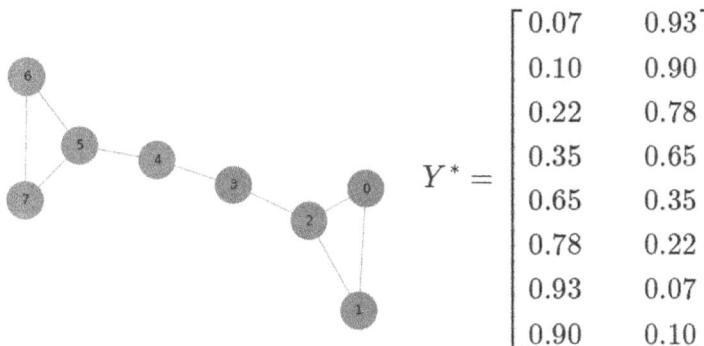

$$Y^* = \begin{bmatrix} 0.07 & 0.93 \\ 0.10 & 0.90 \\ 0.22 & 0.78 \\ 0.35 & 0.65 \\ 0.65 & 0.35 \\ 0.78 & 0.22 \\ 0.93 & 0.07 \\ 0.90 & 0.10 \end{bmatrix}$$

Figure 5.8: The final labeled graph (left) and the final probability assignment matrix (right) generated after applying the label propagation algorithm on the graph shown in Figure 5.2

The result visible in the diagram shown in *Figure 5.8* looks similar to the one obtained using the label propagation algorithm.

In the next section, we will continue our description of supervised graph embedding methods. We will describe how network-based information helps regularize the training and create more robust models.

Graph regularization methods

The shallow embedding methods described in the previous section show how topological information and relations between data points can be encoded and leveraged in order to build more robust classifiers and address semi-supervised tasks. In general terms, network information can be extremely useful in constraining models and enforcing the output to be smooth within neighboring nodes. As we have already seen in previous sections, this idea can be efficiently used in semi-supervised tasks, when propagating the information on neighbor unlabeled nodes.

On the other hand, this can also be used to regularize the learning phase in order to create more robust models that tend to better generalize to unseen examples. Both the label propagation and the label spreading algorithms we have seen previously can be implemented as a cost function to be minimized when we add an additional regularization term. Generally, in supervised tasks, we can write the cost function $\mathcal{L}(x)$ to be minimized in the following form:

$$\mathcal{L}(x) = \sum_{i \in S} \mathcal{L}_s(y_i, f(x_i)) + \sum_{i,j \in S,U} \mathcal{L}_g(f(x_i), f(x_j), G)$$

Let's break down the equation. We have two terms in the sum: the first term focuses on fitting the model to the labeled data, while the second term ensures that the model takes advantage of the graph structure to regularize predictions. Let's look at this in more detail.

$\mathcal{L}_s(y_i, f(x_i))$ is the supervised loss term. It measures how well the predicted label $f(x_i)$ matches the true label y_i for each labeled sample $i \in S$. Its exact form depends on the problem; it can be a common loss function like mean **squared error** (**MSE**) (for regression tasks) or cross-entropy loss (for classification tasks).

$\mathcal{L}_g(f(x_i), f(x_j), G)$ is the graph-based regularization term. It uses the graph structure G to enforce consistency in predictions. Specifically, it ensures that connected nodes x_i and x_j in the graph have similar predicted labels $f(x_i)$ and $f(x_j)$ for each labeled sample $i \in S$ and unlabeled sample $j \in U$. This term depends on the graph topology and the relationship between node features.

In this section, we will further describe such an idea and see how this can be very powerful, especially when regularizing the training of neural networks, which—as you might know—naturally tend to overfit and/or need large amounts of data to be trained efficiently.

Manifold regularization and semi-supervised embedding

Manifold regularization (Belkin et al., 2006) extends the label propagation framework by parametrizing the model function in the **reproducing kernel Hilbert space (RKHS)** and using as a supervised loss function (first term in the previous equation) the **MSE** or the hinge loss. In other words, when training an SVM or a least squares fit manifold regularization applies, a graph regularization term based on the Laplacian matrix L, as follows:

$$\sum_{i,j \in S, U} W_{ij} ||f(x_i) - f(x_j)||_2^2 = \bar{f} L \bar{f}$$

Here, W_{ij} represents the weight of the edge between nodes i and j, while $f(x_i)$ and $f(x_j)$ are the predicted labels. The equation enforces smoothness in the learned prediction function by penalizing differences in function values between connected nodes. Essentially, the term ensures that neighboring nodes in the graph have similar representations, reinforcing the manifold assumption that similar data points lie close to each other in the feature space.

For this reason, these methods are generally labeled as **Laplacian regularization**, and such a formulation leads to **Laplacian regularized least squares (LapRLS)** and **LapSVM** classifications. Label propagation and label spreading can be seen as a special case of manifold regularization. Besides, these algorithms can also be used in the case of no-labeled data (first term in the equation disappearing) reducing to **Laplacian eigenmaps**.

On the other hand, they can also be used in the case of a fully labeled dataset, in which case the preceding terms constrain the training phase to regularize the training and achieve more robust models. Moreover, being the classifier parametrized in the RKHS, the model can be used on unobserved samples and does not require test samples to belong to the input graph. In this sense, it is therefore an *inductive* model.

Manifold learning still represents a form of shallow learning, whereby the parametrized function does not leverage on any form of intermediate embeddings. **Semi-supervised embedding** (Weston et al., 2012) extends the concepts of graph regularization to deeper architectures by imposing the constraint and the smoothness of the function on intermediate layers of a neural network. Let's define g_{h_k} as the intermediate output of the kth hidden layer. The regularization term proposed in the semi-supervised embedding framework reads as follows:

$$\mathcal{L}_G^{h_k} = \sum_{i,j \in S, U} \mathcal{L}(W_{ij}, g_{h_k}(x_i), g_{h_k}(x_j))$$

Here, the function enforces similarity between representations of connected nodes i and j at a specific layer h_k. This encourages smoothness in the learned embeddings, ensuring that the network learns meaningful representations that align with the graph structure.

Depending on where the regularization is imposed, three different configurations (shown in *Figure 5.9*) can be achieved, as follows:

- Regularization is applied to the final output of the network. This corresponds to a generalization of the manifold learning technique to multilayer neural networks.

- Regularization is applied to an intermediate layer of the network, thus regularizing the embedding representation.

- Regularization is applied to an auxiliary network that shares the first k-1 layers. This basically corresponds to training an unsupervised embedding network while simultaneously training a supervised network. This technique basically imposes a derived regularization on the first k-1 layers that are constrained by the unsupervised network as well and simultaneously promotes an embedding of the network nodes.

The following diagram shows an illustration of the three different configurations—with their similarities and differences—that can be achieved using a semi-supervised embedding framework:

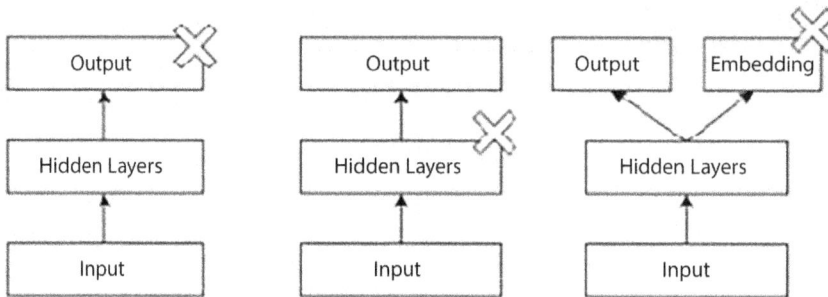

Figure 5.9: Semi-supervised embedding regularization configurations: graph regularization, indicated by the cross, can be applied to the output (left), to an intermediate layer (center), or to an auxiliary network (right)

In its original formulation, the loss function used for the embeddings is the one derived from the Siamese network formulation, as follows:

$$\mathcal{L}\left(W_{ij}, g_{h_k}^{(i)}, g_{h_k}^{(j)}\right) = \begin{cases} ||g_{h_k}^{(i)} - g_{h_k}^{(j)}||^2 \ if \ W_{ij} = 1 \\ \max\left(0, m - ||g_{h_k}^{(i)} - g_{h_k}^{(j)}||^2\right) if \ W_{ij} = 0 \end{cases}$$

As can be seen by this equation, the loss function ensures the embeddings of neighboring nodes stay close. On the other hand, non-neighbors are instead pulled apart to a distance (at least) specified by the threshold m. As compared to the regularization based on the Laplacian $\bar{f}L\bar{f}$ (although for neighboring points, the penalization factor is effectively recovered), the one shown here is generally easier to optimize with gradient descent.

The best choice among the three configurations presented in *Figure 5.9* is largely influenced by the data at your disposal as well as on your specific use case—that is, whether you need a regularized model output or to learn a high-level data representation. However, you should always keep in mind that when using softmax layers (usually done at the output layer), the regularization based on the hinge loss may not be very appropriate or suited for log probabilities. In such cases, regularized embeddings and relative loss should instead be introduced at intermediate layers. However, be aware that embeddings that reside in deeper layers are generally harder to train and require careful tuning of the learning rate and margins to be used. In the next section, we will learn about a generalization of manifold learning. While we have not yet discussed a practical example of manifold learning, we will cover it as part of this generalization.

Neural graph learning

Neural graph learning (NGL) basically generalizes the previous formulations and, as we will see, makes it possible to seamlessly apply graph regularization to any form of a neural network, including CNNs and **recurrent neural networks** (RNNs). In particular, there exists an extremely powerful framework named **neural structured learning** (NSL) that allows us to extend in a very few lines of code a neural network implemented in TensorFlow with graph regularization. The networks can be of any kind: natural or synthetic.

When synthetic, graphs can be generated in several ways. For instance, one approach involves constructing embeddings of raw input features (e.g., using feature vectors from images, text, or tabular data) in an unsupervised manner, then using a similarity or distance metric to define edges between nodes. You can also generate synthetic graphs using adversarial examples. Adversarial examples are artificially generated samples obtained by perturbing actual (real) examples in such a way that we confound the network, trying to force a prediction error. These very carefully designed samples (obtained by perturbing a given sample in the gradient-descent direction in order to maximize errors) can be connected to their related samples, thus generating a graph. These connections can then be used to train a graph-regularized version of the network, allowing us to obtain models that are more robust against adversarially generated examples.

NGL extends the regularization by augmenting the tuning parameters for graph regularization in neural networks, decomposing the contribution of labeled-labeled, labeled-unlabeled, and unlabeled-unlabeled relations using three parameters, α_1, α_2, and α_3, respectively, as follows:

$$L = L_s + \alpha_1 \sum_{i,j \in LL} W_{ij}\, d\left(g_{h_k}^{(i)}, g_{h_k}^{(j)}\right) + \alpha_2 \sum_{i,j \in LU} W_{ij}\, d\left(g_{h_k}^{(i)}, g_{h_k}^{(j)}\right) + \alpha_3 \sum_{i,j \in UU} W_{ij}\, d\left(g_{h_k}^{(i)}, g_{h_k}^{(j)}\right)$$

The function d represents a generic distance between two vectors—for instance, the L2 norm $\|\cdot\|_2$. By varying the coefficients and the definition of g_{h_k}, we can arrive at the different algorithms seen previously as limiting behavior, as follows:

- When $\alpha_1 \neq 0$ $\forall i$, we retrieve the non-regularized version of a neural network.
- When only $\alpha_1 \neq 0$, we recover a fully supervised formulation where relationships between nodes act to regularize the training.
- When we substitute g_{h_k} (which is parametrized by a set of alpha coefficients) with a set of values Y_i^* (to be learned) that map each sample to its instance class, we recover the label propagation formulation.

Loosely speaking, the NGL formulations can be seen as non-linear versions of the label propagation and label spreading algorithms, or as a form of a graph-regularized neural network for which the manifold learning or semi-supervising embedding can be obtained.

We will now apply NGL to a practical example, where you will learn how to use graph regularization in neural networks. To do so, we will use the NLS framework (https://github.com/tensorflow/neural-structured-learning), which is a library built on top of TensorFlow that makes it possible to implement graph regularization with only a few lines of codes on top of standard neural networks.

For our example, we will be using the Cora dataset, which is a labeled dataset that consists of 2,708 scientific papers in computer science that have been classified into seven classes. Each paper represents a node that is connected to other nodes based on citations. In total, there are 5,429 links in the network.

Moreover, each node is further described by a 1,433-long vector of binary values (0 or 1) that represent a dichotomic **bag-of-words (BOW)** representation of the paper: a one-hot-encoding algorithm indicating the presence/absence of a word in a given vocabulary made up of 1,433 terms.

The Cora dataset can be downloaded directly from the stellargraph library with a few lines of code, as follows:

```
from stellargraph import datasets
dataset = datasets.Cora()
dataset.download()
G, labels = dataset.load()
```

This returns two outputs, outlined as follows:

- G, which is the citation network containing the network nodes, edges, and the features describing the BOW representation.

- labels, which is a pandas Series that maps each paper ID (node) to one of the following classes:

```
['Neural_Networks', 'Rule_Learning', 'Reinforcement_Learning',
'Probabilistic_Methods', 'Theory', 'Genetic_Algorithms', 'Case_
Based']
```

Starting from this information, we will create a training set and a validation set. In the training samples, we will include information relating to neighbors (which may or may not belong to the training set and therefore have a label), and this will be used to regularize the training.

Validation samples, on the other hand, will not have neighbor information and the predicted label will only depend on the node features—namely, the BOW representation. Therefore, we will leverage both labeled and unlabeled samples (semi-supervised task) in order to produce an inductive model that can also be used against unobserved samples.

1. As a preparatory step, we conveniently structure the node features as a DataFrame, whereas we store the graph as an adjacency matrix, as follows:

```
adjMatrix = pd.DataFrame.sparse.from_spmatrix(
        G.to_adjacency_matrix(),
        index=G.nodes(), columns=G.nodes()
)
features = pd.DataFrame(G.node_features(), index=G.nodes())
```

Using adjMatrix, we implement a helper function that is able to retrieve the closest topn neighbors of a node, returning the node ID and the edge weight, as illustrated in the following code snippet:

```
def getNeighbors(idx, adjMatrix, topn=5):
    weights = adjMatrix.loc[idx]
    neighbors = weights[weights>0]\
```

```
            .sort_values(ascending=False)\
            .head(topn)
        return [(k, v) for k, v in neighbors.iteritems()]
```

Using the preceding information together with the helper function, we can merge the information into a single DataFrame, as follows:

```
dataset = {
    index: {
        "id": index,
        "words": [float(x)
                    for x in features.loc[index].values],
        "label": label_index[label],
        "neighbors": getNeighbors(index, adjMatrix, topn)
    }
    for index, label in labels.items()
}
df = pd.DataFrame.from_dict(dataset, orient="index")
```

This DataFrame represents the node-centric feature space. This would suffice if we were to use a regular classifier that does not exploit the information of the relationships between nodes. However, in order to allow the computation of the graph regularization term, we need to join the preceding DataFrame with information relating to the neighborhood of each node. We then define a function able to retrieve and join the neighborhood information, as follows:

```
def getFeatureOrDefault(ith, row):
    try:
        nodeId, value = row["neighbors"][ith]
        return {
            f"{GRAPH_PREFIX}_{ith}_weight": value,
            f"{GRAPH_PREFIX}_{ith}_words": df.loc[nodeId]["words"]
        }
    except:
        return {
            f"{GRAPH_PREFIX}_{ith}_weight": 0.0,
            f"{GRAPH_PREFIX}_{ith}_words": [float(x) for x in
np.zeros(1433)]
        }
```

```
def neighborsFeatures(row):
    featureList = [getFeatureOrDefault(ith, row) for ith in
range(topn)]
    return pd.Series(
        {k: v
         for feat in featureList for k, v in feat.items()}
    )
```

As shown in the preceding code snippet, when the neighbors are less than topn, we set the weight and the one-hot encoding of the words to 0. The GRAPH_PREFIX constant is a prefix that is to be prepended to all features that will later be used by the nsl library to regularize the training. Although it can be changed, in the following code snippet we will keep its value equal to the default value: "NL_nbr".

This function can be applied to the DataFrame in order to compute the full feature space, as follows:

```
neighbors = df.apply(neighborsFeatures, axis=1)
allFeatures = pd.concat([df, neighbors], axis=1)
```

We now have in allFeatures all the ingredients we need to implement our graph-regularized model.

2. We can now start the training process by splitting our dataset into a training set and a validation set, as follows:

```
n = int(np.round(len(labels)*ratio))
labelled, unlabelled = model_selection.train_test_split(
    allFeatures, train_size=n, test_size=None, stratify=labels
)
```

Here, ratio is the proportion of the dataset that will be labeled. As the ratio decreases, we expect the performance of standard, non-regularized classifiers to reduce. However, such a reduction can be compensated by leveraging network information provided by unlabeled data. We thus expect graph-regularized neural networks to provide better performance thanks to the augmented information they leverage. For the following code snippet, we will assume a ratio value equal to 0.2.

Before feeding this data into our neural network, we convert the DataFrame into a TensorFlow tensor and dataset, which is a convenient representation that will allow the model to refer to feature names in its input layers.

Since the input features have different data types, it is best to handle the dataset creation separately for the weights, words, and labels values, as follows:

```
train_base = {
    "words": tf.constant([
        tuple(x) for x in labelled["words"].values
    ]),
    "label": tf.constant([
        x for x in labelled["label"].values
    ])
}
train_neighbor_words = {
    k: tf.constant([tuple(x) for x in labelled[k].values])
    for k in neighbors if "words" in k
}
train_neighbor_weights = {
    k: tf.constant([tuple([x]) for x in labelled[k].values])
    for k in neighbors if "weight" in k
}
```

Now that we have the tensor, we can merge all this information into a TensorFlow dataset, as follows:

```
trainSet = tf.data.Dataset.from_tensor_slices({
    k: v
    for feature in [train_base, train_neighbor_words,
                    train_neighbor_weights]
    for k, v in feature.items()
})
```

We can similarly create a validation set. As mentioned previously, since we want to design an inductive algorithm, the validation dataset does not need any neighborhood information. The code is illustrated in the following snippet:

```
validSet = tf.data.Dataset.from_tensor_slices({
    "words": tf.constant([
        tuple(x) for x in unlabelled["words"].values
    ]),
    "label": tf.constant([
        x for x in unlabelled["label"].values
```

```
    ])
})
```

Before feeding the dataset into the model, we split the features from the labels, as follows:

```
def split(features):
    labels=features.pop("label")
    return features, labels
trainSet = trainSet.map(f)
validSet = validSet.map(f)
```

That's it! We have generated the inputs to our model. We could also inspect one sample batch of our dataset by printing the values of features and labels, as shown in the following code block:

```
for features, labels in trainSet.batch(2).take(1):
    print(features)
    print(labels)
```

3. It is now time to create our first model. To do this, we start from a simple architecture that takes as input the one-hot representation and has two hidden layers, composed of a Dense layer plus a Dropout layer with 50 units each, as follows:

```
inputs = tf.keras.Input(
    shape=(vocabularySize,), dtype='float32', name='words'
)
cur_layer = inputs
for num_units in [50, 50]:
    cur_layer = tf.keras.layers.Dense(
        num_units, activation='relu'
    )(cur_layer)
    cur_layer = tf.keras.layers.Dropout(0.8)(cur_layer)
outputs = tf.keras.layers.Dense(
    len(label_index), activation='softmax',
    name="label"
)(cur_layer)
model = tf.keras.Model(inputs, outputs=outputs)
```

Indeed, we could also train this model without graph regularization by simply compiling the model to create a computational graph, as follows:

```
model.compile(
    optimizer='adam',
    loss='sparse_categorical_crossentropy',
    metrics=['accuracy']
)
```

And then, we could run it as usual, also allowing the history file to be written to disk in order to be monitored using TensorBoard, as illustrated in the following code snippet:

```
from tensorflow.keras.callbacks import TensorBoard
model.fit(
    trainSet.batch(128), epochs=200, verbose=1,
    validation_data=validSet.batch(128),
    callbacks=[TensorBoard(log_dir='/tmp/base')]
)
```

At the end of the process, we should have something similar to the following output:

```
Epoch 200/200
loss: 0.7798 - accuracy: 06795 - val_loss: 1.5948 - val_accuracy:
0.5873
```

4. With a top performance around 0.6 in accuracy, we now need to create a graph-regularized version of the preceding model. For this step, it is important to recreate our model from scratch, especially when comparing the results: if we were to use layers already initialized and used in the previous model, the layer weights would not be random but would be used with the ones already optimized in the preceding run.

    Once a new model has been created, adding a graph regularization technique to be used at training time can be done in just a few lines of code, as follows:

```
import neural_structured_learning as nsl
graph_reg_config = nsl.configs.make_graph_reg_config(
    max_neighbors=2,
    multiplier=0.1,
    distance_type=nsl.configs.DistanceType.L2,
    sum_over_axis=-1)
graph_reg= nsl.keras.GraphRegularization(
     model, graph_reg_config)
```

Let's analyze the different hyperparameters of the regularization, as follows:

- max_neighbors tunes the number of neighbors that ought to be used for computing the regularization loss for each node.

- multiplier corresponds to the coefficients that tune the importance of the regularization loss. Since we only consider labeled-labeled and labeled-unlabeled, this effectively corresponds to α_1 and α_2.

- distance_type represents the pairwise distance d to be used.

- sum_over_axis sets whether the weighted average sum should be calculated with respect to features (when set to None) or to samples (when set to -1).

The graph-regularized model can be compiled and run in the same way as before with the following commands:

```
graph_reg.compile(
    optimizer='adam',
    loss='sparse_categorical_crossentropy',
    metrics=['accuracy']
)
model.fit(
    trainSet.batch(128), epochs=200, verbose=1,
    validation_data=validSet.batch(128),
    callbacks=[TensorBoard(log_dir='/tmp/nsl')]
)
```

Note that the loss function now also accounts for the graph regularization term, as defined previously. Therefore, we now also introduce information coming from neighboring nodes that regularizes the training of our neural network. The preceding code, after about 200 iterations, provides the following output:

```
Epoch 200/200
loss: 0.9136 - accuracy: 06405 - scaled_graph_loss: 0.0328 - val_
loss: 1.2526 - val_accuracy: 0.6320
```

As you can see, graph regularization, when compared to the vanilla version, has allowed us to boost the performance in terms of validation accuracy by about 7%. Not bad at all! This is because the model is now able to exploit relational information between data points. While the non-regularized version relies just on individual feature representations, the graph-regularized version now incorporates neighborhood information, which can reduce overfitting and improve generalization, particularly in cases where the data contains inherent structure or relationships.

You can perform several experiments, changing the ratio of labeled/unlabeled samples, the number of neighbors to be used, the regularization coefficient, the distance, and more. We encourage you to play around with the notebook that is provided with this book to explore the effect of different parameters yourself.

In the right panel of the following screenshot, we show the dependence of the performance measured by the accuracy as the supervised ratio increases. As expected, performance increases as the ratio increases. On the left panel, we show the accuracy increments on the validation set for various configurations of neighbors and supervised ratio, defined by:

$$\Delta a = \text{accuracy}_{reg} - \text{accuracy}_{no\ reg}$$

As can be seen in *Figure 5.10*, almost all graph-regularized versions outperform the vanilla models:

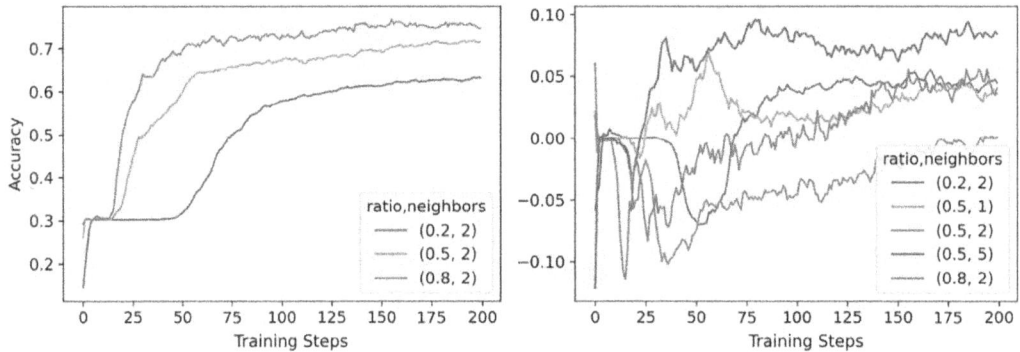

Figure 5.10: Accuracy on the validation set for the graph-regularized neural networks with neighbors = 2 and various supervised ratios (left); accuracy increments on the validation set for the graph-regularized neural networks compared to the vanilla version (right)

The only exceptions are configuration neighbors = 2 and ratio = 0.5, for which the two models perform very similarly. However, the curve has a clear positive trend and we reasonably expect the graph-regularized version to outperform the vanilla model for a larger number of epochs.

In the notebook, we also use another interesting feature of TensorFlow for creating the datasets. Instead of using a pandas DataFrame, as we did previously, we will create a dataset using the TensorFlow Example, Features, and Feature classes, which, besides providing a high-level description of samples, also allow us to serialize the input data (using protobuf) to make them compatible across platforms and programming languages.

If you are interested in further using TensorFlow both for prototyping models and deploying them into production via data-driven applications (maybe written in other languages), we strongly advise you to dig further into these concepts.

Planetoid

The methods discussed so far provide graph regularization that is based on the Laplacian matrix. As we have seen in previous chapters, enforcing constraints based on W_{ij} ensures that first-order proximity is preserved. Yang et al. (2016) proposed a method to extend graph regularization in order to also account for higher-order proximities. Their approach, which they named **Planetoid** (short for **Predicting Labels And Neighbors with Embeddings Transductively Or Inductively from Data**), extends skip-gram methods used for computing node embeddings to incorporate node-label information.

As we have seen in the previous chapter, skip-gram methods are based on generating random walks through a graph and then using the generated sequences to learn embeddings via a skip-gram model. The following diagram shows how the unsupervised version is modified to account for the supervised loss:

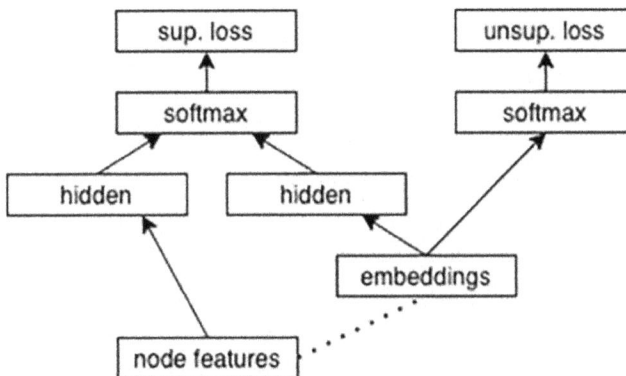

Figure 5.11: Sketch of the Planetoid architecture: the dotted line represents a parametrized function that allows the method to extend from transductive to inductive

As shown in *Figure 5.11*, embeddings are fed to both of the following:

- A softmax layer to predict the graph context of the sampled random-walk sequences
- A set of hidden layers that combine with the hidden layers derived from the node features in order to predict the class labels

The cost function to be minimized to train the combined network is composed of a supervised and an unsupervised loss — $\mathcal{L}_s$ and $\mathcal{L}_u$, respectively. The unsupervised loss is analogous to the one used with skip-gram with negative sampling, whereas the supervised loss minimizes the conditional probability of predicting the label y and can be written as follows:

$$\mathcal{L}_s = -\sum_{i \in S} \log p\left(y_i | x_i, e_i\right)$$

Where S is the set of labeled points and x_i and e_i are the input feature and the learned embeddings for the ith data point, respectively. The goal, then, is to minimize the loss by encouraging the model to make predictions that are closely aligned with the ground truth. The preceding formulation is *transductive* as it requires samples to belong to the graph in order to be applied. In a semi-supervised task, this method can be efficiently used to predict labels for unlabeled examples. However, it cannot be used for unobserved samples. As shown by the dotted line in *Figure 5.11*, an inductive version of the Planetoid algorithm can be obtained by parametrizing the embeddings as a function of the node features, via dedicated connected layers.

As you can see, Planetoid represents a simple extension of the example in the previous section. Therefore, we will not implement it here in the book. Nevertheless, you can find a possible implementation in our GitHub repository.

> All the techniques described so far remain valuable for their simplicity, interpretability, and efficiency. Nevertheless, nowadays, with the rise of neural network-based techniques, it is common to solve such problems using modern techniques such as GNNs. It is important to remark, however, that simpler methods can often achieve comparable results in specific scenarios, offering computational advantages and complementing modern techniques. Highlighting these foundational methods underscores their relevance and prevents them from being overshadowed by the drive for innovation.

Graph CNNs

In *Chapter 4*, *Unsupervised Graph Learning*, we learned the main concepts behind GNNs and **graph convolutional networks (GCNs)**. We also learned the difference between spectral graph convolution and spatial graph convolution. More precisely, we saw that GCN layers can be used to encode graphs or nodes under unsupervised settings by learning how to preserve graph properties such as node similarity.

In this chapter, we will explore such methods under supervised settings. This time, our goal is to learn graphs or node representations that can accurately *predict node or graph labels*. It is indeed worth noting that the encoding function remains the same. What will change is the objective!

Graph classification using GCNs

Let's consider again our PROTEINS dataset. Let's load the dataset as follows:

```
import pandas as pd
from stellargraph import datasets
dataset = datasets.PROTEINS()
graphs, graph_labels = dataset.load()
# necessary for converting default string labels to int
labels = pd.get_dummies(graph_labels, drop_first=True)
```

In the following example, we are going to use (and compare) one of the most widely used GCN algorithms for graph classification—*GCN* by Kipf and Welling:

1. stellargraph, which we are using to build the model, uses tf.Keras as the backend. According to its specific criteria, we need a data generator to feed the model. More precisely, since we are addressing a supervised graph classification problem, we can use an instance of the PaddedGraphGenerator class of stellargraph, which automatically resolves differences in the number of nodes by using padding. Here is the code required for this step:

    ```
    from stellargraph.mapper import PaddedGraphGenerator
    generator = PaddedGraphGenerator(graphs=graphs)
    ```

2. We are now ready to actually create our first model. We will create and stack together four GCN layers through the utility function of stellargraph, as follows:

    ```
    from stellargraph.layer import DeepGraphCNN
    from tensorflow.keras import Model
    ```

```
from tensorflow.keras.optimizers import Adam
from tensorflow.keras.layers import Dense, Conv1D, MaxPool1D,
Dropout, Flatten
from tensorflow.keras.losses import binary_crossentropy
import tensorflow as tf
nrows = 35  # the number of rows for the output tensor
layer_dims = [32, 32, 32, 1]
# backbone part of the model (Encoder)
dgcnn_model = DeepGraphCNN(
    layer_sizes=layer_dims,
    activations=["tanh", "tanh", "tanh", "tanh"],
    k=nrows,
    bias=False,
    generator=generator,
)
```

3. This backbone will be concatenated to **one-dimensional (1D)** convolutional layers and
    fully connected layers using tf.Keras, as follows:

```
# necessary for connecting the backbone to the head
gnn_inp, gnn_out = dgcnn_model.in_out_tensors()
# head part of the model (classification)
x_out = Conv1D(filters=16, kernel_size=sum(layer_dims),
strides=sum(layer_dims))(gnn_out)
x_out = MaxPool1D(pool_size=2)(x_out)
x_out = Conv1D(filters=32, kernel_size=5, strides=1)(x_out)
x_out = Flatten()(x_out)
x_out = Dense(units=128, activation="relu")(x_out)
x_out = Dropout(rate=0.5)(x_out)
predictions = Dense(units=1, activation="sigmoid")(x_out)
```

4. Let's create and compile a model using tf.Keras utilities. We will train the model with a
    binary_crossentropy loss function (to measure the difference between predicted labels
    and ground truth) with the Adam optimizer and a **learning rate** of 0.0001. We will also
    monitor the accuracy metric while training. The code is illustrated in the following snippet:

```
model = Model(inputs=gnn_inp, outputs=predictions)
model.compile(optimizer=Adam(lr=0.0001), loss=binary_crossentropy,
metrics=["acc"])
```

5. We can now exploit scikit-learn utilities to create train and test sets. In our experiments, we will be using 70% of the dataset as a training set and the remainder as a test set. In addition, we need to use the `flow` method of the generator to supply them to the model. The code to achieve this is shown in the following snippet:

```
from sklearn.model_selection import train_test_split
train_graphs, test_graphs = train_test_split(
graph_labels, test_size=.3, stratify=labels,)
gen = PaddedGraphGenerator(graphs=graphs)
train_gen = gen.flow(
    list(train_graphs.index - 1),
    targets=train_graphs.values,
    symmetric_normalization=False,
    batch_size=50,
)
test_gen = gen.flow(
    list(test_graphs.index - 1),
    targets=test_graphs.values,
    symmetric_normalization=False,
    batch_size=1,
)
```

6. It's now time for training. We train the model for 100 epochs. However, feel free to play with the hyperparameters to gain better performance. Here is the code for this:

```
epochs = 100
history = model.fit(train_gen, epochs=epochs, verbose=1,
validation_data=test_gen, shuffle=True,)
```

After 100 epochs, this should be the output:

```
Epoch 100/100
loss: 0.5121 - acc: 0.7636 - val_loss: 0.5636 - val_acc: 0.7305
```

Here, we are achieving about 76% accuracy on the training set and about 73% accuracy on the test set. These results indicate that the model has learned meaningful patterns from the graph structure and node features, which it is exploiting effectively to generalize reasonably well.

Node classification using GraphSAGE

In the next example, we will train GraphSAGE to classify nodes of the Cora dataset.

Let's first load the dataset using stellargraph utilities, as follows:

```
dataset = datasets.Cora()
G, nodes = dataset.load()
```

Follow this list of steps to train GraphSAGE to classify nodes of the Cora dataset:

1. As in the previous example, the first step is to split the dataset. We will be using 90% of the dataset as a training set and the remainder for testing. Here is the code for this step:

    ```
    train_nodes, test_nodes = train_test_split(nodes, train_
    size=0.1,test_size=None, stratify=nodes)
    ```

2. This time, we will convert labels using **one-hot representation**. This representation is often used for classification tasks and usually leads to better performance. Specifically, let c be the number of possible targets (seven, in the case of the Cora dataset), and each label will be converted to a vector of size c, where all the elements are 0 except for the one corresponding to the target class. The code is illustrated in the following snippet:

    ```
    from sklearn import preprocessing
    label_encoding = preprocessing.LabelBinarizer()
    train_labels = label_encoding.fit_transform(train_nodes)
    test_labels = label_encoding.transform(test_nodes)
    ```

3. Let's create a generator to feed the data into the model. We will be using an instance of the GraphSAGENodeGenerator class of stellargraph. We will use the flow method to feed the model with the train and test sets, as follows:

    ```
    from stellargraph.mapper import GraphSAGENodeGenerator
    batchsize = 50
    n_samples = [10, 5, 7]
    generator = GraphSAGENodeGenerator(G, batchsize, n_samples)
    train_gen = generator.flow(train_nodes.index, train_labels,
    shuffle=True)
    test_gen = generator.flow(test_labels.index, test_labels)
    ```

4. Finally, let's create the model and compile it. For this exercise, we will be using a GraphSAGE encoder with three layers of 32, 32, and 16 dimensions, respectively. The encoder will then be connected to a dense layer with softmax activation to perform the classification. We will use an Adam optimizer with a learning rate of 0.03 and categorical_crossentropy as the loss function. The code is illustrated in the following snippet:

```
from stellargraph.layer import GraphSAGE
from tensorflow.keras.losses import categorical_crossentropy
graphsage_model = GraphSAGE(layer_sizes=[32, 32, 16],
generator=generator, bias=True, dropout=0.6,)
gnn_inp, gnn_out = graphsage_model.in_out_tensors()
outputs = Dense(units=train_labels.shape[1], activation="softmax")
(gnn_out)
# create the model and compile
model = Model(inputs=gnn_inp, outputs=outputs)
model.compile(optimizer=Adam(lr=0.003), loss=categorical_
crossentropy, metrics=["acc"],)
```

5. It's now time to train the model. We will train the model for 20 epochs, as follows:

```
model.fit(train_gen, epochs=20, validation_data=test_gen, verbose=2,
shuffle=False)
```

6. This should be the output:

```
Epoch 20/20
loss: 0.8252 - acc: 0.8889 - val_loss: 0.9070 - val_acc: 0.8011
```

We achieved about 89% accuracy over the training set and about 80% accuracy over the test set, indicating that GraphSAGE has strong capability to learn meaningful information from the graph while avoiding excessive overfitting.

Summary

In this chapter, we have learned how supervised ML can be effectively applied on graphs to solve real problems such as node and graph classification. In particular, we first analyzed how graph and node properties can be directly used as features to train classic ML algorithms. We have seen shallow methods and simple approaches to learning node, edge, or graph representations for only a finite set of input data. We then learned how regularization techniques can be used during the learning phase in order to create more robust models that tend to generalize better. Finally, we have seen how GNNs can be applied to solve supervised ML problems on graphs.

But what can these algorithms be useful for? In the next chapter, we will explore common problems that can be solved using graph-based ML techniques.

Get This Book's PDF Version and Exclusive Extras

UNLOCK NOW

Scan the QR code (or go to packtpub.com/unlock).
Search for this book by name, confirm the edition,
and then follow the steps on the page.

Note: Keep your invoice handy. Purchases made
directly from Packt don't require one.

6

Solving Common Graph-Based Machine Learning Problems

Graph **machine learning (ML)** approaches can be useful for a wide range of tasks, with applications ranging from drug design to recommender systems in social networks. Furthermore, given the fact that such methods are *general by design* (meaning that they are not tailored to a specific problem), the same algorithm can be used to solve different problems.

There are common problems that can be effectively solved using graph-based learning techniques, as they excel at capturing relationships and structures in data. In this chapter, we will mention some of the most well studied of these by providing details about how a specific algorithm, among the ones we have already learned about in *Chapters 4, Unsupervised Graph Learning,* and *Chapters 5, Supervised Graph Learning*, can be used to solve a task. After reading this chapter, you will be aware of the formal definition of many common problems you may encounter when dealing with graphs. In addition, you will learn about useful ML pipelines that you can reuse on future real-world problems you will deal with.

More precisely, the following topics will be covered in this chapter:

- Predicting missing links in a graph
- Detecting meaningful structures such as communities
- Detecting graph similarities and graph matching

Technical requirements

All code files relevant to this chapter are available at https://github.com/PacktPublishing/ Graph-Machine-Learning/tree/main/Chapter06. Please refer to the *Practical exercises* section of *Chapters 1, Getting Started with Graphs,* for guidance on how to set up the environment to run the examples in this chapter, either using Poetry, pip, or docker.

Predicting missing links in a graph

Link prediction, also known as **graph completion**, is a common problem when dealing with graphs. More precisely, from a partially observed graph—that is a graph for which only a portion of the existing edges are known—we want to predict whether or not an edge exists between any given node pairs, as seen in *Figure 6.1.* Formally, let $G = (V, E)$ be a graph where V is its set of nodes and $E = E_0 \cup E_u$ is its set of edges. The set of edges E_0 are known as *observed links*, while the set of edges E_u are known as *unknown links*. The goal of the link prediction problem is to exploit the information of V and E_0 to estimate E_u. The partially observed graph can be seen here:

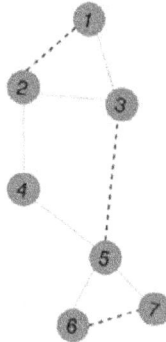

$$E_o = \{\{1, 3\}, \{2, 3\}, \{2, 4\}, \{4, 5\}, \{5, 6\}, \{5, 7\}\}$$
$$E_u = \{\{1, 2\}, \{3, 5\}, \{6, 7\}\}$$

Figure 6.1: Partially observed graph with observed link E_0 (solid lines) and unknown link E_u (dashed lines)

The link prediction problem is widely used in different domains, such as a recommender system, in order to propose friendships in social networks or items to purchase on e-commerce websites. It is also used in criminal network investigations in order to find hidden connections between criminal clusters, as well as in bioinformatics for the analysis of protein-protein interactions. In the next sections, we will discuss two families of approaches to solve the link prediction problem—namely, **similarity-based** and **embedding-based** methods.

Similarity-based methods

In this subsection, we show several simple algorithms to solve the link prediction problem. The main shared idea behind all these algorithms is to estimate a similarity function between each couple of nodes in a graph. If, according to the function, the nodes *look similar*, they will have a higher probability of being connected by an edge. We will divide these algorithms into two sub-families: **index-based** and **community-based** methods. The former contains all the methods through a simple calculation of an index based on the neighbors of a given couple of nodes. The latter contains more sophisticated algorithms, whereby the index is computed using information about the community to which a given pair of nodes belong. In order to give a practical example of these algorithms, we will use the standard implementation available in the networkx library in the networkx.algorithms.link_prediction package.

Index-based methods

In this section, we will show some algorithms available in networkx to compute the probability of an edge between two disconnected nodes. These algorithms are based on the calculation of a simple index through information obtained by analyzing the neighbors of the two disconnected nodes. Note that they are not implemented for DiGraph or MultiGraph and do not take relationship weights into account, as implementing such approaches would require additional theoretical considerations on directionality, edge multiplicity, and weight impact on indices.

Resource allocation index

The resource allocation index method estimates the probability that two nodes v and u are connected by estimating the *resource allocation index* for all node pairs according to the following formula:

$$Resource\ Allocation\ Index(u, v) = \sum_{w \in N(u) \cap N(v)} \frac{1}{|N(w)|}$$

Here, $N(u)$ and $N(v)$ computes the neighbors of nodes u and v, respectively, and w is a node that is a neighbor of both u and v. Thus, for each node w in the common neighborhood $N(u) \cap N(v)$, the inverse of its degree $1/|N(w)|$ captures the idea that nodes with fewer neighbors contribute more to the probability of an edge between u and v.

This index can be computed in `networkx` using the following code:

```
import networkx as nx
edges = [[1,3],[2,3],[2,4],[4,5],[5,6],[5,7]]
G = nx.from_edgelist(edges)
preds = nx.resource_allocation_index(G,[(1,2),(2,5),(3,4)])
```

The first parameter for the `resource_allocation_index` function is an input graph (depicted in *Figure 6.1*), while the second parameter is a list of possible edges. We want to compute the probability of a connection. As a result, we get the following output:

```
[(1, 2, 0.5), (2, 5, 0.5), (3, 4, 0.5)]
```

The output is a list containing couples of nodes such as $(1,2)$, $(2,5)$, and $(3,4)$, which form the resource allocation index. According to this output, the probability of having an edge between those couples of nodes is 0.5.

Jaccard coefficient

The algorithm computes the probability of a connection between two nodes u and v, according to the *Jaccard coefficient*, which measures the similarity between their neighborhoods as the ratio of the size of their intersection to the size of their union, computed as follows:

$$Jaccard\ Coefficient(u, v) = \frac{|N(u) \cap N(v)|}{|N(u) \cup N(v)|}$$

The function can be used in `networkx` using the following code:

```
import networkx as nx
edges = [[1,3],[2,3],[2,4],[4,5],[5,6],[5,7]]
G = nx.from_edgelist(edges)
preds = nx.resource_allocation_index(G,[(1,2),(2,5),(3,4)])
```

The `resource_allocation_index` function has the same parameters as the previous function. The result of the code is shown here:

```
[(1, 2, 0.5), (2, 5, 0.25), (3, 4, 0.3333333333333333)]
```

According to this output, the probability of having an edge between nodes $(1,2)$ is 0.5, while between nodes $(2,5)$ this is 0.25, and between nodes $(3,4)$ this is 0.333.

In networkx, other methods to compute the probability of a connection between two nodes based on their similarity score are nx.adamic_adar_index and nx.preferential_attachment, based on *Adamic/Adar index* and *preferential attachment index* calculations respectively. Those functions have the same parameters as the others, and accept a graph and a list of a pair of nodes where we want to compute the score. In the next section, we will show another family of algorithms based on community detection.

Summarizing, index-based methods are well suited for social networks, recommendation systems with dense graphs, and applications where computational efficiency is a priority. Indeed, their simplicity and efficiency is one of their main strengths. They also work well in networks where local neighborhood information is sufficient to infer missing links. However, they may struggle in case of sparse networks where nodes have few connections. Furthermore, they do not consider higher-order structural properties of the graph.

Community-based methods

As with index-based methods, the algorithms belonging to this family also compute an index representing the probability of the disconnected nodes being connected. The main difference between index-based and community-based methods is related to the logic behind them. Indeed, community-based methods, before generating the index, need to compute information about the community belonging to those nodes. In this subsection, we will show—also providing several examples—some common community-based methods.

Community common neighbor

In order to estimate the probability of two nodes being connected, this algorithm computes the number of common neighbors and adds to this value the number of common neighbors belonging to the same community. Formally, for two nodes v and u, the community common neighbor value is computed as follows:

$$Community\ Common\ Neighbor(u,v) = |N(v) \cup N(u)| + \sum_{w \in N(v) \cap N(u)} f(w)$$

In this formula, $f(w) = 1$ if w belongs to the same community of u and v; otherwise, this is 0. The function can be computed in networkx using the following code:

```
import networkx as nx
edges = [[1,3],[2,3],[2,4],[4,5],[5,6],[5,7]]
G = nx.from_edgelist(edges)
G.nodes[1]["community"] = 0
```

```
G.nodes[2]["community"] = 0
G.nodes[3]["community"] = 0
G.nodes[4]["community"] = 1
G.nodes[5]["community"] = 1
G.nodes[6]["community"] = 1
G.nodes[7]["community"] = 1
preds = nx.cn_soundarajan_hopcroft(G,[(1,2),(2,5),(3,4)])
```

From the preceding code snippet, it is possible to see how we need to assign the community property to each node of the graph. This property is used to identify nodes belonging to the same community when computing the function $f(w)$ defined in the previous equation. The community value, as we will see in the next section, can also be automatically computed using specific algorithms. As the example code above shows, the cn_soundarajan_hopcroft function takes the input graph and a couple of nodes for which we want to compute the score. As a result, we get the following output:

```
[(1, 2, 2), (2, 5, 1), (3, 4, 1)]
```

The output indicates the pairs of nodes and their corresponding scores, where the score reflects the number of shared neighbors between the nodes, adjusted by their community membership. The main difference from the previous function is in the index value. Indeed, we can easily see that the output is not in the range (0,1).

Community resource allocation

As with the previous method, the community resource allocation algorithm merges information obtained from the neighbors of the nodes with the community, as shown in the following formula:

$$Community\ Resource\ Allocation(u, v) = \sum_{w \in N(v) \cap N(u)} \frac{f(w)}{|N(w)|}$$

Here, $f(w) = 1$ if w belongs to the same community of u and v; otherwise, this is 0. The function can be computed in networkx using the following code:

```
import networkx as nx
edges = [[1,3],[2,3],[2,4],[4,5],[5,6],[5,7]]
G = nx.from_edgelist(edges)
G.nodes[1]["community"] = 0
G.nodes[2]["community"] = 0
G.nodes[3]["community"] = 0
G.nodes[4]["community"] = 1
```

```
G.nodes[5]["community"] = 1
G.nodes[6]["community"] = 1
G.nodes[7]["community"] = 1
preds = nx. ra_index_soundarajan_hopcroft(G,[(1,2),(2,5),(3,4)])
```

From the preceding code snippet, it is possible to see how we need to assign the community property to each node of the graph. This property is used to identify nodes belonging to the same community when computing the function $f(w)$ defined in the previous equation. The community value, as we will see in the next section, can also be automatically computed using specific algorithms. As we saw, the ra_index_soundarajan_hopcroft function takes the input graph and a couple of nodes for which we want to compute the score. As a result, we get the following output:

```
[(1, 2, 0.5), (2, 5, 0), (3, 4, 0)]
```

From the preceding output, it is possible to see the influence of the community in the computation of the index. Since nodes 1 and 2 belong to the same community, they have a higher value in the index. On the contrary, edges (2,5) and (3,4) feature the value 0, since they belong to a different community from each other.

In networkx, two other methods to compute the probability of a connection between two nodes based on their similarity score merged with community information are nx.within_inter_cluster and nx.common_neighbor_centrality.

Community-based methods, therefore, are best suited for applications where structural information is crucial, such as criminal network analysis and protein-protein interaction networks, since they leverage community structure to improve accuracy. Furthermore, they are effective in detecting hidden connections in sparse networks. However, they are more computationally expensive than index-based methods and may require community detection as a preprocessing step.

In the next section, we will describe a more complex method based on machine learning techniques and edge embedding to perform prediction of unknown edges.

Embedding-based methods

In this section, we describe a more advanced way to perform link prediction. The idea behind this approach is to solve the link prediction problem as a supervised classification task. More precisely, for a given graph, each couple of nodes is represented with a feature vector (x), and a class label (y) is assigned to each of those node couples. Formally, let $G = (V, E)$ be a graph, and for each couple of nodes i, j, we build the following formula:

$$x = \left[f_{0,0}, \dots, f_{i,j}, \dots, f_{n,n}\right] \quad y = \left[y_{0,0}, \dots, y_{i,j}, \dots, y_{n,n}\right]$$

Here, $f_{ij} \in x$ is the *feature vector* representing the couple of nodes i, j, and $y_{i,j} \in y$ is their *label*. The value for $y_{i,j}$ is defined as follows: $y_{i,j} = 1$ if, in the graph **G**, the edge connecting node i, j exists; otherwise, $y_{i,j} = 0$. Using the feature vector and the labels, we can then train an ML algorithm in order to predict whether a given couple of nodes constitutes a plausible edge for the given graph.

If it is easy to build the label vector for each couple of nodes, it is not so straightforward to build the feature space. In order to generate the feature vector for each couple of nodes, we will use some embedding techniques, such as node2vec and edge2vec, already discussed in *Chapter 4, Unsupervised Graph Learning*. Using those embedding algorithms, the generation of the feature space will be greatly simplified. Indeed, the whole process can be summarized in two main steps, outlined as follows:

1. For each node of the graph G, its embedding vector is computed using a node2vec algorithm.
2. For all the possible pairs of nodes in the graph, the embedding is computed using an edge2vec algorithm.

We can now apply a generic ML algorithm to the generated feature vector in order to solve the classification problem.

In order to give you a practical explanation of this procedure, we will provide an example in the following code snippet. More precisely, we will describe the whole pipeline (from graph construction to link prediction) using the networkx, stellargraph, and node2vec libraries. We will split the whole process into different steps in order to simplify our understanding of the different parts. The link prediction problem was applied to the citation network dataset described in *Chapters 1, Getting Started with Graphs*, available at the following link: https://linqs-data. soe.ucsc.edu/public/lbc/cora.tgz.

Have a look at the steps mentioned ahead.

We will build a networkx graph using the citation dataset, as follows:

```
import networkx as nx
import pandas as pd
edgelist = pd.read_csv("cora.cites", sep='\t', header=None,
names=["target", "source"])
G = nx.from_pandas_edgelist(edgelist)
```

Since the dataset is represented as an edge list (see *Chapters 1, Getting Started with Graphs*), we used the from_pandas_edgelist function to build the graph.

Next, we need to create, from the graph G, training and test sets. More precisely, our training and test sets should contain not only a subset of real edges of the graph G but also couples of nodes that do not represent a real edge in **G**. The couples representing real edges will be *positive instances* (class label 1), while the couples that do not represent real edges will be *negative instances* (class label 0). This process can be easily performed as follows:

```
from stellargraph.data import EdgeSplitter
edgeSplitter = EdgeSplitter(G)
graph_test, samples_test, labels_test = edgeSplitter.train_test_
split(p=0.1, method="global")
```

We used the EdgeSplitter class available in stellargraph. The main constructor parameter of the EdgeSplitter class is the graph (G) we want to use to perform our split. The real splitting is performed using the train_test_split function, with p being the percentage of total edges to be returned (the percentage that will be allocated to the test set), which will generate the following outputs:

- graph_test is a subset of the original graph *G* containing all the nodes but just a selected subset of edges.
- samples_test is a vector containing in each position a couple of nodes. This vector will contain couples of nodes representing real edges (positive instance) but also couples of nodes that do not represent real edges (negative instance).
- labels_test is a vector having the same length as samples_test. It contains only 0 or 1. The value of 0 is present in the position representing a negative instance in the samples_ test vector, while the value of 1 is present in the position representing a positive instance in samples_test.

By following the same procedure used to generate the test set, it is possible to generate the training set, as illustrated in the following code snippet:

```
edgeSplitter = EdgeSplitter(graph_test, G)
graph_train, samples_train, labels_train = edgeSplitter.train_test_
split(p=0.1, method="global")
```

The main difference in this part of code is related to the initialization of EdgeSplitter. In this case, we also provide graph_test in order to not repeat positive and negative instances generated for the test set.

At this point, we have our training and testing datasets with negative and positive instances, where 81% of the edges are in graph_train and will be used to calculate the node2vec embeddings. 9% of the edges are in samples_train, paired with an equal number of negative samples, while 10% of the edges are in samples_test, also matched with an equal number of negative samples. For each of those instances, we now need to generate their feature vectors. In this example, we used the node2vec library to generate the node embedding. In general, every node embedding algorithm can be used to perform this task. For the training set, we can thus generate the feature vector with the following code:

```python
from node2vec import Node2Vec
from node2vec.edges import HadamardEmbedder
node2vec = Node2Vec(graph_train)
model = node2vec.fit()
edges_embs = HadamardEmbedder(keyed_vectors=model.wv)
train_embeddings = [edges_embs[str(x[0]),str(x[1])] for x in samples_
train]
```

From the previous code snippet, it is possible to see the following:

- We generate the embedding for each node in the training graph using the node2vec library.
- We use the HadamardEmbedder class to generate the embedding of each couple of nodes contained in the training set. Those values will be used as feature vectors to perform the training of our model.

It is worth noticing that we generate the embeddings only for the training graph, since calculating embeddings or other node properties on the full graph, including test data, is a common mistake that leads to data leakage. Furthermore, in this example, we used the HadamardEmbedder algorithm, but in general, other embedding algorithms can be used, such as the ones described in *Chapter 4, Unsupervised Graph Learning*.

The previous step needs to also be performed for the test set, with the following code:

```python
edges_embs = HadamardEmbedder(keyed_vectors=model.wv)
test_embeddings = [edges_embs[str(x[0]),str(x[1])] for x in samples_test]
```

The only difference here is given by the samples_test array used to compute the edge embeddings. Indeed, in this case, we use the data generated for the test set. Moreover, it should be noted that the node2vec algorithm was not recomputed for the test set.

Indeed, given the stochastic nature of node2vec, it is not possible to ensure that the two learned embeddings are "comparable" and therefore node2vec embeddings will change between runs.

Everything is set now. We can finally train—using the train_embeddings feature space and the train_labels label assignment—an ML algorithm to solve the label prediction problem, as follows:

```
from sklearn.ensemble import RandomForestClassifier
rf = RandomForestClassifier(n_estimators=1000)
rf.fit(train_embeddings, labels_train);
```

In this example, we used a simple RandomForestClassifier class, but every ML algorithm can be used to solve this task. We can then apply the trained model on the test_embeddings feature space in order to quantify the quality of the classification, as shown in the following code block:

```
from sklearn import metrics
y_pred = rf.predict(test_embeddings)
print('Precision:', metrics.precision_score(labels_test, y_pred))
print('Recall:', metrics.recall_score(labels_test, y_pred))
print('F1-Score:', metrics.f1_score(labels_test, y_pred))
```

As a result, we get the following output:

```
Precision: 0.8557114228456913
Recall: 0.8102466793168881
F1-Score: 0.8323586744639375
```

The methods we just described are just a general schema; each piece of the pipeline—such as the train/test split, the node/edge embedding, and the ML algorithm—can be changed according to the specific problem we are facing. For example, the choice of train/test split may vary depending on the size and structure of the graph: for large graphs, techniques like stratified sampling or temporal splits (that we will see in *Chapter 11, Temporal Graph Machine Learning*) may be more appropriate, while smaller graphs might allow for simpler random splits. Similarly, the embedding method, whether using node2vec or other algorithms, should be selected based on the graph's characteristics (e.g., if the graph is dense or sparse, directed or undirected), as well as the specific downstream task.

This simple yet powerful pipeline can capture complex patterns beyond local neighborhood similarity. However, it requires high computational costs (to train and tune the ML algorithm) and performances, as in other supervised tasks, may depend on the quality and quantity of labels.

In this section, we introduced the link prediction problem. We enriched our explanation by providing a description, with several examples, of different techniques used to find a solution to the link prediction problem. We showed that different ways to tackle the problem are available, from simple index-based techniques to more complex embedding-based techniques. However, it is worth noticing that the scientific literature is full of algorithms to solve the link prediction task. In the paper by Mutlu et al., *Review on Learning and Extracting Graph Features for Link Prediction* (https://arxiv.org/pdf/1901.03425.pdf), a good overview of different techniques used to solve the link prediction problem is available. In the next section, we will investigate the community detection problem.

Detecting meaningful structures such as communities

One common problem data scientists face when dealing with networks is how to identify clusters and communities within a graph. This often arises when graphs are derived from social networks for which communities are known to exist. However, the underlying algorithms and methods can also be used in other contexts, representing another option to perform clustering and segmentation. For example, these methods can effectively be used in text mining to identify emerging topics and to cluster documents that refer to single events/topics. A community detection task consists of partitioning a graph such that nodes belonging to the same community are tightly connected with each other and are weakly connected with nodes from other communities. There exist several strategies to identify communities. In general, we can define them as belonging to one of two categories, outlined as follows:

- **Non-overlapping** community detection algorithms that provide a one-to-one association between nodes and communities, thus with no overlapping nodes between communities
- **Overlapping** community detection algorithms that allow a node to be included in more than one community—for instance, reflecting the natural tendencies of social networks to develop overlapping communities (for example, friends from school, neighbors, playmates, people being in the same football team, and so on), or in biology, where a single protein can be involved in more than one process and bioreaction.

In the following section, we will review some of the most used techniques in the context of community detection.

Embedding-based community detection

One class of methods that allow us to partition nodes into communities can be simply obtained by applying standard shallow clustering techniques on the node embeddings, computed using the methods described in *Chapter 4, Unsupervised Graph Learning*. The embedding methods in fact allow us to project nodes into a vector space where a distance measure that represents a similarity between nodes can be defined. As we have shown in *Chapter 4, Unsupervised Graph Learning*, embedding algorithms are very effective in separating nodes with similar neighborhood and/or connectivity properties. Then, standard clustering techniques can be used, such as distance-based clustering (K-means), connectivity clustering (hierarchical clustering), distribution clustering (Gaussian mixture), and density-based clustering (**Density-Based Spatial Clustering of Applications with Noise (DBSCAN)**). Depending on the algorithm, these techniques may both provide a single-association community detection or a soft cluster assignment. We will showcase how they would work on a simple barbell graph. We start by creating a simple barbell graph using the networkx utility function, as follows:

```
import networkx as nx
G = nx.barbell_graph(m1=10, m2=4)
```

We can then first get the reduced dense node representation using one of the embedding algorithms we have seen previously (for instance, HOPE), shown as follows:

```
from gem.embedding.hope import HOPE
gf = HOPE(d=4, beta=0.01)
gf.learn_embedding(G)
embeddings = gf.get_embedding()
```

We can finally run a clustering algorithm on the resulting vector representation provided by the node embeddings, like this:

```
from sklearn.mixture import GaussianMixture
gm = GaussianMixture(n_components=3, random_state=0)
labels = gm.fit_predict(embeddings)
```

We can plot the network with the computed communities highlighted in different colors, like this:

```
colors = ["blue", "green", "red"]
nx.draw_spring(G, node_color=[colors[label] for label in labels])
```

By doing so, you should obtain the output shown in the following screenshot:

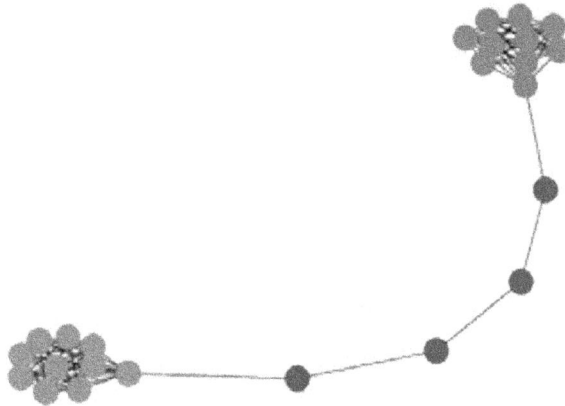

Figure 6.2: Barbell graph where the community detection algorithm has been applied using embedding-based methods

The two clusters, as well as the connecting nodes, have been correctly grouped into three different communities, reflecting the internal structure of the graph.

Spectral methods and matrix factorization

Another way to achieve a graph partition is to process the adjacency matrix or the Laplacian matrix that represents the connectivity properties of the graph. For instance, spectral clustering can be obtained by applying standard clustering algorithms on the eigenvectors of the Laplacian matrix. In some sense, spectral clustering can also be seen as a special case of an embedding-based community detection algorithm where the embedding technique is so-called spectral embedding, obtained by considering the first k-eigenvectors of the Laplacian matrix. By considering different definitions of the Laplacian as well as different similarity matrices, variations to this method can be obtained. A convenient implementation of this method can be found within the communities Python library and can be used on the adjacency matrix representation easily obtained from a networkx graph, as illustrated in the following code snippet:

```
from communities.algorithms import spectral_clustering
adj=np.array(nx.adjacency_matrix(G).todense())
communities = spectral_clustering(adj, k=2)
```

Where k is the number of communities to cluster nodes into. Moreover, the adjacency matrix (or the Laplacian) can also be decomposed using matrix factorization techniques such as **non-negative matrix factorization** (**NMF**)—that allow similar descriptions, as illustrated in the following code snippet:

```
from sklearn.decomposition import NMF
nmf = NMF(n_components=2)
score = nmf.fit_transform(adj)
communities = [set(np.where(score [:,ith]>0)[0])
                 for ith in range(2)]
```

The threshold for belonging to the community was set in this example to 0, although other values can also be used to retain only the community cores. Note that these methods are overlapping community detection algorithms, and nodes might belong to more than one community.

Probability models

Community detection methods can also be derived from fitting the parameters of generative probabilistic graph models. Examples of generative models were already described in *Chapter 1, Getting Started with Graphs*. However, they did not assume the presence of any underlying community, unlike the so-called **stochastic block model** (**SBM**). In fact, this model is based on the assumption that nodes can be partitioned into K disjoint communities and each community has a defined probability of being connected to another. Imagine we want to generate graphs with community structure. For a network of n nodes and K communities, the generative model can be parametrized by the following:

- **Membership matrix**: M, which is an $n \times K$ matrix, represents the probabilities of each node belonging to each community. For a node i, the row M_i contains the probabilities for i being in each of the K communities. Sampling from this distribution gives the community assignment g_i.

- **Probability matrix**: B, which is a $K \times K$ matrix and represents the probability of an edge existing between two nodes based on their community assignment. For communities g_i and g_j, the edge probability is $B_{g_i g_j}$.

The adjacency matrix A of the graph is then generated probabilistically by the following formula:

$$f(A_{ij}) = \begin{cases} Bernoulli\left(B_{g_i, g_j}\right), & if\ i < j \\ 0, & if\ i = 0 \\ A_{j,i}, & if\ i > j \end{cases}$$

For community detection, the goal is reversed. Instead of generating A from M and B, we estimate the community membership (M) and the connection probabilities (B) *given* the observed adjacency matrix A. This can be done via maximum likelihood estimation, which identifies the M and B that best explain A.

Note that the SBM in the limit of the constant probability matrix (that is, $B_{ij} = p$) corresponds to the Erdős-Rényi model. These models have the advantage of also describing a relation between communities, identifying community-community relationships. This is particularly useful in large-scale-graphs for tasks like social network analysis or biological network modeling.

It is worth noticing that we explained here the idea of the method, but the actual implementation would be more complex. A more detailed mathematical formulation can be found in the paper *A review of stochastic block models and extensions for graph clustering* (Lee and Wilkinson, 2019).

Cost function minimization

Community detection in graphs can be approached by optimizing a cost function that evaluates the structure of the graph. Such cost functions typically reward edges within the same community while penalizing edges between different communities. One popular approach involves defining a measure of community quality (such as modularity, for example) and optimizing the assignment of nodes to communities to maximize this measure.

For a binary community structure, each node i is assigned a community label S_i, a variable that takes the values 1 or -1 depending on whether the node belongs to one or two communities. The association between two nodes i and j can then be expressed using their labels, S_i and S_j. Therefore, a cost function to evaluate the presence of edges between different communities can be defined as:

$$\sum_{i,j \in E} A_{i,j}(1 - S_i S_j)$$

Where A is the adjacency matrix of the graph. When two connected nodes $A_{ij} > 0$ belong to a different community $S_i S_j = -1$, the contribution provided by the edge is positive. On the other hand, the contribution is 0, both when two nodes are not connected ($A_{ij} = 0$) and when two connected nodes belong to the same community ($S_i S_j = 0$). Therefore, the problem is to find the best community assignment (S_i and S_j) in order to minimize the preceding function. This method, however, applies only to binary community detection and is therefore rather limited in its application.

Another very popular algorithm belonging to this class is the Louvain method, which takes its name from the university where it was invented. This algorithm aims to maximize the modularity, defined as follows:

$$Q = \frac{1}{2m} \sum_{i,j \in E} \left(A_{ij} - \frac{k_i k_j}{2m} \right) \delta(c_i, c_j)$$

Here, m represents the number of edges, k_i and k_j represent the degree of the i-th and j-th node respectively, and $\delta(c_i, c_j)$ is the Kronecker delta function, which is 1 when c_i and c_j belong to the same community and 0 otherwise.

By subtracting from the adjacency matrix the expected number of edges that i and j would have in a random graph, $k_i k_j / 2m$, we can estimate *how much the actual graph is deviating from the randomness* (the Kronecker delta function ensures contribution from only pairs in the same community). The summation, then, aggregates this information across all node pairs, and the division by $2m$ normalizes modularity, keeping it between –1 and 1. A higher Q indicates that the detected communities have more internal edges than expected by chance, highlighting meaningful structure.

To maximize this modularity efficiently, the Louvain methods iteratively compute the following steps, until no further improvement in modularity can be achieved:

1. **Modularity optimization:** Nodes are swept iteratively, and for each node we compute the change of modularity Q there would be if the node were to be assigned to each community of its neighbors. Once all the ΔQ values are computed, the node is assigned to the community that provides the largest increase. If there is no increase obtained by placing the node in any other community than the one it is in, the node remains in its original community. This optimization process continues until no changes are induced.

2. **Node aggregation:** In the second step, we build a new network by grouping all the nodes in the same community and connecting the communities using edges that result from the sum of all edges across the two communities. Edges within communities are accounted for as well by means of self-loops that have weights resulting from the sum of all edge weights belonging to the community.

A Louvain implementation can already be found in the `communities` library, as can be seen in the following code snippet:

```
from communities.algorithms import louvain_method
communities = louvain_method(adj)
```

Another method to maximize the modularity is the Girvan-Newman algorithm, which is based on iteratively removing edges that have the highest betweenness centrality (and thus connect two separate clusters of nodes) to create connected component communities. Here is the code related to this:

```
from communities.algorithms import girvan_newman
communities = girvan_newman(adj, n=2)
```

Detecting graph similarities and graph matching

Learning a quantitative measure of the *similarity* among graphs is considered a key problem. Indeed, it is a critical step for network analysis and can also facilitate many ML problems, such as classification, clustering, and ranking. Many clustering algorithms, for example, use the concept of similarity for determining if an object should or should not be a member of a group.

In the graph domain, finding an effective similarity measure constitutes a crucial problem for many applications. Consider, for instance, the *role* of a node inside a graph. This node might be very important for spreading information across a network or guaranteeing network robustness: for example, it could be the center of a star graph or it could be a member of a clique. In this scenario, it would be very useful to have a powerful method for comparing nodes according to their roles. For example, you might be interested in searching for individuals showing similar roles or presenting similar unusual and anomalous behaviors. You might also use it for searching similar subgraphs or to determine network compatibility for *knowledge transfer*. For example, if you find a method for increasing the robustness of a network and you know that such a network is very similar to another one, you may apply the same solution that worked well for the first network directly to the second one:

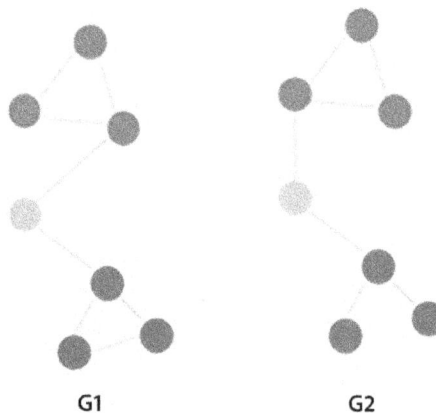

Figure 6.3: Example of differences between two graphs

Several metrics can be used for measuring the similarity (distance) between two objects. Some examples include the *Euclidean distance, Manhattan distance, cosine similarity*, and so on. However, these metrics might fail to capture the specific characteristics of the data being studied, especially on non-Euclidean structures such as graphs. Take the setup shown in *Figure 6.3*: how "distant" are **G1** and **G2**? They indeed look pretty similar. But what if the missing connection in the red community of **G2** causes a severe loss of information? Do they still look similar?

Several algorithmic approaches and heuristics have been proposed for measuring similarity among graphs, based on mathematical concepts such as *graph isomorphism, edit distance*, and *common subgraphs* (we suggest reading `https://link.springer.com/article/10.1007/s10044-012-0284-8` for a detailed review). Many of these approaches are currently used in practical applications, even if they often require exponentially high computational time to provide a solution to **NP-complete** problems in general (where **NP** stands for **nondeterministic polynomial time**). Therefore, it is essential to find or learn a metric for measuring the similarity of data points involved in the specific task. Here is where ML comes to our aid.

Many algorithms among the ones we have already seen in *Chapter 4, Unsupervised Graph Learning*, and *Chapter 5, Supervised Graph Learning*, might be useful for learning an effective similarity metric. According to the way they are used, a precise taxonomy can be defined. Here, we provide a simple overview of graph similarity techniques. A more comprehensive list can be found in the paper *Deep Graph Similarity Learning: A Survey* (`https://arxiv.org/pdf/1912.11615.pdf`). They can be essentially divided into three main categories, even if sophisticated combinations can also be developed. **Graph embedding-based methods** use embedding techniques to obtain an embedded representation of the graphs and exploit such a representation to learn the similarity function; **graph kernel-based methods** define the similarity between graphs by measuring the similarity of their constituting substructures; **graph neural network-based methods** use **graph neural networks (GNNs)** to jointly learn an embedded representation and a similarity function. Let's see all of them in more detail.

Graph embedding-based methods

Such techniques seek to apply graph embedding techniques to obtain node-level or graph-level representations and further use the representations for similarity learning. For example, *DeepWalk* and *Node2Vec* can be used to extract meaningful embedding that can then be used to define a similarity function or to predict similarity scores. For example, in Tixier et al. (2015), **node2vec** was used for encoding node embeddings for representing a graph as an image.

Specifically, **two-dimensional** (**2D**) histograms obtained from those node embeddings were passed to a classical 2D **convolutional neural network** (**CNN**) architecture designed for images. Such a simple yet powerful approach enabled good results to be obtained for many benchmark datasets.

Let's see an example of how such methods could work. First, let's define a toy dataset with simple graphs and let's define random labels for each graph:

```
# Create toy dataset with simple graphs
num_graphs = 10
graphs = [nx.erdos_renyi_graph(10, np.random.rand()) for _ in range(num_
graphs)]


# Generate random labels
labels = [np.random.choice([0,1]) for _ in range(num_graphs)]
```

We can now define a proper function to generate the 2D histograms and use it on node embeddings:

```
# Function to generate 2D histogram from node embeddings
def generate_2d_histogram(node_embeddings, bins=16):
    # Flatten embeddings to create histograms
    embeddings = np.vstack(node_embeddings)
    histogram, xedges, yedges = np.histogram2d(embeddings[:, 0],
embeddings[:, 1], bins=bins)
    return histogram
# Prepare graph-level 2D histograms from node embeddings
graph_histograms = []
for i, graph in enumerate(graphs):
    node2vec = Node2Vec(graph, dimensions=64, walk_length=10, num_
walks=80, workers=4)
    model = node2vec.fit()
    node_embeddings = [model.wv.get_vector(str(node)) for node in graph.
nodes()]
    histogram = generate_2d_histogram(node_embeddings)
    graph_histograms.append(histogram)
```

We can now split our dataset into training and testing sets:

```
# Split histograms into training and testing sets
train_histograms, test_histograms, train_labels, test_labels = train_test_
split(graph_histograms, labels, test_size=0.5, random_state=42)
```

Finally, let's define and train our GraphCNN model. We will be using PyTorch for this:

```python
# Initialize and train the CNN model
num_classes = 2  # Binary classification
input_channels = 1  # Single channel for histogram
bins = 16  # Same as used in generate_2d_histogram function

train_histograms = train_histograms.unsqueeze(1)  # Add channel dimension
test_histograms = test_histograms.unsqueeze(1)

cnn_model = GraphCNN(input_channels, num_classes)
criterion = nn.CrossEntropyLoss()
optimizer = optim.Adam(cnn_model.parameters(), lr=0.001)

# Training loop
cnn_model.train()
epochs = 20
for epoch in range(epochs):
    optimizer.zero_grad()
    outputs = cnn_model(train_histograms)
    loss = criterion(outputs, torch.tensor(train_labels, dtype=torch.
long))
    loss.backward()
    optimizer.step()
    print(f"Epoch {epoch+1}/{epochs}, Loss: {loss.item()}")

# Evaluate on test set
cnn_model.eval()
with torch.no_grad():
    test_outputs = cnn_model(test_histograms)
    predictions = torch.argmax(test_outputs, axis=1)
    accuracy = accuracy_score(test_labels, predictions.numpy())
    print(f"Test Accuracy: {accuracy}")
```

The above code is an example based on a toy random dataset. For this reason, the accuracy would probably be low. However, it effectively shows how such a technique can be used on real graphs and we will invite the reader to explore this approach as an exercise.

Graph kernel-based methods

Graph kernel-based methods have generated a lot of interest in terms of capturing the similarity between graphs. These approaches compute the similarity between two graphs as a function of the similarities between some of their substructures. Different graph kernels exist based on the substructures they use, which include random walks, shortest paths, and subgraphs.

As an example, a method called **Deep Graph Kernels** (**DGK**) (Yanardag et al., 2015) decomposes graphs into substructures that are viewed as "words." Then, **natural language processing** (**NLP**) approaches such as **continuous bag of words** (**CBOW**) and **skip-gram** are used to learn latent representations of the substructures. This way, the kernel between two graphs is defined based on the similarity of the substructure space.

Let's see how such an approach can be used on the toy dataset generated in the previous example. First, let's define a proper function to decompose graphs into substructures using random walks. Then, let's use it to create a "document" for each graph:

```python
# Decompose graphs into substructures using random walks
def random_walks_as_words(graph, walk_length=6, num_walks=10):
    """Generates 'words' from random walks on the graph."""
    walks = []
    for _ in range(num_walks):
        for node in graph.nodes():
            walk = [str(node)]
            for _ in range(walk_length - 1):
                neighbors = list(graph.neighbors(int(walk[-1])))
                if neighbors:
                    walk.append(str(np.random.choice(neighbors)))
                else:
                    break
            walks.append(" ".join(walk))
    return walks


# Create a "document" for each graph using its random walks
graph_documents = []
for graph in graphs:
    walks = random_walks_as_words(graph)
    # Combine all walks into a single document
    graph_documents.append(" ".join(walks))
```

We can now use the `CountVectorizer` function to create a numerical representation of each graph based on the frequency of specific patterns or features. This function processes a collection of textual data and generates a matrix where each entry indicates how often a particular pattern or feature appears in the graph:

```
# Generate "bag of words" representation for each graph
vectorizer = CountVectorizer()
# Sparse matrix of shape (num_graphs, num_features)
graph_bow = vectorizer.fit_transform(graph_documents)
```

Then, we can compute pairwise similarity out of it:

```
# Compute pairwise similarities between graphs
graph_similarity_matrix = cosine_similarity(graph_bow)
```

We now have all the ingredients to train a machine learning algorithm. As usual, we will now split the dataset into training and test sets and use the similarity matrix as input features for a classification task:

```
# Split dataset into training and testing sets
train_indices, test_indices = train_test_split(range(len(graphs)), test_
size=0.5, random_state=42)

train_similarity = graph_similarity_matrix[np.ix_(train_indices, train_
indices)]
test_similarity = graph_similarity_matrix[np.ix_(test_indices, train_
indices)]  # Test against training similarities

train_labels = [labels[i] for i in train_indices]
test_labels = [labels[i] for i in test_indices]

# Train a classifier using the training similarity matrix
svm = SVC(kernel="precomputed")  # Precomputed kernel
svm.fit(train_similarity, train_labels)

# Predict on the test set
test_predictions = svm.predict(test_similarity)
```

```
# Evaluate the classification accuracy
accuracy = accuracy_score(test_labels, test_predictions)
print(f"Test Accuracy using Deep Graph Kernels (DGK): {accuracy}")
```

Also, in this case, the output is not meaningful, since we are training the algorithm on random labels for demonstration purposes. Feel free to use the approach on a real dataset!

It is worth noticing that we have introduced some NLP concepts here. We will cover the topic in more detail in *Chapter 8, Text Analytics and Natural Language Processing Using Graphs.*

GNN-based methods

With the emergence of **deep learning** (**DL**) techniques, GNNs have become a powerful new tool for learning representations on graphs. Such powerful models can be easily adapted to various tasks, including graph similarity learning. Furthermore, they present a key advantage with respect to other traditional graph embedding approaches. Indeed, while the latter generally learn the representation in an isolated stage, in this kind of approach, the representation learning and the target learning task are conducted jointly. Therefore, the GNN deep models can better leverage the graph features for the specific learning task. We have already seen an example of similarity learning using GNNs in *Chapter 4, Unsupervised Graph Learning*, where a two-branch network was trained to estimate the proximity distance between two graphs.

Applications

Similarity learning on graphs has already achieved promising results in many domains. Important applications may be found in chemistry and bioinformatics—for example, for finding the chemical compounds that are most similar to a query compound, as illustrated on the left-hand side of the following diagram. In neuroscience, similarity learning methods have started to be applied to measure the similarity of brain networks among multiple subjects, allowing the novel clinical investigation of brain diseases.

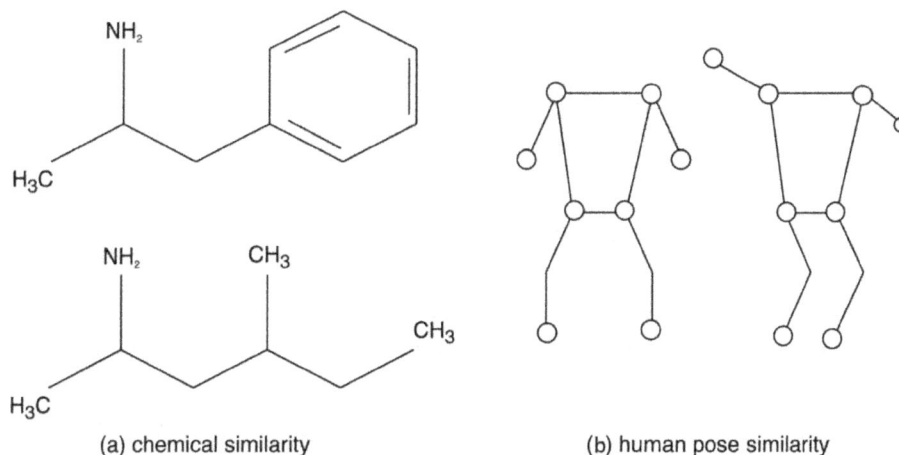

(a) chemical similarity   (b) human pose similarity

Figure 6.4: Example of how graphs can be useful for representing various objects: (a) differences between two chemical compounds; (b) differences between two human poses

Graph similarity learning has also been explored in computer security, where novel approaches have been proposed for the detection of vulnerabilities in software systems as well as hardware security problems (a survey of these approaches can be found at `https://eprint.iacr.org/2019/983.pdf`). Recently, a trend for applying such solutions to solve computer vision problems has been observed (you may check `https://ieeexplore.ieee.org/document/9263681` for further reading). Once the challenging problem of converting images into graph data has been solved, interesting solutions can indeed be proposed for human action recognition in video sequences and object matching in scenes, among other areas. In the context of human action recognition, graphs provide an intuitive way to model the human body. Here, body joints (e.g., elbows, knees, shoulders) are represented as **nodes**, while the connections between them (e.g., bones) serve as **edges**. This creates a graph that encodes the **spatial relationships and hierarchical structure** of the human body (as shown on the right-hand side of *Figure 6.4*). Therefore, such graphs can be studied using flexible techniques such as GCN, which can capture both local and global patterns in the data. Notice also that temporal relationships can be exploited (which are critical for understanding actions) by creating spatio-temporal graphs where multiple body pose graphs are connected over time. These approaches enable effective recognition of complex human activities by leveraging both spatial and temporal dynamics. We will learn more about temporal graphs in *Chapter 11, Temporal Graph Machine Learning*.

Summary

In this chapter, we have learned how graph-based ML techniques can be used to solve many different problems. In particular, we have seen that the same algorithm (or a slightly modified version) can be adapted to solve apparently very different tasks such as link prediction, community detection, and graph similarity learning. We have also seen that each problem has its own peculiarities, which have been exploited by researchers in order to design more sophisticated solutions.

In the next chapter, we will explore real-world problems related to social networking that can be solved using graph-based ML.

Get This Book's PDF Version and Exclusive Extras

UNLOCK NOW

Scan the QR code (or go to packtpub.com/unlock).
Search for this book by name, confirm the edition,
and then follow the steps on the page.

*Note: Keep your invoice handy. Purchases made
directly from Packt don't require one.*

Part 3

Practical Applications of Graph Machine Learning

In this part, you will acquire a more practical knowledge of methods outlined in previous chapters by applying them to real-world use cases and learn how to scale out the approaches to structured and unstructured datasets.

This part comprises the following chapters:

- *Chapter 7, Social Network Graphs*
- *Chapter 8, Text Analytics and Natural Language Processing Using Graphs*
- *Chapter 9, Graphs Analysis for Credit Card Transactions*
- *Chapter 10, Building a Data-Driven Graph-Powered Application*

7
Social Network Graphs

The growth of social networking sites has been one of the most active trends in digital media over recent years. Since the late 1990s, when the first social applications were published, they have attracted billions of active users worldwide, many of whom have integrated digital social interactions into their daily lives. New ways of communication are being driven by social networks such as Facebook, Twitter, and Instagram, among others. Users can share ideas, post updates and feedback, or engage in activities and events while sharing their broader interests on social networking sites.

Besides, social networks constitute a huge source of information for studying user behaviors, interpreting the interaction among people, and predicting their interests. Structuring them as graphs, where a vertex corresponds to a person and an edge represents the connection between them, provides a powerful tool to extract useful knowledge.

However, understanding the dynamics that drive the evolution of the social network is a complex problem due to a large number of variable parameters.

In this chapter, we will talk about how we can analyze the Facebook social network using graph theory and how we can solve useful problems such as link prediction and community detection using machine learning.

The following topics will be covered in this chapter:

- Overview of the dataset
- Network topology and community detection
- Embedding for supervised and unsupervised tasks

Technical requirements

All code files relevant to this chapter are available at https://github.com/PacktPublishing/
Graph-Machine-Learning/tree/main/Chapter07. Please refer to the *Practical exercises* section of
Chapter 1, Getting Started with Graphs, for guidance on how to set up the environment to run the
examples in this chapter, using either Poetry, pip, or docker.

Overview of the dataset

We will be using a **SNAP public dataset, social circles: Facebook,** from Stanford University
(https://snap.stanford.edu/data/ego-Facebook.html).

The dataset was created by collecting Facebook user information from survey participants. In
more detail, ego networks were created for 10 users. Each user was asked to identify all the **circles**
(list of friends) to which their friends belong. On average, each user identified 19 circles in their
ego networks, where each circle has on average 22 friends.

For each user, the following information was collected:

- **Edges:** An edge exists if two users are friends on Facebook.
- **Node features:** Features are scored as **1** if the user has this property in their profile and **0**
  otherwise. Features have been anonymized since the names of the features would reveal
  private data.

The 10 ego networks were then unified in a single graph that we are going to study.

Dataset download

The dataset can be retrieved using the following URL: https://snap.stanford.edu/data/
ego-Facebook.html. In particular, three files can be downloaded: facebook.tar.gz, facebook_
combined.txt.gz, and redme-Ego.txt. Let's inspect each file separately:

- facebook.tar.gz: This is an archive containing four files for each **ego user** (40 files in
  total). Each file is named using the format *nodeId.extension*, where nodeId is the node
  ID of the ego user and extension is one of *edges, circles, feat, egofeat,* or *featnames*. This is
  explained in more detail here:

  a. **nodeId.edges:** Contains a list of edges for the network of node nodeId.

  b. **nodeId.circles:** Contains several lines (one for each circle). Each line consists of
      a name (the circle name) followed by a series of node IDs.

c. **nodeId.feat**: Contains the features (0 if nodeId has that feature, 1 otherwise) for each node in the ego network.

d. **nodeId.egofeat**: Contains the features of the ego user.

e. **nodeId.featname**: Contains the names of the features.

- `facebook_combined.txt.gz`: This is an archive containing a single file, `facebook_combined.txt`, which is a list of edges from all the ego networks combined.

- `readme-Ego.txt`: This contains a description of the above-mentioned files.

Take a look at those files by yourself. It is strongly suggested to explore and become as comfortable as possible with the dataset before starting any machine learning task.

Loading the dataset using networkx

The first step of our analysis will be loading the aggregated ego networks using networkx. As we have seen in the previous chapters, networkx is powerful for graph analysis and, given the size of the datasets, will be the perfect tool for the analysis that we will be doing in this chapter. However, for larger social network graphs with billions of nodes and edges, more specific tools might be required for loading and processing them. We will cover the tools and technologies used for scaling out the analysis in *Chapter 10, Building a Data-Driven Graph-Powered Application*.

As we have seen above, the combined ego network is represented as a list of edges. We can create an undirected graph from a list of edges using networkx as follows:

```
G = nx.read_edgelist("facebook_combined.txt", create_using=nx.Graph(),
nodetype=int)
```

Notice that here we are using an undirected graph because connections on Facebook are undirected: when you accept someone's friend request, you are added to their friend list and vice versa (that is, an undirected edge is added to the graph).

Let's print some basic information about the graph:

```
print(nx.info(G))
```

The output should be as follows:

```
Name:
Type: Graph
Number of nodes: 4039
Number of edges: 88234
Average degree:  43.6910
```

As we can see, the aggregated network contains 4039 nodes and 88234 edges. This is a fairly connected network with the number of edges being more than 20 times the number of nodes. Due to the nature of the network, it is expected that several clusters will emerge, particularly because of the small-world properties of the individual ego networks. In a small-world network, nodes are highly clustered, meaning that individuals (or nodes) are more likely to be connected to their immediate neighbors, and groups of tightly connected nodes (clusters) form naturally. These clusters represent subgroups of ego users, each with their own localized connections, which are typically observed in social network structures.

Drawing the network will also help with better understanding what we are going to analyze. We can draw the graph using networkx as follows:

```
nx.draw_networkx(G, pos=spring_pos, with_labels=False, node_size=35)
```

The output should be as follows:

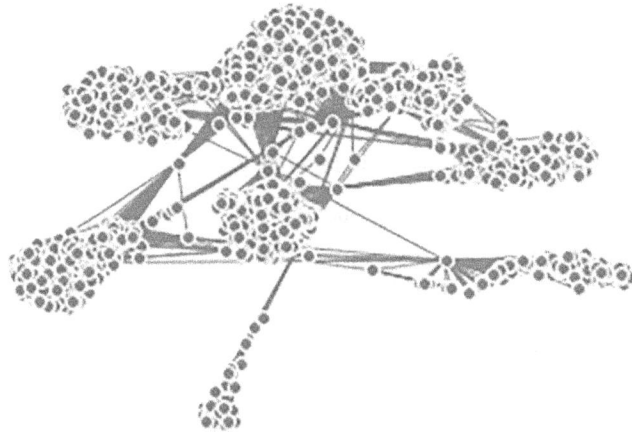

Figure 7.1 – The aggregated Facebook ego network

We can observe the presence of highly interconnected hubs. This is interesting from a social network analysis point of view since these hubs may result from underlying social mechanisms, such as the presence of influential individuals or key opinion leaders. For example, highly interconnected hubs could represent influential people in a community who have strong relationships with a large number of others, often acting as connectors between different groups. These hubs might also emerge due to mechanisms like preferential attachment (where new nodes are more likely to connect to already well-connected nodes) or homophily (where individuals tend to connect with others who are similar to themselves). These mechanisms can be further investigated to better understand the structure of an individual's relationships within their social world.

Before continuing our analysis, let's save the IDs of the ego-user nodes inside the network. We can retrieve them from the files contained in the `facebook.tar.gz` archive.

First, unpack the archive. The extracted folder will be named `facebook`. Let's run the following Python code for retrieving the IDs by taking the first part of each filename:

```
ego_nodes = set([int(name.split('.')[0]) for name in
os.listdir("facebook/")])
```

We are now ready to analyze the graph. In particular, in the next section, we will better understand the structure of the graph by inspecting its properties. This will help us to have a clearer idea of its topology and its relevant characteristics.

Analyzing the graph structure

Understanding the topology of the network, the role of its nodes, and the presence of communities is a crucial step in the analysis of the social network. It is important to keep in mind that, in this context, nodes are actually users, each with its own interests, habits, and behaviors. Such knowledge will be extremely useful when performing predictions and/or finding insights.

We will be using `networkx` to compute a few useful metrics we have seen in *Chapter 1, Getting Started with Graphs*. We will try to give them an interpretation to collect insight into the graph. Let's begin as usual, by importing the required libraries and defining some variables that we will use throughout the code:

```
import os
import math
import numpy as np
import networkx as nx
import matplotlib.pyplot as plt
default_edge_color = 'gray'
default_node_color = '#407cc9'
enhanced_node_color = '#f5b042'
enhanced_edge_color = '#cc2f04'
```

We can now proceed to the analysis.

Topology overview

As we have already seen before, our combined network has 4,039 nodes and more than 80,000 edges. The next metric we will compute is assortativity. It will reveal information about the tendency of users to be connected with users with a similar degree. We can do that as follows:

```
assortativity = nx.degree_pearson_correlation_coefficient(G)
```

The output should be as follows:

```
0.06357722918564912
```

The assortativity coefficient ranges from -1 to 1, where a value of 1 indicates a perfect positive correlation (users tend to connect with others of similar degree) and -1 indicates a perfect negative correlation (users tend to connect with others of different degree). Here, we can observe a positive assortativity, likely showing that well-connected individuals associate with other well-connected individuals (as we have seen in *Chapter 1, Getting Started with Graphs*). This is expected since inside each circle, users tend to be highly connected to each other.

Transitivity could also help with better understanding how individuals are connected. Recall transitivity indicates the mean probability that two people with a common friend are themselves friends. Transitivity can range from 0 (no probability) to 1 (certainty):

```
t = nx.transitivity(G)
```

The output should be as follows:

```
0.5191742775433075
```

Here we have the half probability that two friends can or cannot have common friends.

The observation is also confirmed by computing the average clustering coefficient. The average clustering coefficient ranges from 0 (no clustering) to 1 (perfect clustering). Indeed, it can be considered as an alternative definition of transitivity:

```
aC = nx.average_clustering(G)
```

The output should be as follows:

```
0.6055467186200876
```

Notice that the clustering coefficient tends to be higher than the transitivity. Indeed, by definition, it puts more weight on vertices with a low degree, since they have a limited number of possible pairs of neighbors (the denominator of the local clustering coefficient).

Node centrality

Once we have a clearer idea of what the overall topology looks like, we can proceed by investigating the importance of each individual inside the network. In *Chapter 1*, *Getting Started with Graphs*, we understood the significance of betweenness centrality. It measures how many shortest paths pass through a given node, giving an idea of how *central* that node is for the spreading of information inside the network. The betweenness centrality can range from 0 (no shortest path passes through the node) to a higher value, with larger values indicating more central nodes. We can compute it using:

```
bC = nx.betweenness_centrality(G)
np.mean(list(bC.values()))
```

The output should be as follows:

```
0.0006669573568730229
```

The average betweenness centrality is pretty low, which is understandable given the large amount of non-bridging nodes inside the network. In other words, many nodes have few or no shortest paths passing through them, which reduces their centrality. This suggests that the network may have many nodes that are more peripheral or isolated in terms of information flow. However, we could collect better insights by visually inspecting the graph. In particular, we will draw the combined ego network by enhancing nodes with the highest betweenness centrality. Let's define a proper function for this:

```
def draw_metric(G, dct, spring_pos):
    top = 10
    max_nodes = sorted(dct.items(), key=lambda v: -v[1])[:top]
    max_keys = [key for key,_ in max_nodes]
    max_vals = [val*300 for _, val in max_nodes]
    plt.axis("off")
    nx.draw_networkx(G,
                    pos=spring_pos,
                    cmap='Blues',
                    edge_color=default_edge_color,
                    node_color=default_node_color,
                    node_size=3,
                    alpha=0.4,
                    with_labels=False)
```

```
nx.draw_networkx_nodes(G,
                       pos=spring_pos,
                       nodelist=max_keys,
                       node_color=enhanced_edge_color,
                       node_size=max_vals)
```

Now let's invoke it as follows:

```
draw_metric(G,bC,spring_pos)
```

The output should be as follows:

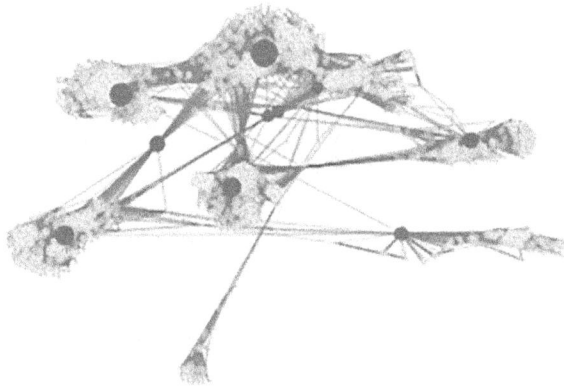

Figure 7.2: Betweenness centrality

Let's also inspect the degree centrality of each node. Since this metric is related to the number of neighbors of a node, we will have a clearer idea of how nodes are well connected to each other. A higher degree centrality means that a node has more connections, which often indicates its importance in the network:

```
deg_C = nx.degree_centrality(G)
np.mean(list(deg_C.values()))
draw_metric(G,deg_C,spring_pos)
```

The output should be as follows:

```
0.010819963503439287
```

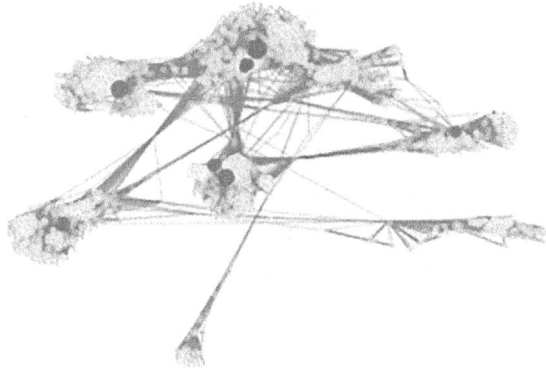

Figure 7.3: Degree centrality

This value represents the average degree centrality across all nodes in the network. Given that degree centrality ranges from 0 (no connections) to 1 (the most connected node), this value suggests that, on average, the nodes in the network have relatively few connections.

Next, let's look at the closeness centrality, which helps us understand how close nodes are to each other in terms of the shortest path. A higher closeness centrality means a node is closer, on average, to all other nodes in the network:

```
clos_C = nx.closeness_centrality(G)
np.mean(list(clos_C.values()))
draw_metric(G,clos_C,spring_pos)
```

The output should be as follows:

```
0.2761677635668376
```

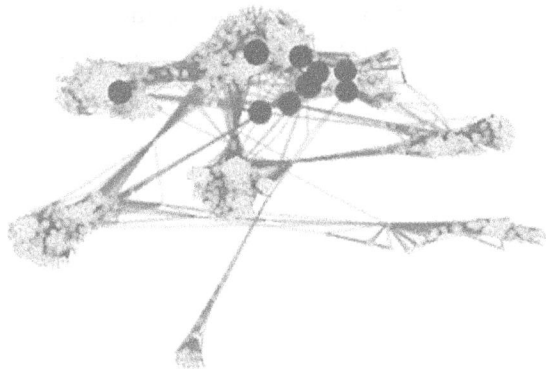

Figure 7.4: Closeness centrality

The closeness centrality ranges from 0 (isolated nodes) to 1 (nodes that can reach all other nodes in the fewest steps). Here, the average closeness centrality suggests that most nodes are somewhat close to each other, but not extremely close on average.

Overall, from the centrality analysis, it is interesting to observe that each central node seems to be part of a sort of community (this is reasonable, since central nodes might correspond to ego nodes of the network). It is also interesting to notice the presence of a bunch of highly interconnected nodes (especially from the closeness centrality analysis). Let's thus identify these communities in the next part of our analysis.

Community detection

Since we are performing social network analysis, it is worth exploring one of the most interesting graph structures for social networks: communities. If you use Facebook, it is very likely that your friends reflect different aspects of your life: friends from the educational environment (high school, college, and so on), friends from your weekly football match, friends you have met at parties, and so on.

An interesting aspect of social network analysis is to automatically identify such groups. This can be done automatically, inferring them from topological properties, or semi-automatically, exploiting some prior insight.

As we have seen in *Chapter 1, Getting Started with Graphs,* and *Chapter 6,* Solving Common Graph-Based Machine Learning Problems, one good criterion is to try to compute the partition of the graph nodes that maximizes the modularity using the Louvain heuristics. Recall that modularity evaluates whether the density of edges within communities is higher than what would be expected in a random graph with the same degree distribution. By aiming to maximize modularity, we ensure that the computed partition reflects significant community structure, meaning the groups are highly interconnected internally but sparsely connected to other groups.

We can do that in networkx with the help of the python-louvain package, as follows:

```
import community
parts = community.best_partition(G)
values = [parts.get(node) for node in G.nodes()]
n_sizes = [5]*len(G.nodes())
plt.axis("off")
```

```
nx.draw_networkx(G, pos=spring_pos, cmap=plt.get_cmap("Blues"), edge_
color=default_edge_color, node_color=values, node_size=n_sizes, with_
labels=False)
```

The output should be as follows:

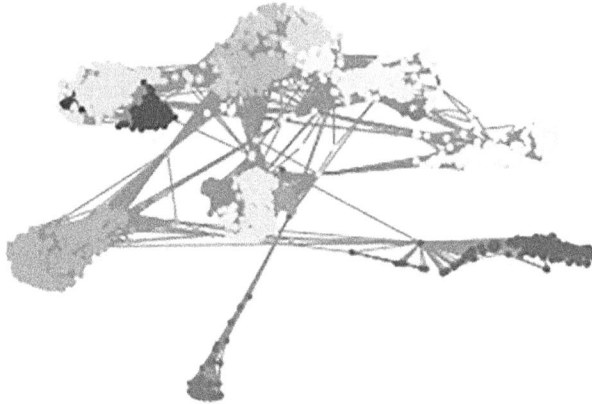

Figure 7.5: Detected communities using networkx

In terms of social network analysis, the detected clusters likely represent groups of tightly connected individuals who share similar interests, affiliations, or social contexts. For example, dense clusters, represented in the figure using different colors, may correspond to groups, such as family units or friend groups. Furthermore, the connections between clusters highlight bridging individuals (e.g., ego users) or relationships, which act as connectors between different communities. These individuals are often influential in information dissemination across the network.

In this context, it is also interesting to investigate whether the ego users occupy some roles inside the detected communities. Let's enhance the size and color of the ego user nodes, as follows:

```
for node in ego_nodes:
    n_sizes[node] = 250
nodes = nx.draw_networkx_nodes(G,spring_pos,ego_nodes,node_color=[parts.
get(node) for node in ego_nodes])
nodes.set_edgecolor(enhanced_node_color)
```

The output should be as follows:

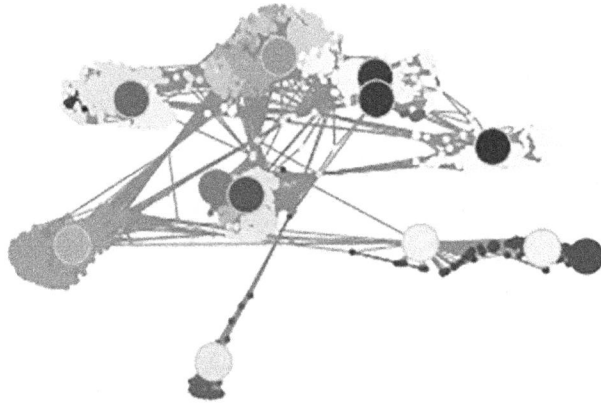

Figure 7.6: Detected communities using networkx with ego user's node size enhanced

The output highlights ego users as larger, distinctively colored nodes within their respective communities, making them easily identifiable. It is interesting to notice that some ego users belong to the same community. It is possible that ego users are actual friends on Facebook, and therefore their ego networks are partially shared. It is also interesting to notice that a few of the high betweenness nodes are not ego nodes and, conversely, some ego nodes are not high betweenness. This occurs because high-betweenness nodes act as connectors between communities, while ego nodes may be deeply embedded within their groups. Thus, structural importance and ego selection criteria are not always aligned.

We have now completed our basic understanding of the graph structure. We now know that some important nodes can be identified inside the network. We have also seen the presence of well-defined communities to which those nodes belong. Keep in mind these observations while performing the next part of the analysis, which is applying machine learning methods for supervised and unsupervised tasks.

Embedding for supervised and unsupervised tasks

Social media nowadays represent one of the most interesting and rich sources of information. Every day, thousands of new connections arise, new users join the communities, and billions of posts are shared. Graphs mathematically represent all those interactions, helping to make order in all such spontaneous and unstructured traffic.

When dealing with social graphs, there are many interesting problems that can be addressed using machine learning. Under the correct settings, it is possible to extract useful insight from this huge amount of data for improving your marketing strategy, identifying users with dangerous behaviors (for example, terrorist networks), and predicting the likelihood that a user will read your new post.

Specifically, link prediction is one of the most interesting and important research topics in this field. Depending on what a *connection* in your social graph represents, by predicting future edges, you will be able to predict your next suggested friend, the next suggested movie, and which product you are likely to buy.

As we have already seen in *Chapter 6, Solving Common Graph-Based Machine Learning Problems*, the link prediction task aims at forecasting the likelihood of a future connection between two nodes, and it can be solved using several machine learning algorithms.

In the next examples, we will be applying machine learning supervised and unsupervised graph embedding algorithms for predicting future connections on the SNAP Facebook social graph. Furthermore, we will evaluate the contribution of node features in the prediction task.

Task preparation

In order to perform the link prediction task, it is necessary to prepare our dataset. The problem will be treated as a supervised task. Pairs of nodes will be provided to each algorithm as input, while the target will be binary, that is, *connected* if the two nodes are actually connected in the network and *not connected* otherwise.

Since we aim to cast this problem as a supervised learning task, we need to create a training and testing dataset. We will, therefore, create two new subgraphs that have the same number of nodes but different numbers of edges (as some edges will be removed and treated as positive samples for training/testing the algorithm).

The stellargraph library provides a useful tool for splitting the data and creating training and test reduced subgraphs. This process is similar to the one we have already seen in *Chapter 6, Solving Common Graph-Based Machine Learning Problems*:

```
from sklearn.model_selection import train_test_split
from stellargraph.data import EdgeSplitter
from stellargraph import StellarGraph
edgeSplitter = EdgeSplitter(G)
graph_test, samples_test, labels_test = edgeSplitter.train_test_
```

```
split(p=0.1, method="global", seed=24)
edgeSplitter = EdgeSplitter(graph_test, G)
graph_train, samples_train, labels_train = edgeSplitter.train_test_
split(p=0.1, method="global", seed=24)
```

We are using the EdgeSplitter class to extract a fraction (p=10%) of all the edges in G, as well as the same number of negative edges, in order to obtain a reduced graph, graph_test. The train_test_split method also returns a list of node pairs, samples_test (where each pair corresponds to an existing or not-existing edge in the graph), and a list of binary targets (labels_test) of the same length of the samples_test list. Then, from such a reduced graph, we repeat the operation to obtain another reduced graph, graph_train, as well as the corresponding samples_train and labels_train lists.

We will be comparing three different methods for predicting missing edges:

- **Method1**: node2vec will be used to learn a node embedding without supervision. The learned embedding will be used as input for a supervised classification algorithm to determine whether the input pair is actually connected.

- **Method2**: The graph neural network-based algorithm GraphSAGE will be used to jointly learn the embedding and perform the classification task.

- **Method3**: Hand-crafted features will be extracted from the graph and used as inputs for a supervised classifier, together with the nodes' IDs.

Let's analyze them in more detail.

Node2Vec-based link prediction

The proposed method involves several steps:

1. We use node2vec to generate node embeddings without supervision from the training graph. This can be done using the node2vec Python implementation, as we have already seen in *Chapter 6, Solving Common Graph-Based Machine Learning Problems*:

    ```
    from node2vec import Node2Vec
    node2vec = Node2Vec(graph_train)
    model = node2vec.fit()
    ```

2. Then, we use HadamardEmbedder to generate an embedding for each pair of embedded nodes. Such feature vectors will be used as input to train the classifier:

```
from node2vec.edges import HadamardEmbedder

edges_embs = HadamardEmbedder(keyed_vectors=model.wv)
train_embeddings = [edges_embs[str(x[0]),str(x[1])] for x in
samples_train]
```

It's time to train our supervised classifier. We will be using the RandomForest classifier, a powerful decision tree-based ensemble algorithm:

```
from sklearn.ensemble import RandomForestClassifier
from sklearn import metrics
rf = RandomForestClassifier(n_estimators=10)
rf.fit(train_embeddings, labels_train);
```

Finally, let's apply the trained model for creating the embedding of the test set:

```
edges_embs = HadamardEmbedder(keyed_vectors=model.wv) test_
embeddings = [edges_embs[str(x[0]),str(x[1])] for x in samples_test]
```

3. Now we are ready to perform the prediction on the test set using our trained model:

```
y_pred = rf.predict(test_embeddings)
print('Precision:', metrics.precision_score(labels_test, y_pred))
print('Recall:', metrics.recall_score(labels_test, y_pred))
print('F1-Score:', metrics.f1_score(labels_test, y_pred))
```

4. The output should be as follows:

```
Precision: 0.9701333333333333
Recall: 0.9162573983125551
F1-Score: 0.9424260086781945
```

Not bad at all! We can observe that the node2vec-based embedding already provides a powerful representation for actually predicting links on the combined Facebook ego network.

GraphSAGE-based link prediction

Next, we will use GraphSAGE to learn node embeddings and classify edges. More specifically, we will build a two-layer GraphSAGE architecture that, given labeled pairs of nodes, outputs a pair of node embeddings. Then, a fully connected neural network will be used to process these embeddings and produce link predictions. Notice that the GraphSAGE model and the fully connected network will be concatenated and trained from end to end so that the embedding learning stage is influenced by the predictions.

A similar example using PyTorch is also available in the GitHub repository (`https://github.com/PacktPublishing/Graph-Machine-Learning/blob/main/Chapter07`) repository for interested readers.

Featureless approach

Before starting, you may recall from *Chapter 4, Unsupervised Graph Learning*, and *Chapter 5, Supervised Graph Learning*, that GraphSAGE needs node descriptors (features). Such features may or may not be available in your dataset. Let's begin our analysis by not considering available node features. In this case, a common approach is to assign to each node a one-hot feature vector of length |V| (the number of nodes in the graph), where only the cell corresponding to the given node is 1, while the remaining cells are 0.

This can be done in Python and `networkx` as follows:

```
eye = np.eye(graph_train.number_of_nodes())
fake_features = {n:eye[n] for n in G.nodes()}
nx.set_node_attributes(graph_train, fake_features, "fake")
eye = np.eye(graph_test.number_of_nodes())
fake_features = {n:eye[n] for n in G.nodes()}
nx.set_node_attributes(graph_test, fake_features, "fake")
```

In the preceding code snippet, we did the following:

1. Created an identity matrix of size |V|. Each row of the matrix is the one-hot vector we need for each node in the graph.

2. Then, we create a Python dictionary where, for each `nodeID` (used as the key), we assign the corresponding row of the above-created identity matrix.

3. Finally, the dictionary is passed to the `networkx` function `set_node_attributes` to assign the "fake" features to each node in the `networkx` graph.

Notice that the process is repeated for both training and test graphs.

The next step will be defining the generator that will be used to feed the model. We will be using the `stellargraph GraphSAGELinkGenerator` for this, which essentially provides the model with pairs of nodes as input:

```
from stellargraph.mapper import GraphSAGELinkGenerator
batch_size = 64
num_samples = [4, 4]
# convert graph_train and graph_test for stellargraph
```

```
sg_graph_train = StellarGraph.from_networkx(graph_train, node_
features="fake")
sg_graph_test = StellarGraph.from_networkx(graph_test, node_
features="fake")
train_gen = GraphSAGELinkGenerator(sg_graph_train, batch_size, num_
samples)
train_flow = train_gen.flow(samples_train, labels_train, shuffle=True,
seed=24)
test_gen = GraphSAGELinkGenerator(sg_graph_test, batch_size, num_samples)
test_flow = test_gen.flow(samples_test, labels_test, seed=24)
```

Note that we also need to define the batch_size (number of inputs per minibatch) and the number of first- and second-hop neighbor samples that GraphSAGE should consider.

Finally, we are ready to create the model:

```
from stellargraph.layer import GraphSAGE, link_classification
from tensorflow import keras
layer_sizes = [20, 20]
graphsage = GraphSAGE(layer_sizes=layer_sizes, generator=train_gen,
bias=True, dropout=0.3)
x_inp, x_out = graphsage.in_out_tensors()
# define the link classifier
prediction = link_classification(output_dim=1, output_act="sigmoid", edge_
embedding_method="ip")(x_out)
model = keras.Model(inputs=x_inp, outputs=prediction)
model.compile(
    optimizer=keras.optimizers.Adam(lr=1e-3),
    loss=keras.losses.mse,
    metrics=["acc"],
)
```

In the preceding snippet, we are creating a GraphSAGE model with two hidden layers of size 20, each with a bias term and a dropout layer for reducing overfitting. Then, the output of the GraphSAGE part of the module is concatenated with a link_classification layer that takes pairs of node embeddings (output of GraphSAGE), uses binary operators (inner product – ip – in our case) to produce edge embeddings, and finally passes them through a fully connected neural network for classification.

The model is optimized via the Adam optimizer (learning rate=1e-3) using the mean squared error as a loss function.

Let's train the model for 20 epochs:

```
epochs = 20
history = model.fit(train_flow, epochs=epochs, validation_data=test_flow)
```

The tail of the output should be as follows:

```
Epoch 18/20
loss: 0.4921 - acc: 0.8476 - val_loss: 0.5251 - val_acc: 0.7884
Epoch 19/20
loss: 0.4935 - acc: 0.8446 - val_loss: 0.5247 - val_acc: 0.7922
Epoch 20/20
loss: 0.4922 - acc: 0.8476 - val_loss: 0.5242 - val_acc: 0.7913
```

As you can see, in the final epochs (18–20), the loss and accuracy values stabilize, suggesting the model has mostly converged. The validation metrics (`val_loss` and `val_acc`) are close to the training metrics, indicating that the model generalizes well and is not overfitting. Let's compute the performance metrics over the test set:

```
from sklearn import metrics
y_pred = np.round(model.predict(train_flow)).flatten()
print('Precision:', metrics.precision_score(labels_train, y_pred))
print('Recall:', metrics.recall_score(labels_train, y_pred))
print('F1-Score:', metrics.f1_score(labels_train, y_pred))
```

The output should be as follows:

```
Precision: 0.7156476303969199
Recall: 0.983125550938169
F1-Score: 0.8283289124668435
```

As we can observe, performances are lower than the ones obtained in the `node2vec`-based approach. A possible reason may be attributed to the particular global structure of the social network: the dataset used here likely benefits from the global structural relationships captured by node2vec. Node2vec relies on the relationships captured through random walks, which emphasize connections between nodes that may not be directly adjacent but share a broader context in the network, while the featureless GraphSAGE relies on aggregating information from immediate neighbors. Without informative node features, this "local" approach may not fully capture the relationships present in the network defined by, for example, common interests. For this reason, real node features may represent a great source of information and we may want to use it. Let's do that in the following test.

In this example, node2vec performs better than the featureless GraphSAGE. However, this may not be the case for other scenarios. The featureless GraphSAGE excels in scenarios dominated by local interactions, such as citation or community-based networks with strong localized patterns, and may be particularly effective when small, connected clusters or highly predictive immediate neighbors are present.

Introducing node features

The process of extracting node features for the combined ego network is quite verbose. This is because, as we explained in the first part of the chapter, each ego network is described using several files, as well as all the feature names and values. We have written useful functions for parsing all the ego networks in order to extract the node features. You can find their implementation in the Python notebook supplied with this book. Here, let's just briefly summarize how they work:

1. The load_features function parses each ego network and creates two dictionaries:

    a. feature_index, which maps numeric indices to feature names

    b. inverted_feature_indexes, which maps names to numeric indices

2. The parse_nodes function receives the combined ego network G and the ego node IDs. Then, each ego node in the network is assigned the corresponding features previously loaded using the load_features function.

Let's invoke them in order to load a feature vector for each node in the combined ego network:

```
load_features()
parse_nodes(G, ego_nodes)
```

We can easily check the result by printing the information of one node in the network (e.g., the node with ID 0):

```
print(G.nodes[0])
```

The output should be as follows:

```
{'features': array([1., 1., 1., ..., 0., 0., 0.])}
```

As we can observe, the node has a dictionary containing a key named features. The corresponding value is the feature vector assigned to this node.

We are now ready to repeat the same steps used before for training the GraphSAGE model, this time using features as the key when converting the networkx graph to the stellargraph format:

```
sg_graph_train = StellarGraph.from_networkx(graph_train, node_
features="features")
sg_graph_test = StellarGraph.from_networkx(graph_test, node_
features="features")
```

Finally, as we have done before, we create the generators, compile the model, and train it for 20 epochs:

```
train_gen = GraphSAGELinkGenerator(sg_graph_train, batch_size, num_samples)
train_flow = train_gen.flow(samples_train, labels_train, shuffle=True,
seed=24)
test_gen = GraphSAGELinkGenerator(sg_graph_test, batch_size, num_samples)
test_flow = test_gen.flow(samples_test, labels_test, seed=24)
layer_sizes = [20, 20]
graphsage = GraphSAGE(layer_sizes=layer_sizes, generator=train_gen,
bias=True, dropout=0.3)
x_inp, x_out = graphsage.in_out_tensors()
prediction = link_classification(output_dim=1, output_act="sigmoid", edge_
embedding_method="ip")(x_out)
model = keras.Model(inputs=x_inp, outputs=prediction)
model.compile(
    optimizer=keras.optimizers.Adam(lr=1e-3),
    loss=keras.losses.mse,
    metrics=["acc"],
)
epochs = 20
history = model.fit(train_flow, epochs=epochs, validation_data=test_flow)
```

Notice that we are using the same hyperparameters (including the number of layers, batch size, and learning rate) as well as the random seed, to ensure a fair comparison between the models.

The tail of the output should be as follows:

```
Epoch 18/20
loss: 0.1337 - acc: 0.9564 - val_loss: 0.1872 - val_acc: 0.9387
Epoch 19/20
```

```
loss: 0.1324 - acc: 0.9560 - val_loss: 0.1880 - val_acc: 0.9340
Epoch 20/20
loss: 0.1310 - acc: 0.9585 - val_loss: 0.1869 - val_acc: 0.9365
```

Let's evaluate the model performance:

```
from sklearn import metrics
y_pred = np.round(model.predict(train_flow)).flatten()
print('Precision:', metrics.precision_score(labels_train, y_pred))
print('Recall:', metrics.recall_score(labels_train, y_pred))
print('F1-Score:', metrics.f1_score(labels_train, y_pred))
```

We'll check the output:

```
Precision: 0.7895418326693228
Recall: 0.9982369978592117
F1-Score: 0.8817084700517213
```

As we can notice, the introduction of real node features has brought a good improvement, even if the best performances are still the ones achieved using the node2vec approach. This can be attributed to the effectiveness of capturing global structural patterns, which seems to be highly advantageous for this social network dataset and more informative than individual node features, according to the observed results. Such global patterns may include relationships between nodes that do not share direct edges but are closely connected within the graph's topology, relationships that the random walks employed by node2vec likely enhance in their identification.

Finally, we will evaluate a shallow embedding approach where hand-crafted features will be used for training a supervised classifier.

Hand-crafted features for link prediction

As we have already seen in *Chapter 5, Supervised Graph Learning*, shallow embedding methods represent a simple yet powerful approach for dealing with supervised tasks. Basically, for each input edge, we will compute a set of metrics that will be given as input to a classifier.

In this example, for each input edge represented as a pair of nodes (u,v), four metrics will be considered, namely:

- **Shortest path**: The length of the shortest path between u and v. If u and v are directly connected through an edge, this edge will be removed before computing the shortest path. The value 0 will be used if u is not reachable from v.

- **Jaccard coefficient**: Given a pair of nodes (u,v), it is defined as the intersection over a union of the set of neighbors of u and v. Formally, let $s(u)$ be the set of neighbors of the node u and $s(v)$ be the set of neighbors of the node v:

$$j(u, v) = \frac{s(u) \cap s(v)}{s(u) \cup s(v)}$$

- **Centrality**: The degree centrality computed for the nodes v and u.
- **u community**: The community ID assigned to the nodes u and v using the Louvain heuristic.

We have written a useful function for computing these metrics using Python and networkx. You can find the implementation in the Python notebook attached to this book.

Let's compute the features for each edge in the training and test sets:

```
feat_train = get_hc_features(graph_train, samples_train, labels_train)
feat_test = get_hc_features(graph_test, samples_test, labels_test)
```

In the proposed shallow approach, these features will be directly used as input for a Random Forest classifier. We will use its scikit-learn implementation as follows:

```
from sklearn.ensemble import RandomForestClassifier
from sklearn import metrics
rf = RandomForestClassifier(n_estimators=10)
rf.fit(feat_train, labels_train);
```

The above lines automatically instantiate and train a Random Forest classifier using the edge features we have computed before. We are now ready to compute the performance as follows:

```
y_pred = rf.predict(feat_test)
print('Precision:', metrics.precision_score(labels_test, y_pred))
print('Recall:', metrics.recall_score(labels_test, y_pred))
print('F1-Score:', metrics.f1_score(labels_test, y_pred))
```

The output will be:

```
Precision: 0.9636952636282395
Recall: 0.9777853337866939
F1-Score: 0.9706891701828411
```

Surprisingly, the shallow method based on hand-crafted features performs better than the others. This result can be attributed to the high expressiveness of the hand-crafted metrics used in this specific dataset.

Metrics such as the shortest path, Jaccard coefficient, centrality, and community information directly capture structural properties of the graph that are particularly predictive for link prediction tasks. Additionally, the Random Forest classifier used in the shallow method may have better exploited these engineered features compared to the more generalized embeddings generated by node2vec and GraphSAGE. However, it is important to note that this observation is dataset-dependent and may not hold true in all cases. The shallow method relies heavily on the quality of the hand-crafted features, which may not always capture all relevant graph properties for more complex or large-scale datasets. In contrast, methods like node2vec and GraphSAGE are designed to generalize across diverse graph structures by learning feature representations directly from the data. These methods often outperform shallow approaches in scenarios where the graph is too complex for simple metrics to capture all predictive patterns.

Summarizing the results

In the above examples, we have trained three algorithms in learning, with and without supervision, useful embeddings for link prediction. In the following table, we summarize the results:

Algorithm	embedding	node features	Precision	Recall	F1-Score
Node2Vec	unsupervised	No	0.97	0.92	0.94
GraphSAGE	supervised	Yes	0.72	0.98	0.83
GraphSAGE	supervised	No	0.79	1.00	0.88
Shallow	manual	No	0.96	0.98	0.97

Table 7.1: Summary of the results achieved for the link prediction task

As shown in *Table 7.1*, the node2vec-based method is already able to achieve a high level of performance without supervision or per-node information. As mentioned earlier, such high results might be related to the particular structure of the combined ego network. Due to the high sub-modularity of the network (since it is composed of several ego networks), predicting whether two users will be connected or not might be highly related to the way the two candidate nodes are connected inside the network. For example, there might be a systematic situation in which two users, both connected to several users in the same ego network, have a high chance of being connected as well. On the other hand, two users belonging to different ego networks, or *very far* from each other, are likely to be not connected, making the prediction task easier. This is also confirmed by the high results achieved using the shallow method.

Such a situation might be confusing, instead, for more complicated algorithms like GraphSAGE, especially when node features are involved. For example, two users might share similar interests, making them very similar. However, they might belong to different ego networks, where the corresponding ego users live in two very different parts of the world. So, similar users that in principle should be connected are not. However, it is also possible that such algorithms are predicting something further in the future. Recall that the combined ego network is a timestamp of a particular situation in a given period of time. Who knows how it might have evolved now!

Interpreting machine learning algorithms is probably the most interesting challenge of machine learning itself. For this reason, we should always interpret results with care. Our suggestion is always to dig into the dataset and try to give an explanation of your results.

Finally, it is important to remark that each of the algorithms was not tuned for the purpose of this demonstration. Different results can be obtained by properly tuning each hyperparameter and we highly suggest you try to do it.

Notice that, in this example, we did not apply any particular feature processing method. However, in other scenarios, we may want to perform feature engineering to improve performances. Skewed data may benefit from transformations, and highly correlated features, though not harmful, could be pruned for efficiency. Addressing outliers enhances model robustness, and dimensionality reduction prevents issues with overly high-dimensional data.

Summary

In this chapter, we have seen how machine learning can be useful for solving practical machine learning tasks on social network graphs. More specifically, we have seen how future connections can be predicted on the SNAP Facebook combined ego network.

We reviewed graph analysis concepts and used graph-derived metrics to collect insights into the social graph. Then, we benchmarked several machine learning algorithms on the link prediction task, evaluating their performance and trying to give them an interpretation.

In the next chapter, we will focus on how similar approaches can be used to analyze a corpus of documents, using text analytics and natural language processing.

8

Text Analytics and Natural Language Processing Using Graphs

Nowadays, a vast amount of information is available in the form of text in natural written language. The very book you are reading right now is one such example. The news you read every morning, the messages or Facebook posts you send or read, the reports you write for school assignments, and the emails you write are all examples of information that is exchanged via written documents and text. It is undoubtedly the most common way of indirect interaction, as opposed to direct interaction, such as talking or gesticulating. It is, therefore, crucial to utilize this kind of information and extract insights from documents and texts. This abundance of textual information has driven significant advancements in **natural language processing (NLP)**. One notable development is ChatGPT, which provides sophisticated conversational capabilities. However, like many other NLP algorithms, ChatGPT is not natively designed to exploit inherent relationships and structures present in text data. As we are learning throughout this book, indeed, such relationships are everywhere and, if properly leveraged, can significantly benefit machine learning. For example, graphs can represent connections between entities, dependencies in sentences, or semantic relationships across documents. By integrating graphs into NLP workflows, we can enhance tasks such as information retrieval, question answering, and recommendation systems. Moreover, graph-based methods often improve the explainability and robustness of models, offering insights into the underlying relationships that would otherwise be overlooked by traditional NLP approaches.

In this chapter, we will show you how to process natural language texts and review some basic models that allow structuring text information. Using the information extracted from a corpus of documents, we will show you how to create networks that can be analyzed using some of the techniques we have seen in previous chapters. In particular, using a tagged corpus, we will show you how to develop both semi-supervised (classification models to classify documents in pre-determined topics) and unsupervised (community detection to discover new topics) algorithms.

The chapter is organized as follows:

- Providing a quick overview of a dataset – the `Reuters-21578` dataset
- Understanding the main concepts and tools used in NLP
- Creating graphs from a corpus of documents
- Building a document topic classifier

Technical requirements

All code files relevant to this chapter are available at `https://github.com/PacktPublishing/` `Graph-Machine-Learning/tree/main/Chapter08`. Please refer to the *Practical exercises* section in *Chapter 1, Getting Started with Graphs,* for guidance on how to set up the environment to run the examples in this chapter, using either Poetry, `pip,` or Docker.

Providing a quick overview of a dataset

In order to show you how to process a corpus of documents with the aim of extracting relevant information, we will be using a dataset derived from a well-known benchmark in the field of NLP: the so-called **Reuters-21578 dataset**. The original dataset includes a set of 21,578 news articles published in the Reuters financial newswire in 1987, which were assembled and indexed in categories. The original dataset has a very skewed distribution, with some categories appearing in only the training set or the test set. For this reason, we use a modified version named **ApteMod**, also referred to as *Reuters-21578 Distribution 1.0*, which has a lesser skew distribution and consistent labels between training and test datasets.

Despite the fact that the news articles in the Reuters financial newswire are a bit outdated, the dataset has been used in a plethora of papers on NLP and still represents a dataset often used for benchmarking algorithms. Nevertheless, the `Reuters-21578` dataset is much smaller in size in comparison to other datasets available today, which can comprise millions or billions of documents. Hence, if you wish to scale your application and analysis, using a dataset with a larger number of documents is advisable (see, for instance, `https://github.com/niderhoff/` `nlp-datasets` for an overview of the most common ones).

It is important to note that these may require larger storage and computational power to allow their processing. In *Chapter 10, Building a Data-Driven Graph-Powered Application*, we will show you some of the tools and libraries that can help you scale your analysis.

Let's overview the dataset: each document of the Reuters-21578 dataset is provided with a set of labels that represent its content and make it a perfect benchmark for testing both supervised and unsupervised algorithms. The Reuters-21578 dataset can be easily download using the nltk library (which is a very useful library for post-processing documents):

```
from nltk.corpus import reuters
corpus = pd.DataFrame([
    {"id": _id,
     "text": reuters.raw(_id).replace("\n", ""),
     "label": reuters.categories(_id)}
    for _id in reuters.fields()
])
```

As you will see by inspecting the DataFrame corpus, the IDs have the form training/{ID} and test/{ID}, making it clear which documents should be used for training and which for testing. To start with, let us list all the topics and see the number of documents per topic using the following code:

```
from collections import Counter
Counter([label for document_labels in corpus["label"] for label in
document_labels]).most_common()
```

The Reuters-21578 dataset includes 90 different topics with a significant degree of unbalance between classes, with almost 37% of the documents in the most common category and only 0.01% in each of the five least common categories. As you can see from inspecting the text, some of the documents have some new-line characters embedded. These artifacts are common in raw text data and can interfere with text processing tasks, such as tokenization, which is a fundamental step in most NLP pipelines (as we will see later). If not removed, such artifacts can lead to tokenization errors, where words or phrases are incorrectly split, resulting in noisy or inconsistent data representations. This can degrade the performance of downstream NLP algorithms. To address this issue, we can easily clean the text by removing newline characters during the preprocessing stage, as follows:

```
corpus["clean_text"] = corpus["text"].apply(
    lambda x: x.replace("\n", "")
)
```

Now that we have loaded the data in memory, we can start analyzing it. In the next section, we will show you some of the main tools that can be used when dealing with unstructured text data to extract structured information that can be more easily used.

Understanding the main concepts and tools used in NLP

When processing documents, the first analytical step is certainly to infer the document language. Most analytical engines used in NLP tasks are in fact trained on documents that have a specific language and should only be used for such a language. Although attempts to build cross-language models (see, for instance, multi-lingual embeddings such as https://fasttext.cc/docs/en/aligned-vectors.html and https://github.com/google-research/bert/blob/master/multilingual.md) have recently gained increasing popularity, these models face challenges, such as lower performance compared to language-specific models and difficulty in handling languages with sparse training data or significantly different syntax and grammatical structures. For these reasons, they still represent a small portion of NLP models. It is, therefore, very common to first infer the language to use the correct downstream analytical NLP pipeline.

In order to infer the language, different methods can be used. One very simple yet effective approach relies on looking for the most common words of a language (so-called stopwords) and building a score based on their frequencies. Its precision, however, tends to be limited for short text and does not make use of the word positioning and context. On the other hand, Python has many libraries that, using more elaborate logic, allow inferring the language in a more precise manner. Some examples of these libraries are fasttext, polyglot, and langdetect, to name just a few.

As an example, we will use langdetect in the following code example, which can be integrated with very few lines and provides support for more than 150 languages. The language can be inferred for all documents using the following snippet:

```
from langdetect import detect
import numpy as np
def getLanguage(text: str):
    try:
        return langdetect.detect(text)
    except:
        return np.nan
corpus["language"] = corpus["text"].apply(langdetect.detect)
```

As you will see in the output, there seem to be documents in languages other than English. Indeed, these documents often are either very short or have a strange structure that does not seem to be actual news. When documents represent text that a human would read and label as news, the model is generally rather precise and accurate.

Now that we have inferred the language, we can continue with the language-dependent steps of the analytical pipeline. Documents in languages other than English will now be filtered out, as these are indeed a very small part of the dataset (also given that the dataset is built by documents in English and that, for the sake of simplicity, the models that we are going to use in this example are specific to English text). Of course, should the number of non-English documents be larger and more significant, and dropping them were not an option, you could either first translate the non-English documents into English using appropriate text-to-text translation models (like Google Translate) or split the analytical pipeline into multiple sub-pipelines depending on the document language.

For the following tasks, we will be using spacy, which is an extremely powerful library that allows us to embed state-of-the- art NLP models with very few lines of code. After installing the library with pip install spacy, language-specific models can be integrated by simply installing them using the spacy download utility. For instance, the following command can be used in order to download and install the English model:

```
python -m spacy download en_core_web_sm
```

Now we should have the language models in English ready to use. Let's then see which information it can provide to us. Using spacy is extremely simple; in just one line of embedded code, a very rich set of information can be embedded. Let us start by applying the model to one of the documents in the Reuters corpus, presented in the following callout box:

SUBROTO SAYS INDONESIA SUPPORTS TIN PACT EXTENSION

Mines and Energy Minister Subroto confirmed Indonesian support for an extension of the sixth **International Tin Agreement (ITA)**, but said a new pact was not necessary. Asked by Reuters to clarify his statement on Monday in which he said the pact should be allowed to lapse, Subroto said Indonesia was ready to back extension of the ITA. "We can support extension of the sixth agreement," he said. "But a seventh accord we believe to be unnecessary." The sixth ITA will expire at the end of June unless a two-thirds majority of members vote for an extension.

spacy can be easily applied by just loading the model and applying it to the text:

```
nlp = spacy.load('en_core_web_md')
parsed = nlp(text)
```

The object parsed, returned by spacy, has several fields that result from the application of many models that are combined into a single pipeline. These provide a different level of text structuring that we will examine one by one:

- **Text segmentation and tokenization**, which is the process aimed at splitting a document into its periods, sentences, and single words (or tokens). This step is generally very important for all subsequent analyses and it usually uses punctuation, blank-spaces, and new lines, in order to infer the best document segmentation. The segmentation engine provided in spacy generally works fairly well. However, please note that depending on the context, a bit of model tuning or rule modification might be necessary. For instance, when dealing with short texts that have slang, emoticons, links, and hashtags, a better choice for text segmentation and tokenization may be the TweetTokenizer included in the nltk library. Depending on the context, we encourage you to explore other possible segmentations available.

  In the document returned by spacy, the segmentation in sentences can be found in the sents attribute of the parsed object. Each sentence can be iterated over its token by simply using:

  ```
  for sent in parsed.sents:
      for token in sent:
          print(token)
  ```

  Each token is a spacy Span object that has attributes that specify the type of token and further characterization that is introduced by the other models.

- **Part-of-speech tagging**. Once the text has been divided into its single words (also referred to as tokens), the next step is to associate each token to a **part-of-speech (PoS)** tag, that is to say, its grammatical type. The inferred tags are usually nouns, verbs, auxiliary verbs, adjectives, and so on. The engines used for PoS tagging are usually models that are trained to classify tokens based on a large, labeled corpus, where each token has an associated PoS tag. Being trained on actual data, they learn to recognize common patterns within a language, for instance, the word "the" (which is a **determinative article (DET)** is usually followed by a noun, and so on. When using spacy, the information about the PoS tagging is usually stored in the label_ attribute of the Span object. The types of tags available can be found at https://spacy.io/models/en. Conversely, you can get a human-readable value for a given type using the function spacy.explain.

- **Named entity recognition.** This analytical step is generally a statistical model that is trained to recognize the type of nouns that appear within the text. Some common examples of entities are organization, person, geographic location and addresses, products, numbers, and currencies. Given the context (the surrounding words) as well as the prepositions that are used, the model infers the most probable type of the entity, if any. As in other steps of the NLP pipeline, these models are also usually trained using a large, tagged dataset on which they learn common patterns and structure. In spacy, the information about the document entities is usually stored in the ents attribute of the parsed object. spacy also provides some utilities to nicely visualize the entities in a text, using the displacy module:

```
displacy.render(parsed, style='ent', jupyter=True)
```

This gives the following output:

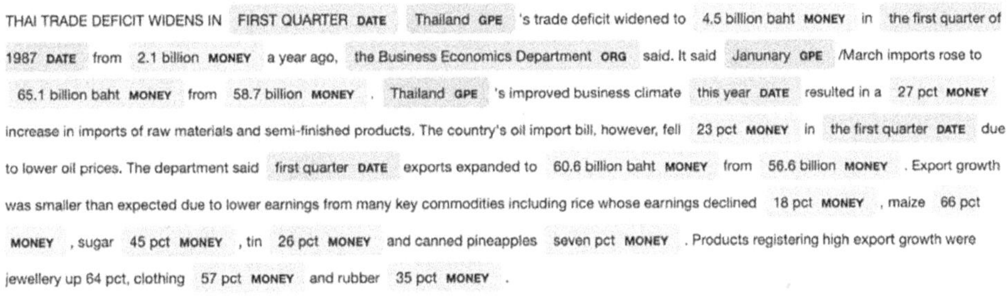

THAI TRADE DEFICIT WIDENS IN FIRST QUARTER DATE Thailand GPE 's trade deficit widened to 4.5 billion baht MONEY in the first quarter of 1987 DATE from 2.1 billion MONEY a year ago, the Business Economics Department ORG said. It said Jananuary GPE /March imports rose to 65.1 billion baht MONEY from 58.7 billion MONEY . Thailand GPE 's improved business climate this year DATE resulted in a 27 pct MONEY increase in imports of raw materials and semi-finished products. The country's oil import bill, however, fell 23 pct MONEY in the first quarter DATE due to lower oil prices. The department said first quarter DATE exports expanded to 60.6 billion baht MONEY from 56.6 billion MONEY . Export growth was smaller than expected due to lower earnings from many key commodities including rice whose earnings declined 18 pct MONEY , maize 66 pct MONEY , sugar 45 pct MONEY , tin 26 pct MONEY and canned pineapples seven pct MONEY . Products registering high export growth were jewellery up 64 pct, clothing 57 pct MONEY and rubber 35 pct MONEY .

Figure 8.1: Example of the spacy output for the named entity recognition engine

- **Dependency parsing.** The dependency parser is an extremely powerful engine that infers the relationships between tokens within a sentence. It basically allows you to build a syntactic tree of how words are related to each other. Let's, for instance, take a simple example taken from the spacy website: *Autonomous cars shift insurance liability towards manufacturers.*

Figure 8.2 shows the dependency tree, where it can be seen that the main verb (or root) "shift" is related by the subject-object relationship to "cars" (subject) and "liability" (object). It also sustains the preposition "towards." In the same way, the remaining nouns/adjectives "Autonomous," "insurance," and "manufacturers" are related to either the subject, the object, or the preposition. Thus, spacy can be used to build a syntactic tree that can be navigated in order to identify relationships between the tokens:

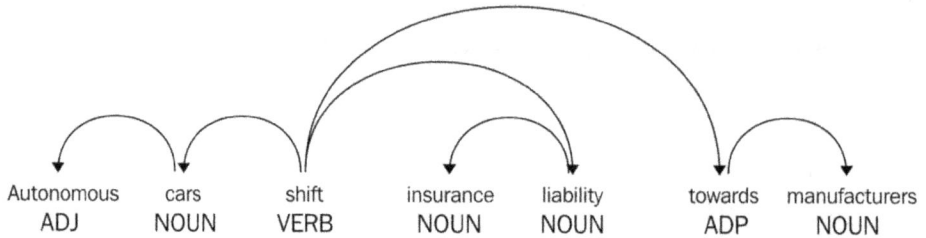

Figure 8.2: Example of a syntactic dependency tree provided by spacy

As we will see in the following section, this information can be crucial when building knowledge graphs.

- **Lemmatization or stemming.** Finally, the very last step of the analytical pipeline is aimed at reducing words to a common root to provide a cleaner version, and limiting the morphological variation of words. Take, for instance, the verb *to be*. It can have many morphological variations, such as "is," "are," "was," etc. which are all different valid forms. Or also consider the difference between "car" and "cars." In most cases, we are not interested in these small differences introduced by morphology. The lemmatizers and stemmers help to reduce tokens to a common, stable form that can be more easily processed. Usually, the lemmatizer is based on a set of rules that associate particular words (with conjugation, plurals, and inflection) to a common root form. More elaborate implementations may also use the context and the *PoS* tagging information in order to be more robust against homonyms. On the other hand, stemmers are generally simpler, and instead of associating words to a common root form, they usually remove the last part of the word to deal with inflectional and derivational variance. Also, stemmers are generally based on a set of rules that remove a certain pattern, rather than taking into account lexica and syntactic information or using extensive vocabulary mappings. In spacy, the lemmatized version of a token can be found in the Span object via the `lemma_` attribute.

As shown below, spacy pipelines can be easily integrated in order to process the entire corpus and store the results in our corpus DataFrame:

```
nlp = spacy.load('en_core_web_md')
sample_corpus["parsed"] = sample_corpus["clean_text"]\
    .apply(nlp)
```

This DataFrame represents the structured information of the documents that will be the base of all our subsequent analysis. In particular, in the next section we will show you how to build graphs starting from such information.

Creating graphs from a corpus of documents

In this section, we will use the information extracted in the previous section using the different text engines to build networks that relate the different information. In particular, we will focus on two kinds of graphs:

- **Knowledge-based graphs,** where we will use the semantic meaning of sentences to infer relationships between the different entities.

- **Bipartite graphs**, connecting the documents with the entities appearing in the text. We will then project the bipartite graph into a homogeneous graph, which might be made of either document or entity nodes only.

Knowledge graphs

A **knowledge graph** is a structured representation of entities, their attributes, and relationships, organized as a graph. It encodes semantic relationships to enable reasoning, querying, and knowledge discovery. It is commonly used in AI and machine learning to power applications like recommendation systems, natural language understanding, and data integration. Knowledge graphs are very interesting as they not only relate entities but also provide direction and meaning to the relationship. For instance, take the following relationship:

I - buy -> book

It is substantially different from the following relationship:

I - sell -> book

Besides the kind of relationship (buying or selling), it is also important to have a direction, where the subject and object are not treated symmetrically, but there is a difference between who is performing the action and who is the target of such an action.

In order to create a knowledge graph, we, therefore, need a function that is able to identify for each sentence the **subject–verb–object** (**SVO**) triplet. This function can then be applied to all sentences in the corpus and all the triplets can be aggregated to generate the corresponding graph.

The SVO extractor can be implemented on top of the enrichment provided by spacy models. Indeed, the tagging provided by the dependency tree parser can be very helpful to separate main sentences and their subordinates, as well as identifying the SOV triplets. The business logic may need to consider a few special cases (such as conjunctions, negations, and preposition handling) but this can be encoded with a set of rules. Moreover, these rules may also change depending on the specific use case, with slight variations to be tuned by the user.

A base implementation of such rules can be found in https://github.com/NSchrading/intro-spacy-nlp/blob/master/subject_object_extraction.py, which has been slightly adopted for our scope and is included in the GitHub repo provided with this book. Using this helper function, we compute all triplets in the corpus and store them in our corpus DataFrame:

```python
from subject_object_extraction import findSVOs
corpus["triplets"] = corpus["parsed"].apply(
    lambda x: findSVOs(x, output="obj")
)
edge_list = pd.DataFrame([
    {
        "id": _id,
        "source": source.lemma_.lower(),
        "target": target.lemma_.lower(),
        "edge": edge.lemma_.lower()
    }
    for _id, triplets in corpus["triplets"].iteritems()
    for (source, (edge, neg), target) in triplets
    # Add stopword filtering
    if not any([source.is_stop, target.is_stop])
    # Add filtering based on PoS tagging
    if (source.pos_ == "PROPN" or source.pos_ == "NOUN")
        and (target.pos_== "PROPN" or target.pos_== "NOUN")
])
```

Note that we have also added stopwords filtering and retention of sources/targets that are nouns (or proper nouns) to only represent relevant edges. Moreover, we reduce the morphological variations of the words using lemmatization.

The type of connection (determined by the sentence main predicate) is stored in the edge column. The first 10 most common relationships can be shown using the following command:

```python
edges["edge"].value_counts().head(10)
```

The most common edge types correspond to very basic predicates. Indeed, together with very general verbs (such as be, have, tell, and give), we can also find predicates more related to a financial context (such as buy, sell, or make). Using all of these edges, we can now create our knowledge base graph using the networkx utility function:

```
G = nx.from_pandas_edgelist(
    edges, "source", "target",
    edge_attr=True, create_using=nx.MultiDiGraph()
)
```

By filtering the edge DataFrame and creating a subnetwork using this information, we can analyze specific relationship types, such as the `lend` edge:

```
G = nx.from_pandas_edgelist(
    edges[edges["edge"]=="lend"], "source", "target",
    edge_attr=True, create_using=nx.MultiDiGraph()
)
```

Figure 8.3 shows the subgraph based on the *lend* relations. As can be seen, it already provides interesting economical insights, such as the economic relationships between countries, such as Venezuela-Ecuador and US-Sudan.

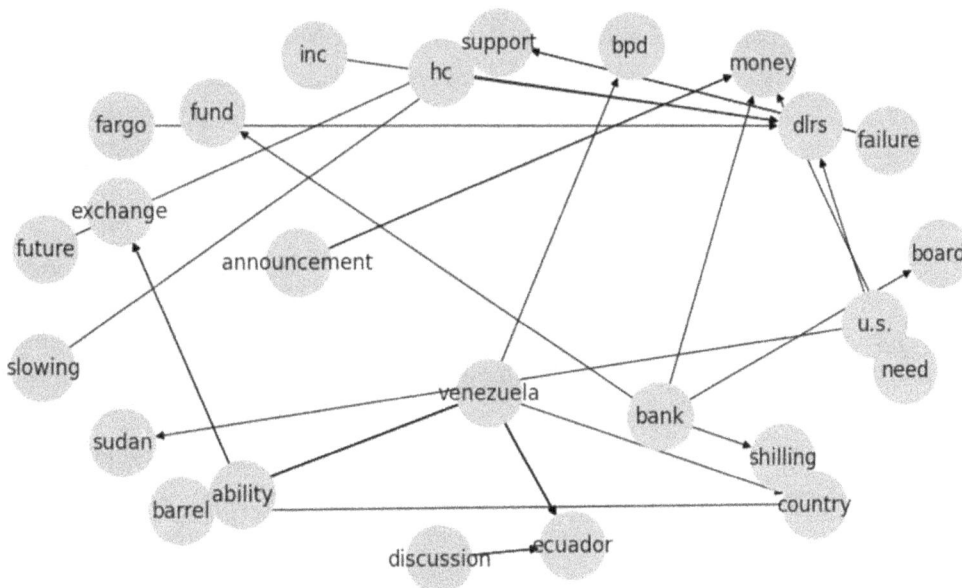

Figure 8.3: Example of a portion of the knowledge graph, for the edges relating to lending relationships

You can indeed play around with the preceding code by filtering the graph based on other relationships, and we definitely encourage you to do so, in order to further unveil interesting insights from the knowledge graphs we just created. In the next section, we will show you another method that allows us to encode the information extracted from the text into a graph structure.

In doing so, we will also make use of a particular type of graph, named *bipartite graphs, which* we introduced in *Chapter 1, Getting Started with Graphs*.

Knowledge graphs are rather interesting for unveiling and querying aggregated information over entities. However, other graph representations are also possible and can be useful in other situations. For example, when you want to cluster documents semantically, the knowledge graph may not be the best data structure to be used and analyzed. Knowledge graphs such as the one presented earlier are also not very effective for finding indirect relationships, such as identifying competitors, similar products, and so on, that do not often occur in the same sentence, but that often occur in the same document.

In order to address these limitations, in the next section, we will encode the information present in the document under the form of a **bipartite graph**.

Bipartite document/entity graphs

Bipartite graphs, a special class of graphs as described in *Chapter 1, Getting Started with Graphs*, are well suited for representing relationships between two distinct sets of entities. Unlike general graphs, where connections can exist between any pair of nodes, bipartite graphs enforce a structure: edges exist only between nodes belonging to different sets. This makes them particularly powerful for modeling systems with naturally dual components, such as users and items in recommendation systems, or, as in our case, documents and entities.

In this section, for each document, we will extract the entities that are most relevant, and connect a node, representing the document, with all the nodes representing the relevant entities in such a document. Each node may have multiple relationships: by definition, each document connects multiple entities. By contract terms, an entity can be referenced in multiple documents. As we will see in the following subsections, multiple cross-referencing in general can be used for creating a measure of similarity between entities and documents. This similarity can also be used to project the bipartite graph into one particular set of nodes, either the document nodes or the entity nodes.

To this aim, in order to build our bipartite graph, we first need to extract the relevant entities of a document. The term *relevant entity* is clearly fuzzy and broad. In the present context, we will consider as a relevant entity either a named entity (such as organization, person, or location recognized by the NER engine) or a keyword, that is to say, a word (or a composition of words) that identifies and generally describes the document and its content. For instance, suitable keywords for this book may be "graph," "network," "machine learning," "supervised model," or "unsupervised model."

There exist many algorithms that extract keywords from a document. One very simple way of doing this is based on the so-called TF-IDF score, which is based on building a score for each token (or group of tokens, often referred to as *grams*). This score is calculated using two components:

- **Term frequency** (**TF**), which measures the frequency of a word *i* in a specific document *j*, normalized by the total number of words in the document. This is given by:

$$TF = c_{i,j} / \sum c_{i,j}$$

- **Inverse document frequency** (**IDF**), which measures how unique a word *i* is across the corpus. It is given by:

$$IDF = \log N / (1 + D_i)$$

where N is the total number of documents, and D_i is the number of documents containing the word *i*.

The TF-IDF score can be computed as:

$$TF \cdot IDF$$

The TF-IDF score therefore promotes words that are repeated many times in the document and penalizes words that are standard and therefore might not be very representative for a document. There also exist more sophisticated algorithms.

One method that is quite powerful and worth mentioning in the context of this book is **TextRank**; it is also based on a graph representation of the document. TextRank creates a network where the nodes are a single token (e.g., a word) and where edges are created between tokens that appear within a specified window of one another in the text. Here, "window" refers to a fixed number of consecutive words in the text; if two tokens occur within this range, they are considered connected.

After creating such a network, **PageRank** is used to compute the centrality for each token, providing a score that allows a ranking within the document based on the centrality score. The most central nodes (up to a certain ratio, generally between 5% and 20% of the document size) are identified as candidate keywords. When two candidate keywords occur close to each other, they get aggregated into composite keywords, made up of multiple tokens. Implementations of TextRank are available in many NLP packages. One such implementation is gensim, which can be used quite straightforwardly:

```
from gensim.summarization import keywords
text = corpus["clean_text"][0]
keywords(text, words=10, split=True, scores=True,
        pos_filter=('NN', 'JJ'), lemmatize=True)
```

This provides an output of this form:

```
[('trading', 0.4615130639538529),
 ('said', 0.3159855693494515),
 ('export', 0.2691553824958079),
 ('import', 0.17462010006456888),
 ('japanese electronics', 0.1360932626379031),
 ('industry', 0.1286043740379779),
 ('minister', 0.12229815662000462),
 ('japan', 0.11434500812642447),
 ('year', 0.10483992409352465)]
```

Where the score represents the centrality, representing the importance for a given token. As you can see, some composite tokens may also occur, such as japanese electronics. Keyword extraction can be implemented to compute the keywords for the entire corpus, storing the information in our corpus DataFrame:

```
corpus["keywords"] = corpus["clean_text"].apply(
    lambda text: keywords(
        text, words=10, split=True, scores=True,
        pos_filter=('NN', 'JJ'), lemmatize=True)
)
```

To create a bipartite graph using both keywords and named entities as relevant entities, we need to extract named entities and encode the information in a data format that is compatible with our representation of keywords. This can be done using a few utility functions for the following tasks:

1. Extracting all the entities of a certain type (e.g., location, organization, and person) and filtering by the number of occurrences in the document (to only retain the most relevant entities), returning a pandas DataFrame:

    ```
    def extractEntities(ents, minValue=1,
                        typeFilters=["GPE", "ORG", "PERSON"]):
        entities = pd.DataFrame([
            {
                "lemma": e.lemma_,
                "lower": e.lemma_.lower(),
                "type": e.label_
            } for e in ents if hasattr(e, "label_")
        ])
    ```

```
        if len(entities)==0:
            return pd.DataFrame()
        g = entities.groupby(["type", "lower"])
        summary = pd.concat({
            "alias": g.apply(lambda x: x["lemma"].unique()),
            "count": g["lower"].count()
        }, axis=1)
        return summary[summary["count"]>1]\
                .loc[pd.IndexSlice[typeFilters, :, :]]
```

2. Reparsing the pandas DataFrame to only extract the entities of a given type into a list (empty if there are no entities of that given type):

```
def getOrEmpty(parsed, _type):
    try:
        return list(parsed.loc[_type]["count"]\
            .sort_values(ascending=False).to_dict().items())
    except:
        return []
```

```
Wrapping up the previous two utilities functions into a more high-
level function:
```

```
def toField(ents):
    typeFilters=["GPE", "ORG", "PERSON"]
    parsed = extractEntities(ents, 1, typeFilters)
    return pd.Series({_type: getOrEmpty(parsed, _type)
                      for _type in typeFilters})
```

With these functions, parsing the spacy tags can be simply done with:

```
entities = corpus["parsed"].apply(lambda x: toField(x.ents))
```

The entities DataFrame can be easily merged with the corpus DataFrame using the pd.concat function, placing all information in a single data structure:

```
merged = pd.concat([corpus, entities], axis=1)
```

Now that we have all the ingredients needed for our bipartite graph, we can create the edge list by looping over all document-entity or document-keyword pairs:

```
edges = pd.DataFrame([
    {"source": _id, "target": keyword, "weight": score, "type": _type}
    for _id, row in merged.iterrows()
    for _type in ["keywords", "GPE", "ORG", "PERSON"]
    for (keyword, score) in row[_type]
])
```

Once the edge list is created, we can produce the bipartite graph using the networkx APIs:

```
G = nx.Graph()
G.add_nodes_from(edges["source"].unique(), bipartite=0)
G.add_nodes_from(edges["target"].unique(), bipartite=1)
G.add_edges_from([
    (row["source"], row["target"])
    for _, row in edges.iterrows()
])
```

We can now start by looking at an overview of our graph by using `nx.info`:

```
Type: Graph
Number of nodes: 25933
Number of edges: 100726
Average degree:   7.7682
```

Next, we will project the bipartite graph into either of the two sets of nodes: **entities** or **documents**. This will allow us to explore the difference between the two graphs we obtain and cluster both the terms and documents using the unsupervised techniques described in *Chapter 4, Unsupervised Graph Learning*.

Entity-entity graph

We start by projecting our graph into the set of entity nodes. In other words, we create a graph with each node representing an entity or a keyword; in this graph, entities are considered connected if they appear in the same document. networkx provides a special submodule to deal with bipartite graphs, `networkx.algorithms.bipartite`, where a number of algorithms have already been implemented. In particular, the submodule `networkx.algorithms.bipartite.projection` provides a number of utility functions to project bipartite graphs on a sub set of nodes.

Before performing the projection, we conveniently categorize the nodes into two distinct sets (either documents or entities), using the "bipartite" property we created when generating the graph:

```
document_nodes = {n
                  for n, d in G.nodes(data=True)
                  if d["bipartite"] == 0}
entity_nodes = {n
                for n, d in G.nodes(data=True)
                if d["bipartite"] == 1}
```

The graph projection basically creates a new graph with the set of selected nodes. Edges are places between the nodes depending on whether two nodes have neighbors in common. The basic function projected_graph creates such a network with unweighted edges. However, it is usually more informative to have edges weighted depending on the number of common neighbors or by summing up the weights of all common neighbors. The projection module provides different functions depending on how the weights are computed. In the following subsections, we will use the overlap_weighted_projected_graph, where the edge weight is computed using the Jaccard similarity based on common neighbors. However, we encourage you to also explore the other options, which, depending on your use case and context, may better suit your aims.

Filtering the graph: Be aware of dimensions

A point of caution when dealing with projections: be aware of the dimension of the projected graph. In certain cases, such as the one we are considering here, projection may create extremely large numbers of edges, which makes the graph hard to be analyzed. In our use case, in fact, following the logic we used to create our network, a document node is connected to at least 10 keywords, plus a few entities. In the resulting entity-entity graph, all these entities will be connected with each other, as they share at least one common neighbor (the document that contains them). Therefore, for just one document, we will be generating around $15 \cdot 14/2 \approx 100$ edges. If we multiply this number by the number of documents, $\sim 10^5$, we end up with a number of edges that, despite the small use case, already becomes almost intractable, with a few million edges. Although this surely represents a conservative upper bound (as some of the co-occurrence between entities will be common in many documents and therefore not repeated), it nevertheless provides an order of magnitude of the complexity that you might expect. We, therefore, encourage you to use some caution before projecting your bipartite graph, depending on the topology of the underlying network and the size of your graph. One trick to reduce this complexity and make the projection feasible is to only consider high-degree entity nodes.

Most of the complexity indeed arises from the presence of entities that appear only once or very few times, but still generate *cliques* within the graph. Such entities are not very informative to capture patterns and provide insights. Besides, they are possibly strongly affected by statistical variability. On the other hand, we should focus on strong correlations that are supported by larger occurrences and provide more reliable statistical results.

We, therefore, only consider high-degree entity nodes. To this aim, we first generate the filtered bipartite subgraph, which excludes nodes with low degree value, namely, smaller than 5:

```
nodes_with_low_degree = {n
    for n, d in nx.degree(G, nbunch=entity_nodes) if d<5}
subGraph = G.subgraph(set(G.nodes) - nodes_with_low_degree)
```

This subgraph can now be projected without generating a graph with an excessive number of edges:

```
entityGraph = overlap_weighted_projected_graph(
    subGraph,
    {n for n in subGraph.nodes() if n in entity_nodes}
)
```

We can check the dimension of the graph with the `nx.info` networkx function, which gives the following outcome:

```
Number of nodes: 2378
Number of edges: 120512
Average degree: 101.3558
```

Despite the filters applied, the number of edges and the average node degree are still quite large. *Figure 8.4* shows the distribution of the degree and the edge weights, where we can observe one peak in the degree distribution at fairly low values, with a fat tail toward large degree values. Also, the edge weight shows a similar behavior, with a peak at rather low values and fat right tails. These distributions indeed suggest the presence of a number of small communities, namely cliques, connected to each other via some central nodes.

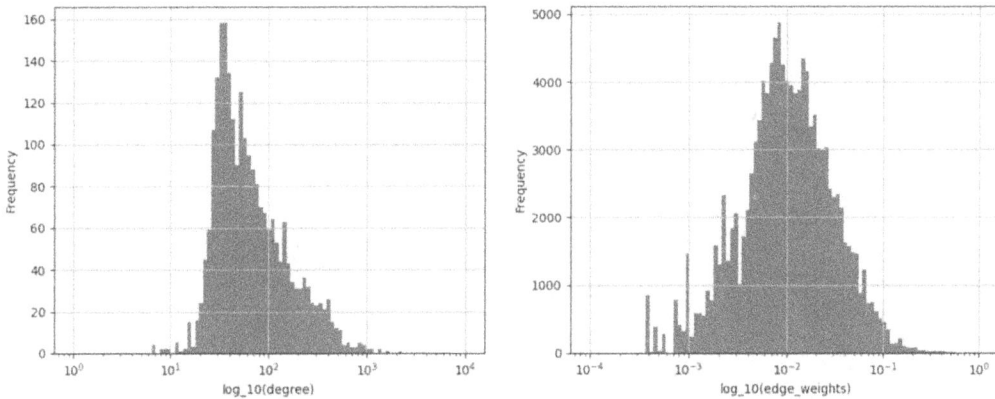

Figure 8.4: Degree (left) and weight (right) distribution for the entity-entity network

The distribution of the edge weights also suggests that a second filter could be applied. The filter on the entity degree that we previously applied on the bipartite graph indeed allowed us to filter out rare entities that appeared on only very few documents. However, the resulting graph could also be affected by an opposite problem: popular entities may be connected just because they tend to appear often in documents, even if there is not an interesting causal connection between them. Consider **U.S.** and **Microsoft**. They are almost surely connected, as it is extremely likely that there will be at least one or few documents where they both appear. However, if there is not a strong and causal connection between them, it is very unlikely that the Jaccard similarity will be large. Considering only the edges with the largest weights allow you to focus on the most relevant and possibly stable relations. The edge weight distribution shown in *Figure 8.4* suggests that a suitable threshold could be 0.05:

```
filteredEntityGraph = entityGraph.edge_subgraph(
    [edge
     for edge in entityGraph.edges
     if entityGraph.edges[edge]["weight"]>0.05])
```

Indeed, such a threshold reduces the number of edges significantly, making the network feasible to be analyzed, as we can see by using nx.info:

```
Number of nodes: 2265
Number of edges: 8114
Average degree:    7.1647
```

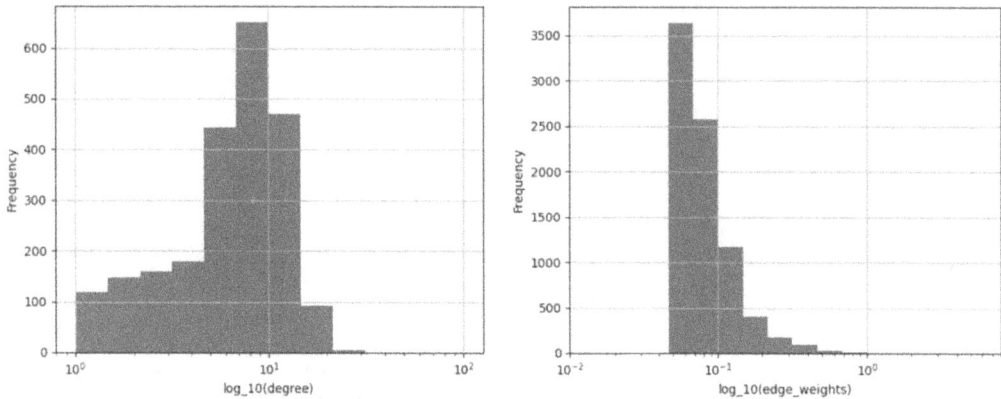

Figure 8.5: Degree distribution (left) and edge weight distribution (right) for the resulting graph, after the filtering based on the edge weight

Figure 8.5 shows the distribution of node degree and edge weights for the filtered graph. The distribution for the edge weights corresponds to the right tail of the distribution in *Figure 8.4*. The relation of degree distribution with *Figure 8.4* is less obvious, and it shows a peak of nodes that has a degree around 10, as opposed to the peak seen in *Figure 8.4* that was observed in the low range, around 100.

Analyzing the graph

To obtain some further insights on the topology of the network, we also compute its connected components alongside some global measures, such as the average shortest path, clustering coefficient, and global efficiency.

As shown in the previous chapters, the connected components can be obtained using networkx's utility functions:

```
components = nx.connected_components(filteredEntityGraph)
pd.Series([len(c) for c in components])
```

In our analysis, the filtered graph is composed of five different connected components, although the largest one almost entirely accounts for the whole graph (having 2,254 out of 2,265 nodes) with the other connected components being very small.

The graph can also be visualized using Gephi, as shown in the previous chapters:

Figure 8.6: Entity-entity network highlighting the presence of multiple small subcommunities

The global properties of the largest components can be found using the following snippet:

```
comp = components[0]
global_metrics = pd.Series({
    "shortest_path": nx.average_shortest_path_length(comp),
    "clustering_coefficient": nx.average_clustering(comp),
    "global_efficiency": nx.global_efficiency(comp)
})
```

Note that the shortest path and global efficiency may require a few minutes of computations. The following results are obtained:

```
{
    'shortest_path':  4.69461456714924,
    'clustering_coefficient':  0.2108929012144935,
    'global_efficiency':  0.22802077231368997
}
```

Already from the magnitude of these metrics (shortest path of about 5 and clustering coefficient around 0.2), together with the degree distribution shown above, we can see a general tendency of the network of having multiple communities of limited size. The use of other interesting local properties, such as degree, PageRank, and betweenness centralities distributions, is shown in *Figure 8.7*, which shows how all these measures tend to correlate and connect to each other. Indeed, as also discussed in *Chapter 1, Getting Started with Graphs*, the correlation between these measures indicates that, despite the different formulation, they are all relevant quantities to identify "central" nodes: nodes with many neighbors also tend to be highly connected (and highly connect with the various nodes), as well as being the nodes that are most important according to the page rank score.

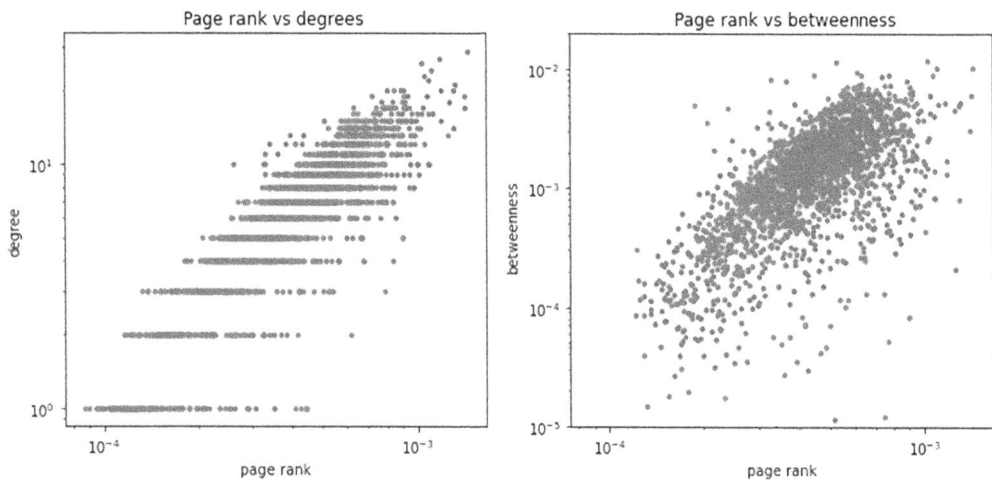

Figure 8.7: Relations and distribution between degree, page rank, and betweenness centrality measures

After providing a description in terms of local/ global measures, as well as a general visualization of the network, we will apply some of the techniques we have seen in previous chapters to identify some insights and information within the network, using the unsupervised techniques described in *Chapter 4, Unsupervised Graph Learning*.

We start by using the Louvain community detection algorithm, which aims at identifying the best partitioning of the nodes into disjointed communities by optimizing the modularity:

```
import community
communities = community.best_partition(filteredEntityGraph)
```

Note that the results might vary between runs because of random seeds. However, a similar partition, with a distribution of cluster memberships, to the one shown in *Figure 8.8* should emerge. We generally observe around 30 communities, with the larger ones having around 130-150 documents:

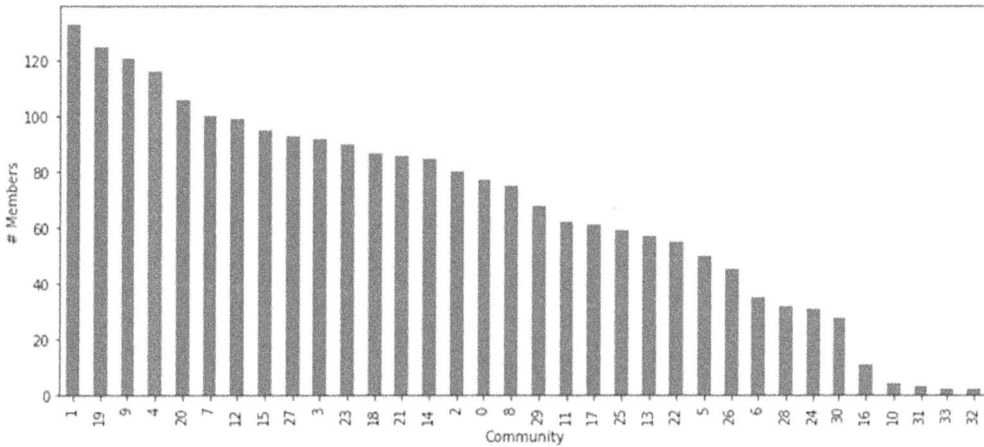

Figure 8.8: Distribution of the size of the detected communities

In *Figure 8.9*, we show a close-up of one of the communities, where one can identify a particular topic/argument related to the word "turkey":

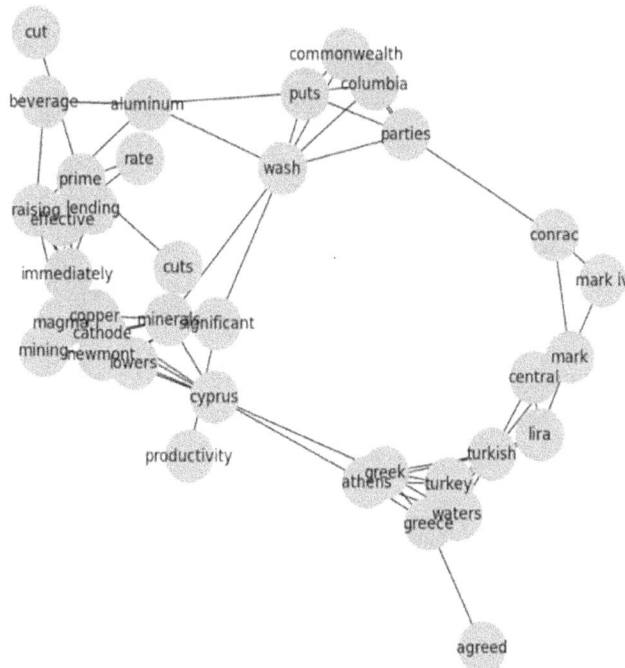

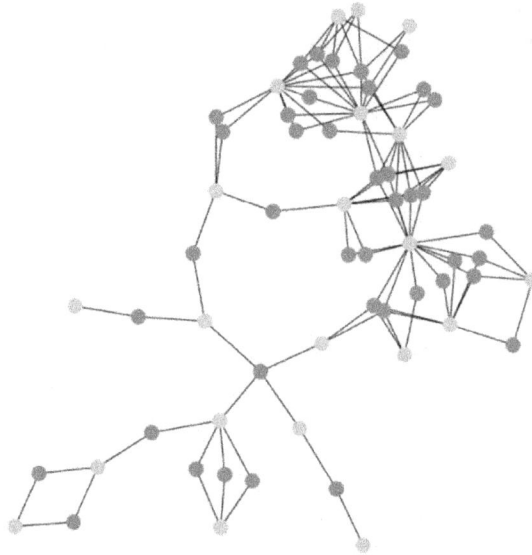

Figure 8.9: Close-up for one of the communities identified. In the top image, beside the entity nodes, we also show the document nodes, thus uncovering the structure of the related bipartite graph

Several analyses are possible using the bipartite graph above. For instance, from *Figure 8.9* we can uncover a close relationship between Turkey and Greece, which, by filtering the documents based on this relationship, we can discover to be due to some tensions between the two countries because of oil drilling in the Aegean Sea (please also refer to the attached Python notebooks for further information or explore other relationships).

As shown in *Chapter 4, Unsupervised Graph Learning*, another way of extracting insightful information on the topology and similarities between entities could also be obtained via node embeddings. In particular, we can use node2vec, which, by feeding a randomly generated walk to a skip-gram model, is able to project the nodes into a vector space, where close-by nodes are mapped into nearby points:

```
from node2vec import Node2Vec
node2vec = Node2Vec(filteredEntityGraph, dimensions=5, num_walks=500)
model = node2vec.fit(window=10)
embeddings = model.wv
```

In the vector space of embeddings, one could apply traditional clustering algorithms, such as *GaussianMixture, K-means,* or *DBSCAN*, for instance. As done in previous chapters, we could also project the embeddings into a 2D plane using t-SNE to visualize clusters and communities. Besides giving another option to identify clusters/communities within the graph, Node2Vec can also be used to also provide similarity between words, as traditionally done by Word2Vec. For instance, we can query the Node2Vec embedding model and find the most similar words to turkey:

```
[('greek', 0.9906390309333801),
 ('workers', 0.9882111549377441),
 ('norwegian', 0.9852849245071411),
 ('agreed', 0.9809741377830505),
 ('greece', 0.9735589623451233),
 ('lme', 0.9709354043006897),
 ('lira', 0.9682418704032898)]
```

The words above indeed show the words that are most similar to the "Turkey" entity, measured as the cosine similarity between embedding vectors for various entities based on their "economic" relation derived from the document, for instance, the aforementioned tension between Turkey and Greece that was also highlighted in the community detection above. Although the two approaches, Node2Vec and Word2Vec, indeed share some methodological similarities, the two embedding schemes different type of information: Word2Vec is built directly from the text and encloses relationships at a sentence/syntactic level. Node2Vec, on the other hand, encodes a description that acts more on a document level, derived from the bipartite entity-document graph and its projection on the entity-entity graph used to train the Node2Vec model. As such, the Node2Vec embedding provides a similarity among entities, is inferred from the corpus, and is not based on syntactic references. For instance, the distance between Turkey and Greece may be comparable/similar to the one between any other pair of countries in a Word2Vec embedding, given that the embedding may be able to generally represent the "nation" concept in the embedded space. However, given that Greece and Turkey often appear together in various documents because of the tension in the Aegean Sea, as also shown by the community detection algorithm, the two nations will be closer in a Node2Vec embedding, which encodes the structure of the entity graph.

Document-document graph

We will now turn to projecting the bipartite graph into a set of document nodes in order to create a document-document network to be analyzed. In other words, we will create a graph where nodes represent documents and edges represent shared entities or keywords.

In a similar way to as what we did when creating an entity-entity network, we use the overlap_ weighted_projected_graph function to obtain a weighted graph that can be filtered to reduce the number of significant edges. Indeed, the topology of the network and the business logic used to build the bipartite graph do not favor clique creation as we have seen with the entity-entity graph: two nodes will be connected only when they share at least one keyword: organization, location, or person. While this connection is certainly possible, it is not very likely within groups of 10-15 nodes, as observed for the entities.

As before, we can easily build our network with the following code:

```
documentGraph = overlap_weighted_projected_graph(
    G,
    document_nodes
)
```

In *Figure 8.10*, we can inspect the distribution of the degree and the edge weight in order to decide the value of the threshold to be used to filter out edges. Interestingly, the node degree distribution shows a clear peak toward large values as compared to the degree distribution observed for the entity-entity graph. This suggests the presence of several *supernodes* (a node with a rather large degree and that is highly connected). Also, the edge weight distribution shows the tendency of the Jaccard index to attain values close to 1, thus much larger than the ones observed in the entity-entity graph.

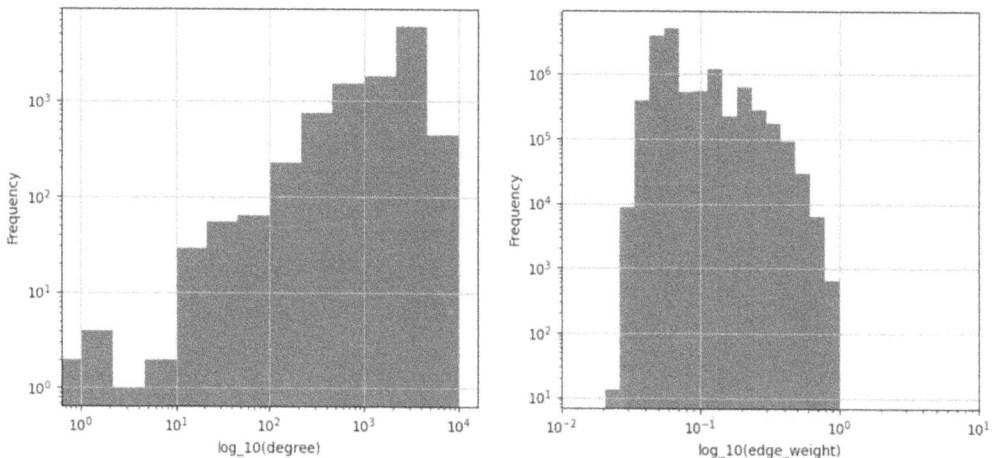

Figure 8.10: Degree and edge weight distribution for the projection of the bipartite graph into the document-document network

These two observations highlight a profound difference between the two networks: whereas the entity-entity graph is characterized by a large number of tightly connected communities (namely cliques), the document-document graph is characterized by a core of tightly connected nodes with large degrees and a periphery of weakly connected or disconnected nodes.

It can be convenient to store all the edges in a DataFrame in order to plot them, and later use the DataFrame to filter them to create a subgraph:

```
allEdgesWeights = pd.Series({
    (d[0], d[1]): d[2]["weight"]
    for d in documentGraph.edges(data=True)
})
```

From *Figure 8.10*, it seems a reasonable choice to set a threshold value for the edge weight of 0.6, which only retains a few thousand edges, therefore allowing us to generate a more tractable network using the networkx function edge_subgraph:

```
filteredDocumentGraph = documentGraph.edge_subgraph(
    allEdgesWeights[(allEdgesWeights>0.6)].index.tolist()
)
```

Figure 8.11 shows the resulting distribution for the degree and for the edge weight for the reduce graph:

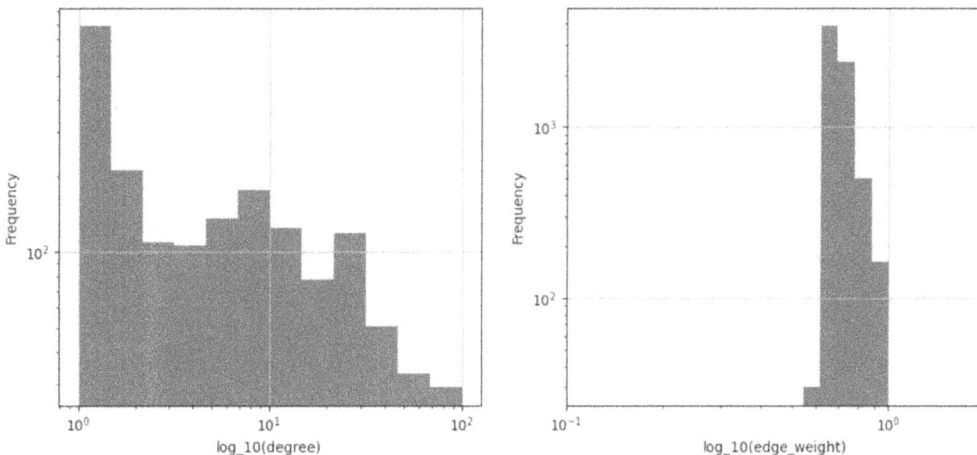

Figure 8.11: Degree and edge weight distribution for the document-document filtered network

The substantial difference in topology of the document-document graph with respect to the entity-entity graph can also be clearly seen in *Figure 8.12* where we offer you a full network visualization. As anticipated by the distributions, the document-document network is characterized by a core network and a number of weakly connected satellites. The satellites represent all the documents that share no or few keywords or entity common occurrences. The number of disconnected documents is actually quite large and accounts for almost 50% of the total.

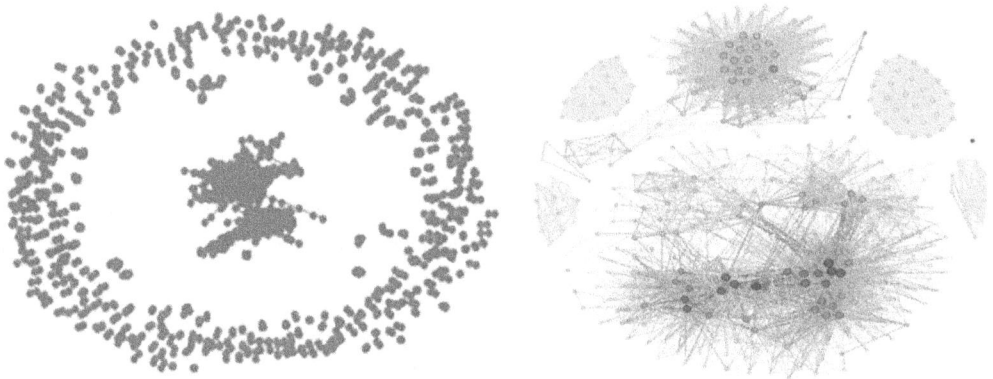

Figure 8.12: (Left) Representation of the document-document filtered network, highlighting the presence of a core and a periphery. (Right) Close-up of the core, with some subcommunities embedded. The node size is proportional to the node degree

It can be of interest to extract the connected components for this network, using the following commands:

```
components = pd.Series({
    ith: component
    for ith, component in enumerate(
        nx.connected_components(filteredDocumentGraph)
    )
})
```

In *Figure 8.13*, we also show the distribution for the connected component sizes, where the presence of a few very large clusters (the cores) can clearly be seen, together with a large number of disconnected or very small components (the periphery or satellites). This structure is strikingly different from the one observed for the entity-entity graph where all nodes generated a very large, connected cluster.

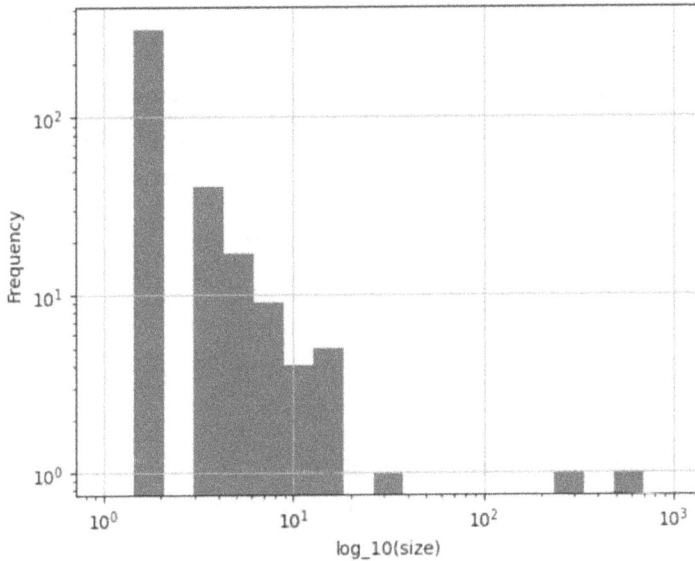

Figure 8.13: Distribution of the connected component sizes, highlighting the presence of many small-size communities (representing the periphery) and a few large communities (representing the core)

It can be interesting to further investigate the structure of the core components. We can extract from the full graph the subgraph composed of the largest components of the network with the following code:

```
coreDocumentGraph = nx.subgraph(
    filteredDocumentGraph,
    [node
    for nodes in components[components.apply(len)>8].values
    for node in nodes]
)
```

We can inspect the properties of the core network using nx.info, which gives the following output:

```
Type: Graph
Number of nodes: 1050
Number of edges: 7112
Average degree:   13.5467
```

The left panel in *Figure 8.12* shows a Gephi visualization of the core. As can be seen, the core is composed of few communities, with nodes with fairly large degrees strongly connected to each other.

As done for the entity-entity network, we can process the network to identify communities embedded in the graph. However, differently from before, the document-document graph now provides a means for judging the clustering using the document labels. Indeed, we expect documents belonging to the same topic to be close and connected to each other. Moreover, as we will see, this will also allow us to identify similarities among topics.

First, let us start by extracting the candidate communities:

```
import community
communities = pd.Series(
    community.best_partition(filteredDocumentGraph)
)
```

Differently from before, the number of detected communities is now very large, around 400, reflecting the core-periphery structure with multiple small communities in the periphery.

We then extract the topic mixture within each community in order to see whether there is homogeneity (all documents belonging to the same class) or some correlation between topics.

We first create a function that returns the number of occurrences for the various topics in a given sub-corpus, represented in a DataFrame, df:

```
from collections import Counter
def getTopicRatio(df):
    return Counter([label
                    for labels in df["label"]
                    for label in labels])
```

We then use this function on each of the communities that the algorithm had extracted:

```
communityTopics = pd.DataFrame.from_dict({
    cid: getTopicRatio(corpus.loc[comm.index])
    for cid, comm in communities.groupby(communities)
}, orient="index")
```

Finally, we normalize the counts to obtain the ratio of topics, summing up to 1:

```
normalizedCommunityTopics = (
    communityTopics.T / communityTopics.sum(axis=1)
).T
```

Therefore, the `normalizedCommunityTopics` is a DataFrame that for each community (row in the DataFrame) provides the topic mixture (in percentage) of the different topics (along the column axis). In order to quantify the heterogeneity of the topic mixture within the clusters/communities, we compute the **Shannon entropy** of each community. Shannon entropy is a measure of diversity in a system, where higher diversity corresponds to greater entropy. It is defined as follows:

$$I_c = -\sum_i \log t_{ci}$$

Where I_c represents the entropy of the cluster c and the t_{ci} corresponds to the percentage of topic i in community c. We compute the empirical Shannon entropy for all communities as follows:

```
normalizedCommunityTopics.apply(
    lambda x: np.sum(-np.log(x)), axis=1)
```

In *Figure 8.14*, we show the entropy distribution across all communities. Most of communities have zero or very low entropy, thus suggesting that documents belonging to the same class (label) tend to cluster together.

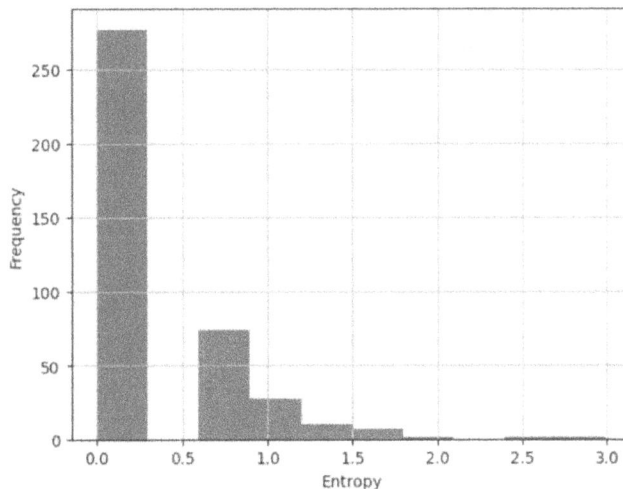

Figure 8.14: Entropy distribution of the topic mixture in each community

Even if most of the communities show zero or low variability around topics, it is interesting to investigate whether there exists a relationship between topics, when communities show some heterogeneity. Namely, we compute the correlation between topic distributions:

```
topicsCorrelation = normalizedCommunityTopics.corr().fillna(0)
```

They can be represented and visualized using a topic-topic network:

```
topicsCorrelation[topicsCorrelation<0.8]=0
topicsGraph = nx.from_pandas_adjacency(topicsCorrelation)
```

In the left panel of *Figure 8.15*, we show the full graph representation for the topics network. As observed for the document-document network, the topic-topic graph also shows a structure organized in a periphery of disconnected nodes and a strongly connected core. The right panel of *Figure 8.15* shows a close-up of the core network that interestingly indicates a correlation that is supported by a semantic meaning with the topics relating to commodities tightly connected to each other.

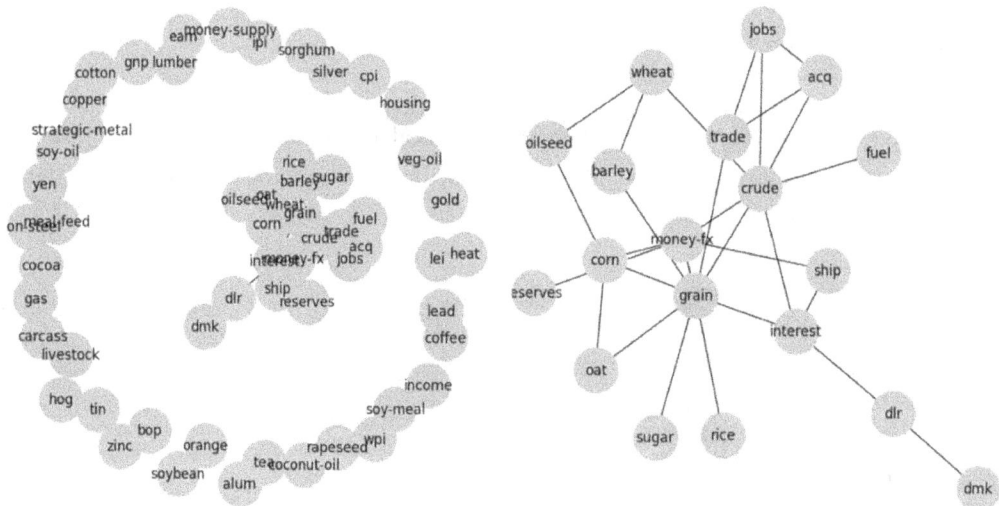

Figure 8.15: (Left) Topic-topic correlation graph, organized with a periphery-core structure.
(Right) Close-up of the core of the network

In this section, we have deeply analyzed the different forms of networks that arise when analyzing documents and, more generally, text sources. In order to do so, we used both global and local properties to statistically describe the networks, as well as some unsupervised algorithms that allowed us to unveil some structure within the graph. In the next section, we will show you how to utilize these graph structures when building a machine learning model.

Building a document topic classifier

In order to show you how to utilize t graph structure, we will focus in the following subsections on using the topological information and the connections between entities and documents provided by the bipartite entity-document graph to train multi-label classifiers that are able to predict the document topics.

In order to do this, we will analyze two different approaches:

- **A shallow machine learning approach,** where we will use the embeddings extracted from the bipartite network to train *traditional* classifiers, such as a RandomForest classifier

- **A more integrated and differentiable approach** using a graph neural network, which is applied on heterogeneous graphs (such as the bipartite graph)

In the following code block, we will consider the 10 most common topics for which we have enough documents to train and evaluate our models:

```python
from collections import Counter
topics = Counter(
    [label
      for document_labels in corpus["label"]
      for label in document_labels]
).most_common(10)
```

The preceding code block produces the output shown below, showing the names of the topics that we will focus on in the following analysis:

```
[('earn', 3964), ('acq', 2369), ('money-fx', 717),
 ('grain', 582), ('crude', 578), ('trade', 485),
 ('interest', 478), ('ship', 286), ('wheat', 283),
 ('corn', 237)]
```

When training topic classifiers, we will restrict our focus only to those documents that belong to the preceding labels. The filtered corpus can be easily obtained by the following code:

```python
topicsList = [topic[0] for topic in topics]
topicsSet = set(topicsList)
dataset = corpus[corpus["label"].apply(
    lambda x: len(topicsSet.intersection(x))>0
)]
```

Now that we have extracted and structured the dataset, we are ready to start training our topic models and evaluating their performances. In the next subsection, we will start by creating a simple model using shallow-learning methods, and then increase the complexity of the model in the following subsection by using graph neural networks.

Shallow-learning methods

We start by implementing a shallow approach for the topic classification tasks, employing the network information. We will show you how to do this, in a step-by-step procedure that you can further customize depending on your use case:

1. First of all, we start by computing the embeddings using Node2Vec on the bipartite graph. The choice of using the bipartite graph is crucial in order to efficiently utilize the topological information and the connection between entities and documents. In fact, filtered document-document networks are characterized by a periphery with many nodes disconnected, which would not benefit from the topological information. On the other hand, the unfiltered document-document network will have a large number of edges, which makes the scalability of the approach an issue:

   ```python
   from node2vec import Node2Vec
   node2vec = Node2Vec(G, dimensions=10)
   model = node2vec.fit(window=20)
   embeddings = model.wv
   ```

   Here, the embedding `dimension` as well as the `window` used for generating the walks can be seen as hyper-parameters to be optimized via cross-validation. Some possible choices are `10`, `20`, and `30` for both variables, although this may vary based on the use case.

2. In order to make it computationally efficient, a set of embeddings can be computed beforehand, saved to disk, and later used in the optimization. This would work under the assumption that we are in a *semi-supervised* setting or in a *transductive* task, where we have all the connection information on the entire dataset, apart from the labels, at training time. Later in this chapter, we will outline another approach, based on the graph neural network, that provides an inductive framework for integrating a topology when training classifiers.

   We therefore store the embeddings in a file:

   ```python
   pd.DataFrame(embeddings.vectors,
               index=embeddings.index2word
   ).to_pickle(f"graphEmbeddings_{dimension}_{window}.p")
   ```

   We can choose a different name for each combination of `dimension` and `window`.

3. These embeddings can be simply integrated into a scikit-learn Transformer in order to be used in a grid-search cross-validation:

```
from sklearn.base import BaseEstimator
class EmbeddingsTransformer(BaseEstimator):

    def __init__(self, embeddings_file):
        self.embeddings_file = embeddings_file

    def fit(self, *args, **kwargs):
        self.embeddings = pd.read_pickle(
            self.embeddings_file)
        return self

    def transform(self, X):
        return self.embeddings.loc[X.index]

    def fit_transform(self, X, y):
        return self.fit().transform(X)
```

4. In order to build a modeling training pipeline, we split our corpus into training and test sets, using the helper function:

```
def train_test_split(corpus):
    indices = [index for index in corpus.index]
    train_idx = [idx
                    for idx in indices
                    if "training/" in idx]
    test_idx = [idx
                    for idx in indices
                    if "test/" in idx]
    return corpus.loc[train_idx], corpus.loc[test_idx]
```

We then use the helper function to split the dataset, as follows:

```
train, test = train_test_split(dataset)
```

5. We also build functions to conveniently extract features and labels:

```
def get_features(corpus):
    return corpus["parsed"]
def get_labels(corpus, topicsList=topicsList):
    return corpus["label"].apply(
```

```
        lambda labels: pd.Series(
            {label: 1 for label in labels}
        ).reindex(topicsList).fillna(0)
    )[topicsList]
def get_features_and_labels(corpus):
    return get_features(corpus), get_labels(corpus)
features, labels = get_features_and_labels(train)
```

6. We can now instantiate the modeling pipeline:

```
from sklearn.pipeline import Pipeline
from sklearn.ensemble import RandomForestClassifier
from sklearn.multioutput import MultiOutputClassifier
pipeline = Pipeline([
    ("embeddings", EmbeddingsTransformer(
        "my-place-holder")
    ),
    ("model", MultiOutputClassifier(
        RandomForestClassifier())
    )
])
```

7. We define the parameter space as well as the configuration for the cross-validated grid search:

```
from glob import glob
param_grid = {
    "embeddings__embeddings_file": glob("graphEmbeddings_*"),
    "model__estimator__n_estimators": [50, 100],
    "model__estimator__max_features": [0.2,0.3, "auto"],
}
grid_search = GridSearchCV(
    pipeline, param_grid=param_grid, cv=5, n_jobs=-1)
```

8. Finally, we train our topic model by using the `fit` method of the `sklearn` API:

```
model = grid_search.fit(features, labels)
```

Great! You have just created your topic model that uses the graph information. Once the best model has been identified, we can also use this model on the test dataset to evaluate its performance. In order to do so, we start by defining the following helper function, which allows us to obtain a set of predictions:

```
def get_predictions(model, features):
    return pd.DataFrame(
        model.predict(features),
        columns=topicsList, index=features.index)
preds = get_predictions(model, get_features(test))
labels = get_labels(test)
```

Using `sklearn`'s functionalities, we can promptly look at the performances of the trained classifier:

```
from sklearn.metrics import classification_report
print(classification_report(labels, preds))
```

The code provides the following output, showing an overall F1-score ranging from 0.6 to 0.8. Such variation may depend on how unbalanced classes are accounted for:

	precision	recall	f1-score	support
0	0.97	0.94	0.95	1087
1	0.93	0.74	0.83	719
2	0.79	0.45	0.57	179
3	0.96	0.64	0.77	149
4	0.95	0.59	0.73	189
5	0.95	0.45	0.61	117
6	0.87	0.41	0.56	131
7	0.83	0.21	0.34	89
8	0.69	0.34	0.45	71
9	0.61	0.25	0.35	56
micro avg	0.94	0.72	0.81	2787
macro avg	0.85	0.50	0.62	2787
weighted avg	0.92	0.72	0.79	2787
samples avg	0.76	0.75	0.75	2787

You can play around with the model type and hyper-parameter of the analytical pipeline, by varying the models as well as experimenting with different values to encode the embeddings. As mentioned before, the approach above is clearly transductive, since it uses an embedding trained on the entire dataset. This is a common situation in semi-supervised tasks, where the labeled information is present only in a small subset of points, and the task is to infer the labels for all unknown samples. In the next sub section, we outline how to build an inductive classifier using graph neural networks that can be used when test samples are not known at training time.

Graph neural network

We will now describe a neural network-based approach that natively integrates and makes use of the graph structure. Graph neural networks have already been introduced in *Chapter 3, Neural Networks and Graphs*, and *Chapter 4, Unsupervised Graph Learning*. However, here we will show you how to apply this framework to heterogeneous graphs, that is, a graph where there is more than one type of node. Each node type might have a different set of features and the training might target only one specific node type over the other.

The approach we present here will make use of `stellargraph` and the GraphSAGE algorithm, that we have already described previously. A similar example using PyTorch is also available in the GitHub repository (`https://github.com/PacktPublishing/Graph-Machine-Learning/blob/main/Chapter08`) for interested readers.

More specifically, given that we are dealing with bipartite graphs, we will be using the HinSAGE algorithm, which generalizes the GraphSAGE algorithm from homogeneous graphs (i.e., with only one node type) to heterogenous graphs (i.e., with multiple node types). These methods also support the use of features for each node, instead of just relying on the topology of the graph. If one does not have node features, the one-hot node representation might be used in its place, as also shown in *Chapter 7, Social Network Graphs*. However, here, in order to make it more general, we will produce a set of node features based on the TF-IDF score (which we saw earlier) for each entity and keyword. Here, we will show you a step-by-step guide to help you train and evaluate a model based on graph neural networks for predicting document topic classification:

1. We start by computing the TF-IDF score for each document. `sklearn` already provides some functionalities that allow us to easily compute the TF-IDF scores from a corpus of documents. The `sklearn` class `TfidfVectorizer` already comes with a tokenizer embedded. However, since we already have a tokenized and lemmatized version extracted with `spacy`, we can also provide an implementation of a custom tokenizer that makes use of `spacy` processing:

```
def my_spacy_tokenizer(pos_filter=["NOUN", "VERB", "PROPN"]):
    def tokenizer(doc):
        return [token.lemma_
                for token in doc
                if (pos_filter is None) or
                   (token.pos_ in pos_filter)]
    return tokenizer
```

That can be used in the `TfidfVectorizer`:

```
cntVectorizer = TfidfVectorizer(
    analyzer=my_spacy_tokenizer(),
    max_df = 0.25, min_df = 2, max_features = 10000
)
```

In order to make the approach truly inductive, the training of the TF-IDF should be done only for the training set and then only applied to the test set:

```
trainFeatures, trainLabels = get_features_and_labels(train)
testFeatures, testLabels = get_features_and_labels(test)

trainedIDF = cntVectorizer.fit_transform(trainFeatures)
testIDF = cntVectorizer.transform(testFeatures)
```

For our convenience, the two TF-IDF representations (for the training and test sets) can now be stacked together into a single data structure representing the features for the document nodes for the whole graph:

```
documentFeatures = pd.concat([trainedIDF, testIDF])
```

2. Besides the feature information for document nodes, we also build a simple feature vector for entities, based on the one-hot-encoding representation of the entity type:

```
entityTypes = {
    entity: ith
    for ith, entity in enumerate(edges["type"].unique())
}
entities = edges\
    .groupby(["target", "type"])["source"]\
    .count()\
    .groupby(level=0).apply(
```

```
        lambda s: s.droplevel(0)\
                 .reindex(entityTypes.keys())\
                 .fillna(0))\
    .unstack(level=1)
entityFeatures = (entities.T / entities.sum(axis=1))
```

3. We now have all the information to create an instance of a `StellarGraph`, by merging the information of the node features, both for documents and for entities, with the connections provided by the edges DataFrame. We should only filter out some of the edges/nodes in order to only include the documents belonging to the targeted topics:

```
from stellargraph import StellarGraph


_edges = edges[edges["source"].isin(documentFeatures.index)]
nodes = {"entity": entityFeatures,
         "document": documentFeatures}
stellarGraph = StellarGraph(
    nodes, _edges,
    target_column="target", edge_type_column="type"
)
```

We have now created our `StellarGraph`. We can inspect the network similar to how we did for `networkx`, with:

```
print(stellarGraph.info())
```

That provides the following overview:

```
StellarGraph: Undirected multigraph
 Nodes: 23998, Edges: 86849

 Node types:
  entity: [14964]
    Features: float32 vector, length 6
    Edge types: entity-GPE->document, entity-ORG->document, entity-
PERSON->document, entity-keywords->document
  document: [9034]
    Features: float32 vector, length 10000
    Edge types: document-GPE->entity, document-ORG->entity,
document-PERSON->entity, document-keywords->entity
```

```
Edge types:
    document-keywords->entity: [78838]
        Weights: range=[0.0827011, 1], mean=0.258464,
std=0.0898612
        Features: none
    document-ORG->entity: [4129]
        Weights: range=[2, 22], mean=3.24122, std=2.30508
        Features: none
    document-GPE->entity: [2943]
        Weights: range=[2, 25], mean=3.25926, std=2.07008
        Features: none
    document-PERSON->entity: [939]
        Weights: range=[2, 14], mean=2.97444, std=1.65956
        Features: none
```

The StellarGraph description is actually very informative. Also, StellarGraph handles natively different types of nodes and edges and provides out-of-the-box segmented statistics for each node/edge type.

4. You may have noticed that the graph we have just created includes both training and test data. In order to truly test the performance of an inductive approach and avoid leakage of information between train and test, we need to create a subgraph with only the data available at training time:

```
targets = labels.reindex(documentFeatures.index).fillna(0)
sampled, hold_out = train_test_split(targets)
allNeighbors = np.unique([n
    for node in sampled.index
    for n in stellarGraph.neighbors(node)
])
subgraph = stellarGraph.subgraph(
    set(sampled.index).union(allNeighbors)
)
```

The considered subgraph has 16,927 nodes and 62,454 edges, as compared to the 23,998 nodes and 86,849 edges of the total graph.

5. Now that we only have the data and the network available at training time, we can build our machine learning model on top of it. To do so, we first split the data into train, validation, and test data. For training, we will only use 10% of the data, also resembling a semi-supervised task:

```
from sklearn.model_selection import train_test_split
train, leftOut = train_test_split(
    sampled,
    train_size=0.1,
    test_size=None,
    random_state=42
)
validation, test = train_test_split(
    leftOut, train_size=0.2, test_size=None, random_state=100,
)
```

6. We can now start to build our graph neural network model using `stallargraph` and the keras API. First of all, we create a generator able to produce the samples that will feed the neural network. Note that, since we are dealing with a heterogeneous graph, we need a generator that will sample examples from nodes that belongs to a specific class only. In the following code block, we will be using the `HinSAGENodeGenerator` class, which generalizes the node generator used for homogeneous graphs to heterogeneous graphs, allowing us to specify the node type we want to target:

```
from stellargraph.mapper import HinSAGENodeGenerator
batch_size = 50
num_samples = [10, 5]
generator = HinSAGENodeGenerator(
    subgraph, batch_size, num_samples,
    head_node_type="document"
)
```

Using this object, we can then create a generator for the train and validation datasets:

```
train_gen = generator.flow(train.index, train, shuffle=True)
val_gen = generator.flow(validation.index, validation)
```

7. We can now create our GraphSAGE model. As already done for the generator, also in this case, we need to use a model that is able to handle heterogenous graph. HinSAGE will be used in place of GraphSAGE:

```
from stellargraph.layer import HinSAGE
from tensorflow.keras import layers
graphsage_model = HinSAGE(
    layer_sizes=[32, 32], generator=generator,
    bias=True, dropout=0.5
)
x_inp, x_out = graphsage_model.in_out_tensors()
prediction = layers.Dense(
    units=train.shape[1], activation="sigmoid"
)(x_out)
```

Note that in the final dense layer, we use a *sigmoid* activation function instead of *softmax*, since the problem at hand is a multi-class, multi-label task. Softmax would in fact normalize all scores to sum to 1, and it is therefore more suited to single-label classification. On the other hand, given that a document may belong to more than one class, the sigmoid activation function is a more sensible choice in this context. As usual, we compile our Keras model:

```
from tensorflow.keras import optimizers, losses, Model
model = Model(inputs=x_inp, outputs=prediction)
model.compile(
    optimizer=optimizers.Adam(lr=0.005),
    loss=losses.binary_crossentropy,
    metrics=["acc"]
)
```

8. Finally, we train the neural network model with:

```
history = model.fit(
    train_gen, epochs=50, validation_data=val_gen,
    verbose=1, shuffle=False
)
```

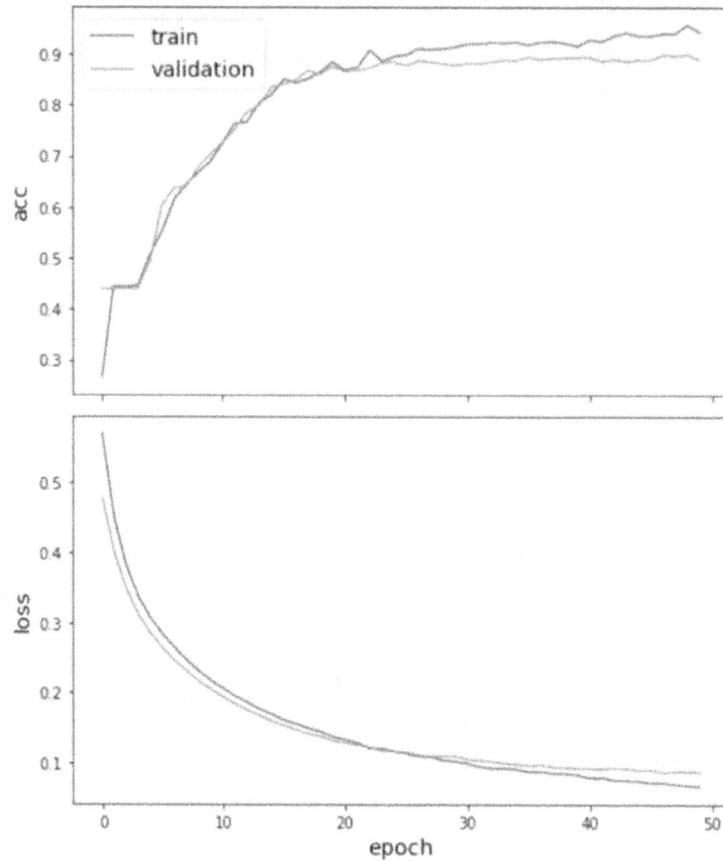

Figure 8.16: (Top) Train and validation accuracy versus the number of epochs. (Bottom) Binary cross-entropy loss for the training and validation dataset versus the number of epochs

Figure 8.16 shows the plots of the evolution for the train and validation losses and accuracy versus the number of epochs. As the figure shows, train and validation accuracy increase consistently up to around 30 epochs. After the accuracy on the validation set settles to a *plateau*, whereas training accuracy continues to increase, indicating a tendency to slight overfitting. Thus, stopping training in the range between 30 and 50 seems a rather legitimate choice.

9. Once the model is trained, we can test its performance on the test set:

```
test_gen = generator.flow(test.index, test)
test_metrics = model.evaluate(test_gen)
```

This should provide the following values:

```
loss: 0.0933
accuracy: 0.8795
```

Note that because of the unbalanced label distribution, accuracy may not be the best choice for assessing performances. Besides, the value of 0.5 generally used for thresholding and providing label assignment may also be sub-optimal in unbalanced settings.

10. To identify the best threshold to be used to classify the documents, we first compute the topic assignment probabilities predicted by the model over all the test samples:

```
test_predictions = pd.DataFrame(
    model.predict(test_gen), index=test.index,
    columns=test.columns)
test_results = pd.concat({
    "target": test,
    "preds": test_predictions
}, axis=1)
```

We then compute the F1-score with a macro average (where the F1-score for the individual classes are averaged) for different threshold choices:

```
thresholds = [0.01,0.05,0.1,0.2,0.3,0.4,0.5]
f1s = {}
for th in thresholds:
    y_true = test_results["target"]
    y_pred = 1.0*(test_results["preds"]>th)
    f1s[th] = f1_score(y_true, y_pred, average="macro")
pd.Series(f1s).plot()
```

As shown in *Figure 8.17*, a threshold value of 0.2 seems to be the best choice that achieves the best performance:

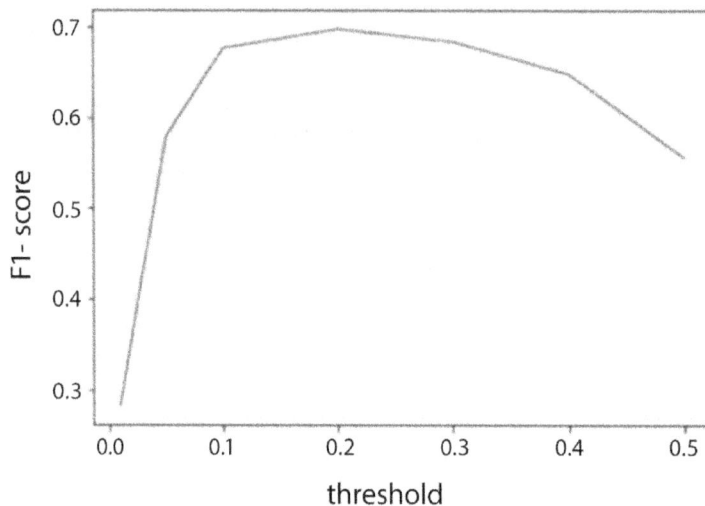

Figure 8.17: Macro-averaged F1-score versus the threshold used for labeling

11. Using the threshold value of 0.2, we can finally extract the classification report on the test set:

```
print(classification_report(
    test_results["target"], 1.0*(test_results["preds"]>0.2))
)
```

This gives the following output:

	precision	recall	f1-score	support
0	0.92	0.97	0.94	2075
1	0.85	0.96	0.90	1200
2	0.65	0.90	0.75	364
3	0.83	0.95	0.89	305
4	0.86	0.68	0.76	296
5	0.74	0.56	0.63	269
6	0.60	0.80	0.69	245
7	0.62	0.10	0.17	150
8	0.49	0.95	0.65	149
9	0.44	0.88	0.58	129

micro avg	0.80	0.89	0.84	5182
macro avg	0.70	0.78	0.70	5182
weighted avg	0.82	0.89	0.84	5182
samples avg	0.83	0.90	0.85	5182

12. At this point, we have trained a graph neural network model and assessed its performance. We now aim to apply this model on a set of unobserved data—the data that was left out at the very beginning—and represent the true test data in an inductive setting. We therefore need to instantiate a new generator:

```
generator = HinSAGENodeGenerator(
    stellarGraph, batch_size, num_samples,
    head_node_type="document")
```

Note that the graph taken as an input of the HinSAGENodeGenerator is now the entire graph (in place of the filtered one used before), with both training and test documents. Using this class, we can create a generator that samples from the test nodes only, filtering out the ones that do not belong to one of our main, selected topics:

```
hold_out = hold_out[hold_out.sum(axis=1) > 0]
hold_out_gen = generator.flow(hold_out.index, hold_out)
```

13. The model can then be evaluated over these samples and labels are predicted using the threshold identified earlier of 0.2:

```
hold_out_predictions = model.predict(hold_out_gen)
preds = pd.DataFrame(1.0*(hold_out_predictions > 0.2),
                     index = hold_out.index,
                     columns = hold_out.columns)
results = pd.concat(
    {"target": hold_out,"preds": preds}, axis=1
)
```

We can finally extract the performance on the inductive test dataset:

```
print(classification_report(
    results["target"], results["preds"])
)
```

This provides the following table:

	precision	recall	f1-score	support
0	0.93	0.99	0.96	1087
1	0.90	0.97	0.93	719
2	0.64	0.92	0.76	179
3	0.82	0.95	0.88	149
4	0.85	0.62	0.72	189
5	0.74	0.50	0.59	117
6	0.60	0.79	0.68	131
7	0.43	0.03	0.06	89
8	0.50	0.96	0.66	71
9	0.39	0.86	0.54	56
micro avg	0.82	0.89	0.85	2787
macro avg	0.68	0.76	0.68	2787
weighted avg	0.83	0.89	0.84	2787
samples avg	0.84	0.90	0.86	2787

As compared to the shallow-learning method, we can see that we have achieved a substantial improvement in performance, between 5 and 10%.

Summary

In this chapter, you have learned how to process unstructured information and how to represent such information by means of a graph. Starting from a well-known benchmark dataset, Reuters-21578, we applied standard NLP engines to tag and structure textual information. These high-level features were then used to create different types of networks: knowledge base networks, bipartite networks, projections of bipartite networks onto each subset of node types, and a topic-topic similarity network. The different graphs also allowed us to use the tools we have presented in previous chapters to extract insights from the network representation.

We used local and global properties in order to show you how these quantities can represent and describe structurally different types of networks. Unsupervised techniques were then used in order to identify semantic communities and cluster together documents belonging to similar subjects/topics. Finally, we used the labeled information provided in the dataset to train supervised multi-class multi-label classifiers, which also utilized the topology of the network.

In particular, we applied supervised techniques to the case of a heterogeneous graph, where two different node types are present: documents and entities. In this setting, we showed you how to implement both transductive and inductive approaches using shallow learning and graph neural networks, respectively.

In the next chapter, we will turn to another domain where graph analytics can be efficiently used to extract insights and/or create machine learning models that leverage network topology: transactional data. The use case we look at in the next chapter will also allow you to take the bipartite graph concepts introduced in this chapter to another level: tripartite graphs.

Get This Book's PDF Version and Exclusive Extras

UNLOCK NOW

Scan the QR code (or go to `packtpub.com/unlock`). Search for this book by name, confirm the edition, and then follow the steps on the page.

Note: Keep your invoice handy. Purchases made directly from Packt don't require one.

9

Graph Analysis for Credit Card Transactions

Analysis of financial data is one of the most common and important domains in big data and data analysis. Due to the increasing number of mobile devices and the introduction of standard platforms for online payments, the amount of transactional data that banks are producing and consuming is increasing exponentially.

As a consequence, new tools and techniques are needed to exploit as much as we can from this huge amount of information in order to better understand customers' behavior and support data-driven decisions in business processes. Data can also be used to build better mechanisms to improve security in the online payment process. Indeed, at the same time as online payment systems are becoming increasingly popular due to e-commerce platforms, cases of fraud are also increasing. An example of a fraudulent transaction is one performed with a stolen credit card. In this case, the fraudulent transaction will be different from the transactions made by the original owner of the credit card. This difference in transactional patterns can serve as a basis for detecting fraud in automated fraud detection systems.

However, building automatic procedures to detect fraudulent transactions could be a complex problem due to the large number of variables involved. Graph machine learning offers a powerful approach to tackle this challenge by representing transactional data as a graph, enabling the detection of complex relationships and patterns, such as communities of fraudulent behavior, that may be missed by traditional methods.

In this chapter, we will describe how we can represent credit card transaction data as a graph in order to automatically detect fraudulent transactions using machine learning algorithms. We will start processing the dataset by applying some of the techniques and algorithms we described in previous chapters to build a fraud detection algorithm.

The following topics will be covered in this chapter:

- Generating a graph from credit card transactions
- Extraction of properties and communities from the graph
- Application of supervised and unsupervised machine learning algorithms to fraud classification

Technical requirements

All code files relevant to this chapter are available at `https://github.com/PacktPublishing/ Graph-Machine-Learning/tree/main/Chapter09`. Please refer to the *Practical exercises* section of *Chapter 1, Getting Started with Graphs,* for guidance on how to set up the environment to run the examples in this chapter, either using Poetry, `pip`, or Docker.

Building graphs from credit card transactions

The dataset used in this chapter is the *Credit Card Transactions Fraud Detection* dataset available on *Kaggle* at the following URL: `https://www.kaggle.com/kartik2112/fraud- detection?select=fraudTrain.csv`. We will build two approaches for fraud detection, based on bipartite and tripartite graphs.

Overview of the dataset

The dataset is made up of simulated credit card transactions containing legitimate and fraudulent transactions for the period January 1, 2019 to December 31, 2020. It includes the credit cards of 1,000 customers performing transactions with a pool of 800 merchants. The dataset was generated using *Sparkov data generation*. More information about the generation algorithm is available at the following URL: `https://github.com/namebrandon/Sparkov_Data_Generation`.

For each transaction, the dataset contains 23 different features. In the following table, we will show only the information that will be used in this chapter:

Column name	Column description	Type
index	Unique identifier for each row	Integer
cc_num	Credit card number of customer	String
merchant	Merchant name	String
amt	Number of transactions (in dollars)	Double
is_fraud	Target variable. It is 0 if it is a genuine transaction, 1 for a fraudulent transaction	Binary

Table 9.1: List of variables used in this chapter

The dataset is indeed a simplified version, where only transaction core data is retained. Despite this, the dataset will nevertheless allow a rich analysis of the behavioral patterns and build a fraud detection model. In real-world scenarios, transactional data is generally complemented by metadata for both customers (like emails, phone numbers, addresses, etc.) and merchants (address, product group, legal entity, retailer, etc.) that can be encoded in node features and is generally extremely critical to allow graph-based entity resolution.

For the purposes of our analysis, we will use the fraudTrain.csv file. As already suggested, take a look at the dataset by yourself. It is strongly suggested to explore and become as comfortable as possible with the dataset before starting any machine learning task.

Loading the dataset

The first step of our analysis will be to load the dataset and build a graph. Since the dataset represents a simple list of transactions, we need to perform several operations to build the final credit card transaction graph. The dataset is a simple CSV file; we can use pandas to load the data as follows:

```
import pandas as pd
df = df[df["is_fraud"]==0].sample(frac=0.20, random_state=42).
append(df[df["is_fraud"] == 1])
```

In order to help you deal with the dataset, we selected 20% of the genuine transactions and all of the fraudulent transactions. As a result, from a total of 1,296,675 transactions, we will only use 265,342. Moreover, we can also investigate the number of fraudulent and genuine transactions in our dataset as follows:

```
df["is_fraud"].value_counts()
```

By way of a result, we get the following:

```
0       257834
1         7506
```

In other words, from a total of 265,342 transactions, only 7506 (2.83 %) are fraudulent transactions, while the others are genuine.

Building the graphs using networkx

The dataset can be represented as a graph using the networkx library. Before starting with the technical description, we will start by specifying how the graph is built from the data. We used two different approaches to build the graph – namely, the bipartite and tripartite approaches, as described in the paper *APATE: A Novel Approach for Automated Credit Card Transaction Fraud Detection Using Network-Based Extensions*, available at https://www.scinapse.io/papers/614715210.

For the **bipartite approach**, we build a weighted bipartite graph $G = (V, E, \omega)$, where $V = V_c \cup V_m$, and where each node $c \in V_c$ represents a customer, and each node $m \in V_m$ represents a merchant. An edge (V_c, V_m) is created if a transaction exists from the customer, V_c, to the merchant, V_m. Finally, to each edge of the graph, we assign an always-positive weight representing the amount (in US dollars) of the transaction. In our formalization, we allow both directed and undirected graphs.

Since the dataset represents temporal transactions, multiple interactions can happen between a customer and a merchant. In both our formalizations, we decided to collapse all that information into a single graph. In other words, if multiple transactions are present between a customer and a merchant, we will build a single edge between the two nodes with its weight given by the sum of all the transaction amounts. A graphical representation of the direct bipartite graph is shown in *Figure 9.1*:

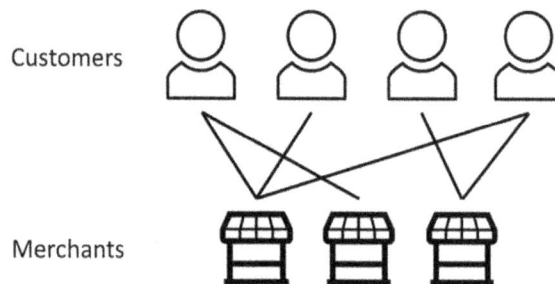

Figure 9.1: Bipartite graph generated from the input dataset

The bipartite graph we defined can be built using the following code:

```
def build_graph_bipartite(df_input, graph_type=nx.Graph()):
    df = df_input.copy()
    mapping = {x:node_id for node_id,x in enumerate(set(df["cc_num"].
values.tolist() + df["merchant"].values.tolist()))}
    df["from"] = df["cc_num"].apply(lambda x: mapping[x])
    df["to"] = df["merchant"].apply(lambda x: mapping[x])
    df = df[['from', 'to', "amt", "is_fraud"]].groupby(['from', 'to']).
agg({"is_fraud": "sum", "amt": "sum"}).reset_index()
    df["is_fraud"] = df["is_fraud"].apply(lambda x: 1 if x>0 else 0)
    G = nx.from_edgelist(df[["from", "to"]].values, create_using=graph_type)
    nx.set_edge_attributes(G, {(int(x["from"]), int(x["to"])):x["is_
fraud"] for idx, x in df[["from","to","is_fraud"]].iterrows()}, "label")
    nx.set_edge_attributes(G,{(int(x["from"]), int(x["to"])):x["amt"] for
idx, x in df[["from","to","amt"]].iterrows()}, "weight")
    return G
```

The code is quite simple. To build the bipartite credit card transaction graph, we use different networkx functions. To go more in-depth, the operations we performed in the code are as follows:

1. We built a map to assign a node_id to each merchant or customer.

2. Multiple transactions are aggregated in a single transaction.

3. The networkx function, nx.from_edgelist, is used to build the networkx graph.

4. Two attributes (namely, weight and label) are assigned to each edge. The former represents the total number of transactions between the two nodes, whereas the latter indicates whether the transaction is genuine or fraudulent.

As we can also see from the code, we can select whether we want to build a directed or an undirected graph. We can build an undirected graph by calling the following function:

```
G_bu = build_graph_bipartite(df, nx.Graph(name="Bipartite Undirect"))))
```

We can instead build a direct graph by calling the following function:

```
G_bd = build_graph_bipartite(df, nx.DiGraph(name="Bipartite Direct"))))
```

The only difference is given by the second parameter we pass in the constructor.

The **tripartite approach** is an extension of the previous one, also allowing the transactions to be represented as a vertex. While, on the one hand, this approach drastically increases network complexity, on the other hand, it allows extra node embeddings to be built for merchants and cardholders and every transaction. Formally, for this approach, we build a weighted tripartite graph, $G = (V, E, \omega)$, where $V = v_t \in v_c \cup v_m$, where each node $c \in V_c$ represents a customer, each node $m \in V_m$ represents a merchant, and each node $t \in V_t$ is a transaction. Two edges $(v_c v_t)$ and $(v_t v_m)$ are created for each transaction, v_t, from the customer, v_c, to the merchant, v_m.

Finally, to each edge of the graph, we assign an always-positive weight representing the amount (in US dollars) of the transaction. Since, in this case, we create a node for each transaction, we do not need to aggregate multiple transactions from a customer to a merchant. Moreover, as for the other approach, in our formalization, we allow both directed and undirected graphs. A graphical representation of the direct bipartite graph is shown in *Figure 9.2*:

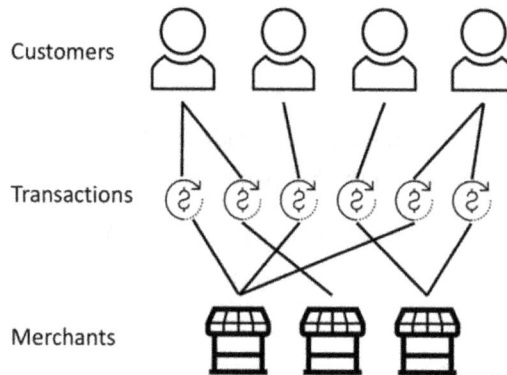

Figure 9.2: Tripartite graph generated from the input dataset

The tripartite graph we defined can be built using the following code:

```python
def build_graph_tripartite(df_input, graph_type=nx.Graph()):
    df = df_input.copy()
    mapping = {x:node_id for node_id,x in enumerate(set(df.index.
values.tolist() + df["cc_num"].values.tolist() + df["merchant"].values.
tolist()))}
    df["in_node"] = df["cc_num"].apply(lambda x: mapping[x])
    df["out_node"] = df["merchant"].apply(lambda x: mapping[x])
    G = nx.from_edgelist([(x["in_node"], mapping[idx]) for idx, x
in df.iterrows()] + [(x["out_node"], mapping[idx]) for idx, x in
df.iterrows()], create_using=graph_type)
```

```
    nx.set_edge_attributes(G,{(x["in_node"], mapping[idx]):x["is_fraud"]
for idx, x in df.iterrows()}, "label")
    nx.set_edge_attributes(G,{(x["out_node"], mapping[idx]):x["is_fraud"]
for idx, x in df.iterrows()}, "label")
    nx.set_edge_attributes(G,{(x["in_node"], mapping[idx]):x["amt"] for
idx, x in df.iterrows()}, "weight")
    nx.set_edge_attributes(G,{(x["out_node"], mapping[idx]):x["amt"] for
idx, x in df.iterrows()}, "weight")
    return G
```

The code is quite simple. To build the tripartite credit card transaction graph, we use different networkx functions. To go more in-depth, the operations we performed in the code are as follows:

1. We built a map to assign a node_id to each merchant, customer, and transaction.
2. The networkx function, nx.from_edgelist, is used to build the networkx graph.
3. Two attributes (namely, weight and label) are assigned to each edge. The former represents the total number of transactions between the two nodes, whereas the latter indicates whether the transaction is genuine or fraudulent.

As we can also see from the code, we can select whether we want to build a directed or an undirected graph. We can build an undirected graph by calling the following function:

```
G_tu = build_graph_tripartite(df, nx.Graph(name="Tripartite Undirect"))
```

We can instead build a direct graph by calling the following function:

```
G_td = build_graph_tripartite(df, nx.DiGraph(name="Tripartite Direct"))
```

The only difference is given by the second parameter we pass in the constructor.

In the formalized graph representation that we introduced, the real transactions are represented as edges. According to this structure for both bipartite and tripartite graphs, the classification of fraudulent/genuine transactions is described as an edge classification task. In this task, the goal is to assign to a given edge a label (0 for genuine, 1 for fraudulent) describing whether the transaction the edge represents is fraudulent or genuine.

In the rest of this chapter, we use both bipartite and tripartite undirected graphs for our analysis, denoted by the Python variables G_bu and G_tu, respectively. We will leave to you, as an exercise, an extension of the analyses proposed in this chapter to direct graphs.

We begin our analysis with a simple check to validate whether our graph is a real bipartite graph using the following code:

```
from networkx.algorithms import bipartite
all([bipartite.is_bipartite(G) for G in [G_bu,G_tu]])
```

As a result, we get True. This check gives us the certainty that the two graphs are actually bipartite/tripartite graphs. Note that networkx does not have an is_tripartite check, but the constraint to be verified here is that a link between nodes of the same type doesn't exist, which would apply to bipartite, tripartite, or any n-partite graphs.

Moreover, using the following command, we can get some basic statistics:

```
for G in [G_bu, G_tu]:
    print(nx.info(G))
```

By way of a result, we get the following:

```
Name: Bipartite Undirect
Type: Graph
Number of nodes: 1676
Number of edges: 201725
Average degree: 240.7220
Name: Tripartite Undirect
Type: Graph
Number of nodes: 267016
Number of edges: 530680
Average degree:    3.9749
```

As we can see, the two graphs differ in the number of nodes and the number of edges. The bipartite undirected graph has 1,676 nodes, equal to the number of customers plus the number of merchants, and 201,725 edges.

The tripartite undirected graph has 267,016 nodes, equal to the number of customers plus the number of merchants plus all the transactions. In this graph, the number of edges, as expected, is higher (530,680) compared to the bipartite graph. The interesting difference in this comparison is given by the average degree of the two graphs: the average degree of the bipartite graph is higher than that of the tripartite graph. This was expected, as in the tripartite graph, the connections are "split" by the presence of the transaction nodes. This makes the connections more "direct," which increases connectivity between nodes. Therefore, the resulting average degree is lower.

In the next section, we will describe how we can now use the transaction graphs that we generated to perform a more complete statistical analysis.

Network topology and community detection

In this section, we are going to analyze some graph metrics to have a clear picture of the general structure of the graph. We will be using networkx to compute most of the useful metrics we saw in *Chapter 1, Getting Started with Graphs*. We will try to interpret the metrics to gain insights into the graph.

Network topology

A good starting point for our analysis is the extraction of simple graph metrics to have a general understanding of the main properties of bipartite and tripartite transaction graphs.

We start by looking at the distribution of the degree for both bipartite and tripartite graphs using the following code:

```
for G in [G_bu, G_tu]:
    plt.figure(figsize=(10,10))
    degrees = pd.Series({k: v for k, v in nx.degree(G)})
    degrees.plot.hist()
    plt.yscale("log")
```

By way of a result, we get the following plots:

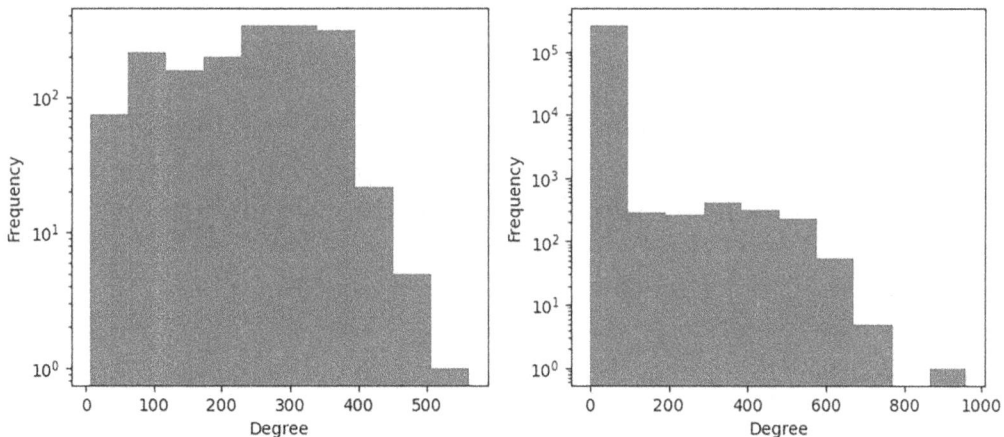

Figure 9.3: Degree distribution for bipartite (left) and tripartite (right) graphs

From *Figure 9.3*, it is possible to see how the distribution of nodes reflects the average degree we previously saw. In greater detail, the bipartite graph has a more variegated distribution, with a peak of around 300. For the tripartite graph, the distribution has a big peak for degree 0, while the other part of the tripartite degree distribution is similar to the bipartite distribution. These distributions completely reflect the differences in how the two graphs were defined. Indeed, if bipartite graphs are made by connections from the customer to the merchant, in the tripartite graph, all the connections pass through the transaction nodes. Those nodes are the majority in the graph, and they all have a degree of 2 (an edge from a customer and an edge to a merchant). As a consequence, the frequency in the bin representing degree 2 is equal to the number of transaction nodes.

In the next subsection, we will continue our investigation by analyzing some key metrics for the graphs:

- Edge weight
- Node centrality
- Assortativity

Edge weight

1. We begin by computing the quantile distribution, as it provides a concise summary of the data by dividing the range of the edge weights into intervals of equal probability:

```
for G in [G_bu, G_tu]:
  allEdgesWeights = pd.Series({(d[0], d[1]): d[2]["weight"] for d in
G.edges(data=True)})
  np.quantile(allEdgesWeights.values,[0.10,0.50,0.70,0.9])
```

2. By way of a result, we get the following:

```
array([  5.03 ,   58.25 ,   98.44 ,  215.656])
array([  4.21,   48.51,   76.4 ,  147.1 ])
```

3. We can also plot (in log scale) the distribution of edges weight, cut to the 90th percentile by using allEdgesWeightsFiltered.plot.hist(bins=40). The result is shown in the following charts:

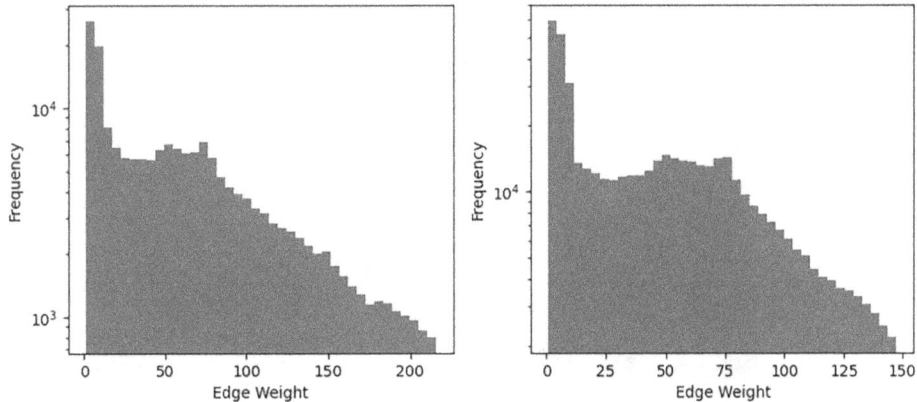

Figure 9.4: Edge weight distribution for bipartite (left) and tripartite (right) graphs

We can see how, due to the aggregation of the transaction having the same customer and merchant, the distribution of the bipartite graph is shifted to the right (high values) compared to the tripartite graph, where edge weights were not computed by aggregating multiple transactions.

Node centrality

1. We will now investigate the node centrality via the betweenness centrality metric. It measures how many shortest paths pass through a given node, giving an idea of how *central* that node is for the spreading of information inside the network. We can compute the distribution of node centrality by using the following command:

```
for G in [G_bu, G_tu]:
  plt.figure(figsize=(10,10))
  bc_distr = pd.Series(nx.betweenness_centrality(G, k=200))
  bc_distr.plot.hist()
  plt.yscale("log")
```

2. As a result, we get the following distributions:

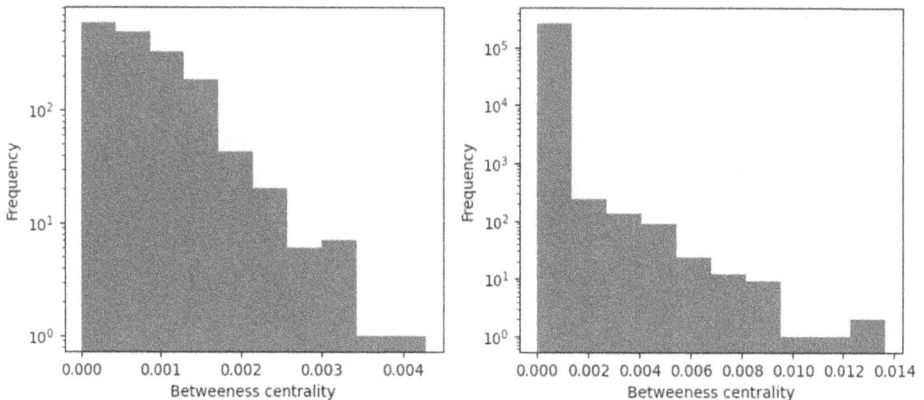

Figure 9.5: Betweenness centrality distribution for bipartite (left) and tripartite (right) graphs

As can be seen, the betweenness centrality is low for both graphs. This can be understood due to the large number of non-bridging nodes inside the network. Similar to what we saw for the degree distribution, the distribution of betweenness centrality values is different in the two graphs. The bipartite graph has a more variegated distribution with a mean of `0.00072`, while in the tripartite graph, the transaction nodes move the distribution values and lower the mean to 1.38e-05. Also, we can see that the distribution for the tripartite graph has a big peak, representing the transaction nodes, and the rest of the distribution is quite similar to the bipartite distribution.

Assortativity

1. We can finally compute the assortativity of the two graphs using the following code:

```
for G in [G_bu, G_tu]:
    print(nx.degree_pearson_correlation_coefficient(G))
```

2. By way of a result, we get the following:

```
-0.1377432041049189
-0.8079472914876812
```

Here, we can observe how both graphs have a negative assortativity, which likely shows that well-connected individuals associate with poor-connected individuals. For the bipartite graph, the value is low (-0.14), since customers who have a low degree are only connected with merchants who have high degrees due to the high number of incoming transactions.

The assortativity is even lower (-0.81) for the tripartite graph. This behavior could be attributed to the presence of the transaction nodes, as these nodes always have a degree of 2, and they are linked to customers and merchants represented by highly connected nodes.

Community detection

Another interesting analysis we can perform is community detection, as communities often correspond to meaningful substructures in the graph, where fraudulent activities may be concentrated. This analysis can help to identify specific fraudulent patterns:

1. The code to perform community extraction is as follows:

```
import community
for G in [G_bu, G_tu]:
    parts = community.best_partition(G, random_state=42,
weight='weight')
    communities = pd.Series(parts)   print(communities.value_
counts().sort_values(ascending=False))
```

   In this code, we simply use the `community` library to extract the communities in the input graph. We then print the communities detected by the algorithms, sorted according to the number of nodes contained.

2. For the bipartite graph, we obtain the following output:

5	546
0	335
7	139
2	136
4	123
3	111
8	83
9	59
10	57
6	48
11	26
1	13

3. For the tripartite graph, we obtain the following output:

11	4828
3	4493

```
26      4313
94      4115
8       4036

        ...

47      1160
103     1132
95       954
85       845
102      561
```

4. Due to the large number of nodes in the tripartite graph, we found 106 communities (we reported just a subset of them), whereas, for the bipartite graph, only 12 communities were found.

5. To have a clear picture of the tripartite graph, it is better to plot the distribution of the nodes contained in the different communities using the following command:

```
communities.value_counts().plot.hist(bins=20)
```

6. By way of a result, we get the following:

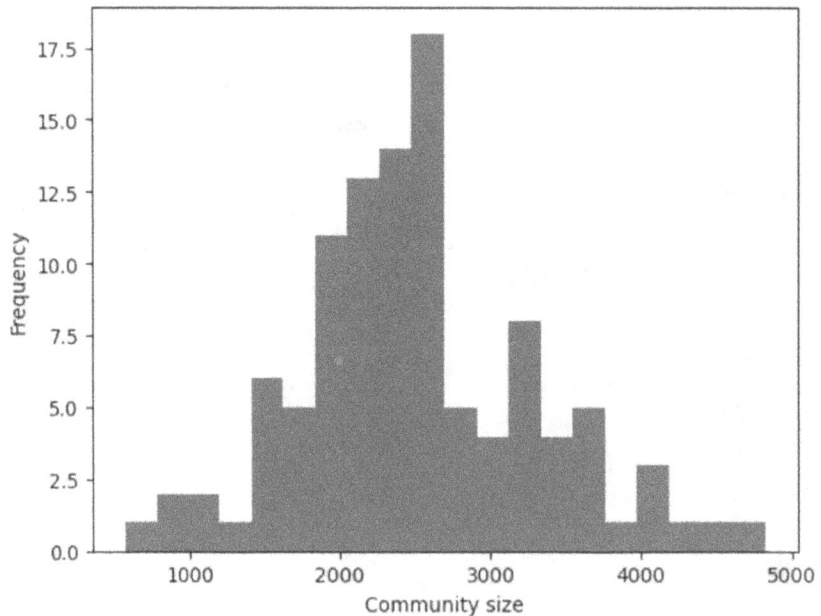

Figure 9.6: Distribution of communities' node size

From the diagram, it is possible to see that the peak is reached around 2,500. This means that more than 30 large communities have more than 2,000 nodes. From the plot, it is also possible to see that a few communities have fewer than 1,000 nodes and more than 3,000 nodes.

7. For each set of communities detected by the algorithm, we can compute the percentage of fraudulent transactions. The goal of this analysis is to identify specific subgraphs where there is a high concentration of fraudulent transactions:

```
graphs = []
d = {}
for x in communities.unique():
    tmp = nx.subgraph(G, communities[communities==x].index)
    fraud_edges = sum(nx.get_edge_attributes(tmp, "label").values())
    ratio = 0 if fraud_edges == 0 else (fraud_edges/tmp.number_of_
edges())*100
    d[x] = ratio
    graphs += [tmp]
print(pd.Series(d).sort_values(ascending=False))
```

8. The code simply generates a node-induced subgraph by using the nodes contained in a specific community. The graph is used to compute the percentage of fraudulent transactions as a ratio of the number of fraudulent edges over the number of all the edges in the graph.

9. For the bipartite graph, we will obtain the following output:

```
9       26.905830
10      25.482625
6       22.751323
2       21.993834
11      21.333333
3       20.470263
8       18.072289
4       16.218905
7        6.588580
0        4.963345
5        1.304983
1        0.000000
```

10. We can also plot a node-induced subgraph detected by the community detection algorithm by using the following code:

```
gId = ...
spring_pos = nx.spring_layout(graphs[gId])
edge_colors = ["r" if x == 1 else "g" for x in nx.get_edge_
attributes(graphs[gId], 'label').values()]
nx.draw_networkx(graphs[gId], pos=spring_pos, node_color=default_
node_color, edge_color=edge_colors, with_labels=False, node_size=15)
```

Given a particular community index, gId, the code extracts the node-induced subgraph, using the node available in the gId community index, and plots the graph obtained.

11. For each community, we have the percentage of its fraudulent edges. To have a better description of the subgraph, we can plot community 10 by executing the previous line of code using gId=10. As a result, we get the following:

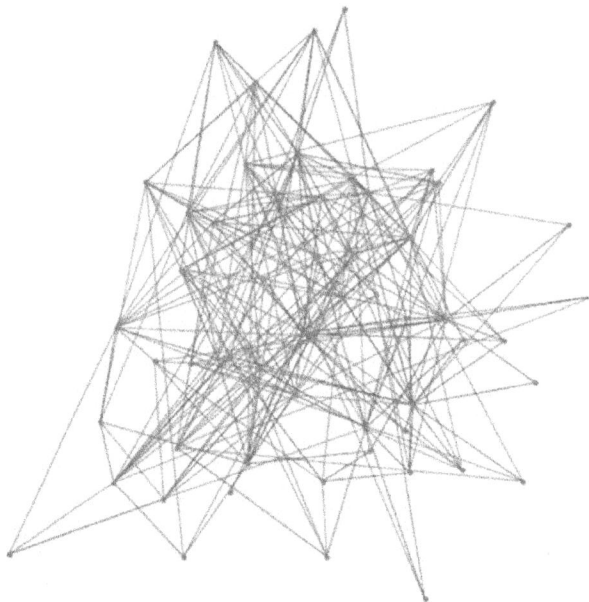

Figure 9.7: Induced subgraph of community 10 for the bipartite graph

12. The image of the induced subgraph allows us to better understand whether specific patterns are visible in the data.

13. Similarly, by running the above code on the tripartite graph, we obtain the following output:

```
6      6.857728
94     6.551151
8      5.966981
1      5.870918
89     5.760271

       ...

102    0.889680
72     0.836013
85     0.708383
60     0.503461
46     0.205170
```

14. Due to the large number of communities, we can plot the distribution of the fraudulent over genuine ratio with the following command:

```
pd.Series(d).plot.hist(bins=20)
```

15. By way of a result, we get the following:

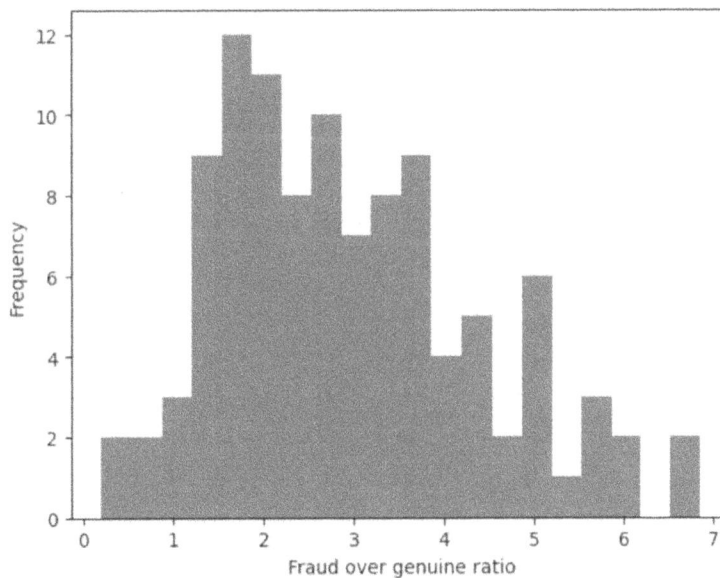

Figure 9.8: Distribution of communities' fraudulent/genuine edge ratio

From the diagram, we can observe that a large part of the distribution is around communities having a ratio of between 2 and 4. There are a few communities with a low ratio (<1) and with a high ratio (>5).

16. Also, for the tripartite graph, we can plot community 6 (with a ratio of 6.86), made by 1,935 nodes, by executing the previous line of code using gId=6:

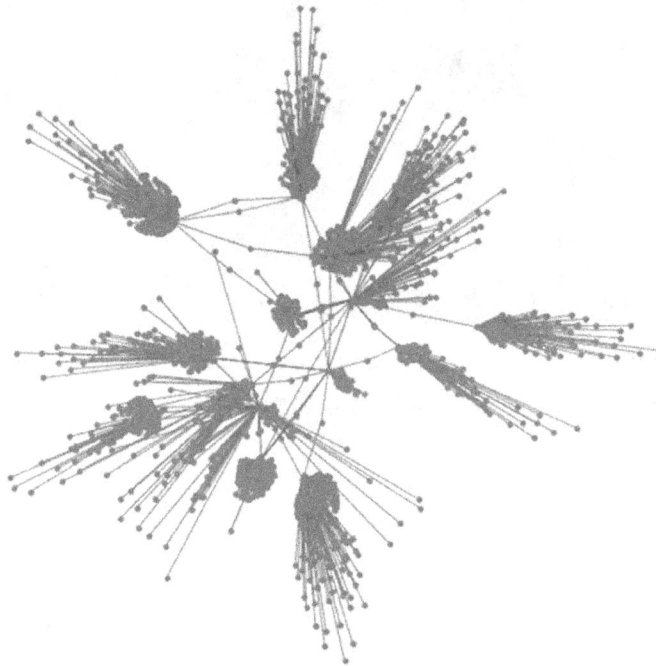

Figure 9.9: Induced subgraph of community 6 for the tripartite graph

As for the bipartite use case, in this image, we can see an interesting pattern that could be used to perform a deeper exploration of some important graph sub-regions.

In this section, we performed some explorative tasks to better understand the graphs and their properties. We also gave an example describing how a community detection algorithm can be used to spot patterns in the data. In the next section, we will describe how the bipartite and tripartite graphs and their properties can be utilized by graph machine learning algorithms to build automatic procedures for fraud detection using supervised and unsupervised approaches.

Applying supervised and unsupervised fraud approaches to fraud detection

As we already discussed at the beginning of this chapter, transactions are represented by edges, and we then want to classify each edge in the correct class: fraudulent or genuine. The pipeline we will use to perform the classification task is the following:

- A sampling procedure for the imbalanced task

- The use of an unsupervised embedding algorithm to create a feature vector for each edge

- The application of supervised and unsupervised machine learning algorithms to the feature space defined in the previous point

Dataset resampling

Since our dataset is strongly imbalanced, with fraudulent transactions representing 2.83% of total transactions, we need to apply some techniques to deal with unbalanced data. In this use case, we will apply a simple random undersampling strategy. Going into more depth, we will take a subsample of the majority class (genuine transactions) to match the number of samples of the minority class (fraudulent transactions). This is just one of the many techniques available in literature. It is also possible to use outlier detection algorithms, such as isolation forests, to detect fraudulent transactions as outliers in the data. We leave it to you, as an exercise, to extend the analyses using other techniques to deal with imbalanced data, such as random oversampling or using cost-sensitive classifiers for the classification task. Specific techniques for node and edge sampling that can be directly applied to the graph will be described in *Chapter 13, Novel Trends on Graphs*. Here are the steps to try this:

1. The code we use for random undersampling is as follows:

```
from sklearn.utils import resample
df_majority = df[df.is_fraud==0]
df_minority = df[df.is_fraud==1]
df_maj_dowsampled = resample(df_majority, n_samples=len(df_
minority), random_state=42)
df_downsampled = pd.concat([df_minority, df_maj_dowsampled])
G_down = build_graph_bipartite(df_downsampled, nx.Graph())
```

2. The code is straightforward. We applied the `resample` function of the `sklearn` package to filter the downsample function of the original data frame. We then build a graph using the `build_graph_bipartite` function defined at the beginning of the chapter. To create the tripartite graph, the `build_graph_tripartite` function should be used.

3. As the next step, we split the dataset into training and validation with a ratio of 80/20:

```
from sklearn.model_selection import train_test_split
train_edges, val_edges, train_labels, val_labels = train_test_
split(list(range(len(G_down.edges))), list(nx.get_edge_attributes(G_
down, "label").values()), test_size=0.20, random_state=42)
```

```
edgs = list(G_down.edges)
train_graph = G_down.edge_subgraph([edgs[x] for x in train_edges]).
copy()
train_graph.add_nodes_from(list(set(G_down.nodes) - set(train_graph.
nodes)))
```

As before, in this case, the code is straightforward since we simply apply the train_test_split function of the sklearn package.

Node feature generation

We can now build the feature space using the Node2Vec algorithm, as follows:

```
from node2vec import Node2Vec
node2vec = Node2Vec(train_graph, weight_key='weight')
model = node2vec_train.fit(window=10)
```

Node2Vec results (which provide each node with a vector) are then aggregated to produce the feature set for the edges, as described in *Chapter 4, Unsupervised Graph Learning*.

Different choices for the aggregation can be used, resulting in slightly different variations for the Edge2Vec algorithm. For instance, using the WeightedL1Embedder, the embedding can be obtained by:

```
from node2vec.edges import WeightedL1Embedder
embeddings = WeightedL1Embedder(keyed_vectors=model.wv)
```

The edge embedding that is obtained in this way will generate the final feature space used by the classifier. The specific type of node feature aggregation algorithm (WeightedL1Embedder, in this case) together with the various parameters used in the Node2Vec algorithm represent the hyperparameters of our machine learning pipeline.

Training and evaluating the model

Finally, we can train and evaluate a machine learning model using the feature set generated in the previous step. Here, we will use a RandomForestClassifier from the sklearn Python library, but of course, other choices could also be valid. Different performance metrics (namely, precision, recall, and F1 score) are computed on the validation test:

1. First, we build the training and validation datasets:

    ```
    # Building training and validation sets
    train_embeddings = [embeddings[str(edgs[x][0]), str(edgs[x][1])] for
    ```

```
    x in train_edges]
        val_embeddings = [embeddings[str(edgs[x][0]), str(edgs[x][1])]
    for x in val_edges]
```

2. Then, we train the classifier on the training set:

```
rf = RandomForestClassifier(n_estimators=1000, random_state=42)
rf.fit(train_embeddings, train_labels)
```

3. Finally, we evaluate the performances on the validation set:

```
y_pred = rf.predict(val_embeddings)
print('Precision:', metrics.precision_score(val_labels, y_pred))
print('Recall:', metrics.recall_score(val_labels, y_pred))
print('F1-Score:', metrics.f1_score(val_labels, y_pred))
```

Hyperparameter tuning

As mentioned in the previous subsection, the Edge2Vec algorithm can be seen as a hyperparameter for the machine learning pipeline. In the following code snippet, we loop over some possible choices for the node feature aggregation function, to identify the best choice for this particular dataset. The code to perform this task is the following:

```
from sklearn import metrics
from sklearn.ensemble import RandomForestClassifier
from node2vec.edges import HadamardEmbedder, AverageEmbedder,
WeightedL1Embedder, WeightedL2Embedder
classes = [HadamardEmbedder, AverageEmbedder, WeightedL1Embedder,
WeightedL2Embedder]
for cl in classes:
    embeddings = cl(keyed_vectors=model.wv)
    train_embeddings = [embeddings[str(edgs[x][0]), str(edgs[x][1])] for x
in train_edges]
    val_embeddings = [embeddings[str(edgs[x][0]), str(edgs[x][1])] for x
in val_edges]
    rf = RandomForestClassifier(n_estimators=1000, random_state=42)
    rf.fit(train_embeddings, train_labels)
    y_pred = rf.predict(val_embeddings)
    print(cl)
    print('Precision:', metrics.precision_score(val_labels, y_pred))
    print('Recall:', metrics.recall_score(val_labels, y_pred))
    print('F1-Score:', metrics.f1_score(val_labels, y_pred))
```

We can apply the preceding code to both bipartite and tripartite graphs to solve the fraud detection task. In the following table, we report the performances for the bipartite graph:

EMBEDDING ALGORITHM	PRECISION	RECALL	F1 SCORE
HADAMARD EMBEDDER	0.73	0.76	0.75
AVERAGE EMBEDDER	0.71	0.79	0.75
WEIGHTED L1 EMBEDDER	0.64	0.78	0.70
WEIGHTED L2 EMBEDDER	0.63	0.78	0.70

Table 9.2: Supervised fraud edge classification performances for a bipartite graph

In the following table, we report the performances for the tripartite graph:

EMBEDDING ALGORITHM	PRECISION	RECALL	F1 SCORE
HADAMARD EMBEDDER	0.89	0.29	0.44
AVERAGE EMBEDDER	0.74	0.45	0.48
WEIGHTED L1 EMBEDDER	0.66	0.46	0.55
WEIGHTED L2 EMBEDDER	0.66	0.47	0.55

Table 9.3: Supervised fraud edge classification performances for a tripartite graph

In *Table 9.2* and *Table 9.3*, we reported the classification performances obtained using bipartite and tripartite graphs. As we can see from the results, in terms of F1 score, precision, and recall, the two methods show significant differences. Since, for both graph types, Hadamard and average edge embedding algorithms give the most interesting results, we are going to focus our attention on those two.

Going into more detail, the tripartite graph has better precision compared to the bipartite graph (0.89 and 0.74 for the tripartite graph versus 0.73 and 0.71 for the bipartite graph). In contrast, the bipartite graph has a better recall (0.76 and 0.79 for the bipartite graph versus 0.29 and 0.45 for the tripartite graph) and F1 score (0.75 and 0.75 for the bipartite graph versus 0.44 and 0.48 for the tripartite graph) than the tripartite graph. We can therefore conclude that, in this specific case, the use of a bipartite graph could be a better choice since it achieves high performances in terms of F1 score and recall with a smaller graph (in terms of nodes and edges) than the tripartite graph.

In the context of fraud detection, it is common to prioritize recall over precision, given the importance of not missing the detection of any fraud, and also at the expense of more false positives.

Unsupervised approach to fraudulent transaction identification

The same approach can also be applied in unsupervised tasks using k-means. The main difference is that the generated feature space will not undergo a train-validation split. Indeed, in the following code, we will compute the Node2Vec algorithm on the entire graph generated following the downsampling procedure:

```
nod2vec_unsup = Node2Vec(G_down, weight_key='weight')
unsup_vals = nod2vec_unsup.fit(window=10)
```

When building the node feature vectors, we can use different Egde2Vec algorithms to run the k-means algorithm, as follows:

```
from sklearn.cluster import KMeans
classes = [HadamardEmbedder, AverageEmbedder, WeightedL1Embedder,
WeightedL2Embedder]
true_labels = [x for x in nx.get_edge_attributes(G_down, "label").
values()]
for cl in classes:
    embedding_edge = cl(keyed_vectors=unsup_vals.wv)
    embedding = [embedding_edge[str(x[0]), str(x[1])] for x in G_down.
edges()]
    kmeans = KMeans(2, random_state=42).fit(embedding)
    nmi = metrics.adjusted_mutual_info_score(true_labels, kmeans.labels_)
    ho = metrics.homogeneity_score(true_labels, kmeans.labels_)
    co = metrics.completeness_score(true_labels, kmeans.labels_)
    vmeasure = metrics.v_measure_score(true_labels, kmeans.labels_)
    print(cl)
    print('NMI:', nmi)
    print('Homogeneity:', ho)
    print('Completeness:', co)
    print('V-Measure:', vmeasure)
```

Different steps are performed in the previous code:

1. For each `Edge2Vec` algorithm, the previously computed `Node2Vec` algorithm on training and validation sets is used to generate the feature space.

    A `KMeans` clustering algorithm from the `sklearn` Python library is fitted on the feature set generated in the previous step. In the previous code snippet, we assumed k=2 as an example, although in general, the number of clusters is one hyperparameter of the modeling pipeline and is to be varied and optimized.

2. Different performance metrics are used – namely, adjusted `mutual information` (MNI), homogeneity, completeness, and v-measure scores.

We can apply the code to both bipartite and tripartite graphs to solve the fraud detection task using the unsupervised algorithm. In the following table, we report the performances for the bipartite graph:

EMBEDDING ALGORITHM	MNI	HOMOGENEITY	COMPLETENESS	V-MEASURE
HADAMARD EMBEDDER	0.34	0.33	0.36	0.34
AVERAGE EMBEDDER	0.07	0.07	0.07	0.07
WEIGHTED L1 EMBEDDER	0.06	0.06	0.06	0.06
WEIGHTED L2 EMBEDDER	0.05	0.05	0.05	0.05

Table 9.4: Unsupervised fraud edge classification performances for the bipartite graph

In the following table, we report the performances for the tripartite graph:

EMBEDDING ALGORITHM	MNI	HOMOGENEITY	COMPLETENESS	V-MEASURE
HADAMARD EMBEDDER	0.44	0.44	0.45	0.44
AVERAGE EMBEDDER	0.06	0.06	0.06	0.06
WEIGHTED L1 EMBEDDER	0.001	0.001	0.00	0.06
WEIGHTED L2 EMBEDDER	0.0004	0.0004	0.0004	0.0004

Table 9.5: Unsupervised fraud edge classification performances for the tripartite graph

In *Table 9.4* and *Table 9.5*, we reported the classification performances obtained using bipartite and tripartite graphs with the application of an unsupervised algorithm. As we can see from the results, the two methods show significant differences. It is also worth noticing that, in this case, the performances obtained with the Hadamard embedding algorithm clearly outperform all other approaches.

As shown by the tables, for this task, the performances obtained with the tripartite graph outstrip those obtained with the bipartite graph. In the unsupervised case, we can see how the introduction of the transaction nodes improves the overall performance. This additional layer of information may enable more accurate embeddings. Thus, in the unsupervised setting, for this specific use case and using as a reference the results obtained in *Table 9.4* and *Table 9.5*, the use of the tripartite graph could be a better choice since it enables the attainment of superior performances compared with the bipartite graph.

Additional resources

On Kaggle, you can find further resources to analyze and explore datasets with financial transactions in which you can similarly apply the frameworks learned here. In particular, we suggest you investigate two other datasets:

The first one is the Czech Bank's Financial Analysis dataset, available at `https://github.com/Kusainov/czech-banking-fin-analysis`. This dataset came from an actual Czech bank in 1999, for the period covering 1993–1998. The data pertaining to clients and their accounts consists of directed relations. The dataset does not come with labels on the transactions. It is therefore not possible to train a fraud detection engine using supervised machine learning techniques, whereas the unsupervised techniques would still apply.

The second dataset is the paysim1 dataset, available at `https://www.kaggle.com/datasets/ealaxi/paysim1`. This dataset comprises simulated mobile money transactions based on a sample of real transactions extracted from one month of financial logs from a mobile money service implemented in an African country. The original logs were provided by a multinational company that is the provider of a mobile financial service and is currently running in more than 14 countries across the globe. This dataset also contains labels on fraudulent/genuine transactions.

Moreover, it is worth pointing out that there can be different types of fraud, depending on the relationship between the fraudster and the entity being defruaded. In general, frauds are generally divided into:

1. First-party fraud, where the fraudster uses their identity and/or fabricated information to deceive a business and obtain personal gain. Some examples of first-party fraud are making unauthorized purchases with stolen or false credit cards, or providing false records to obtain a loan.

2. Second-party fraud, where a person internal to an organization or a business colludes with an external party in order to pursue an advantage against the organization or the business itself. The internal person generally abuses their position, knowledge, and/or access to internal information in order to perpetuate a personal gain.

3. Third-party fraud, where an external entity, unrelated to a business or its customers, commits fraud at the expense of either of the two. Some examples of third-party fraud are phishing scams, counterfeit goods, or identity theft.

Each type of fraud may require a different set of controls and measures, as well as specific analytic algorithms. Indeed, the transaction dataset and the analysis presented in the chapter are more suited to address first-party frauds as, in general, models to spot this type of fraud focus on identifying spending patterns to then capture transactions dissimilar from the expected behavior. Graph analytics is generally very effective in clustering users, merchants, and communities to provide an effective implementation of behavior analytics. On the other hand, second-party fraud can be identified with the implementation of monitoring employee behavior, as well as compliance checks. Graph analytics can indeed be useful for these use cases. Similar to the first-party models, employee behavior can also be analyzed using graph machine learning, although the dataset may need to encode a number of other sources of information besides transactional data. From a compliance standpoint, process mining techniques that still rely on a graph representation of the various procedural steps/pathways can be effective in identifying fraudulent behavior or non-compliant processes. Finally, third-party fraud, especially in the form of phishing attacks, can also be addressed using graph machine learning. In this context, understanding the network from which the phishing attack comes as well as the URLs being used (which can also benefit from a graph representation) can be critical for building an effective phishing classifier.

Summary

In this chapter, we described how a classical fraud detection task can be described as a graph problem and how the techniques described in the previous chapters can be used to tackle the problem. Going into more detail, we introduced the dataset we used and described the procedure to transform the transactional data into two types of graph – namely, bipartite and tripartite undirected graphs.

We then computed local (along with their distributions) and global metrics for both graphs, comparing the results. Moreover, a community detection algorithm was applied to the graphs in order to spot and plot specific regions of the transaction graph where the density of fraudulent transactions is higher compared to the other communities.

Finally, we solved the fraud detection problem using supervised and unsupervised algorithms, comparing the performance of the bipartite and tripartite graphs. As the first step, since the problem was unbalanced with a higher presence of genuine transactions, we performed simple downsampling. We then applied different Edge2Vec algorithms in combination with a random forest for the supervised task, and k-means for the unsupervised task, achieving good classification performances.

This chapter concludes the series of examples that are used to show how graph machine learning algorithms can be applied to problems belonging to different domains, such as social network analysis, text analytics, and credit card transaction analysis.

In the next chapter, we will describe some practical uses for graph databases and graph processing engines that are useful for scaling out the analysis to large graphs.

10

Building a Data-Driven Graph-Powered Application

So far, we have provided you with both theoretical and practical ideas to allow you to design and implement machine learning models that leverage graph structures. Besides designing the algorithm, it is often very important to embed the modeling/analytical pipeline into a robust and reliable end-to-end application. This is especially true in industrial applications, where the goal is usually to design and implement production systems that support data-driven decisions and/or provide users with timely information. However, creating a data-driven application that resorts to graph representation/modeling is indeed a challenging task that requires a proper design that is a lot more complicated than simply importing networkx. This chapter aims to provide you with a general overview of the key concepts and frameworks that are used when building graph-based, scalable, data-driven applications.

We will start by providing an overview of the so-called **Lambda architectures**, which provide a framework to structure scalable applications that require large-scale processing and real-time updates. We will then continue by applying this framework in the context of *graph-powered applications*, that is, applications that leverage graph structures using techniques such as the ones described in this book. We will describe their two main analytical components: **graph processing engines** and **graph querying engines**. We'll present some of the technologies used, both in shared memory machines and distributed memory machines, outlining the similarities and differences. The following topics will be covered in this chapter:

- Overview of Lambda architectures
- Lambda architectures for graph-powered applications

- Technologies and examples of graph processing engines
- Graph querying engines and graph databases

Technical requirements

All the code files relevant to this chapter are available at https://github.com/PacktPublishing/ Graph-Machine-Learning/tree/main/Chapter10. Please refer to the *Practical exercises* section in *Chapter 1, Getting Started with Graphs,* for guidance on how to set up the environment to run the examples in this chapter, either using Poetry, pip, or Docker.

Overview of Lambda architecture

In recent years, great focus has been given to designing scalable architectures that will allow, on the one hand, the *processing of a large amount of data,* and, on the other, *providing answers/alerts/ actions in real time, using the latest available information.* Additionally, these systems need to also be able to scale out seamlessly to a larger number of users or a larger amount of data by increasing resources horizontally (adding more servers) or vertically (using servers that are more powerful).

Lambda architecture is a particular data-processing architecture that is designed to process massive quantities of data and ensure large throughput in a very efficient manner, preserving reduced latency and ensuring fault tolerance and negligible errors.

The Lambda architecture is composed of three different layers:

- **The batch layer:** This layer sits on top of the storage system – either local or distributed – and can handle and store all historical data, as well as performing **online analytical processing (OLAP)** computation on the entire dataset. New data is continuously ingested and stored, as it would be traditionally done in data warehouse systems. Large-scale processing is generally achieved via massively parallel jobs, which aim to produce aggregation, structuring, and computation of relevant information. In the context of machine learning, model training that relies on historical information is generally done in this layer, thus producing a trained model to be used either in a batch prediction job or in real-time execution.

- **The speed layer:** This is a low-latency layer that allows the real-time processing of the information to provide timely updates and information. It is generally fed by a streaming process, usually involving fast computation that does not require long computational time or load. It produces an output that is integrated with the data generated by the batch layer in (near) real time, providing support for **online transaction processing (OLTP)** operations.

The speed layer might also very well use some outputs of the OLAP computations, such as a trained model. Oftentimes, applications that use machine learning modeling in real time (for example, fraud detection engines used in credit card transactions) embed trained models in their speed layers that provide prompt predictions and trigger real-time alerts of potential fraud. Libraries may operate at an event level (such as Apache Storm) or over mini-batches (such as Spark Streaming), providing, depending on the use case, slightly different requirements for latency, fault tolerance, and computational speed.

- **The serving layer**: The serving layer is responsible for organizing, structuring, and indexing information in order to allow the fast retrieval of data coming from the batch and speed layers. The serving layer thus integrates the outputs of the batch layer with the most updated and real-time information of the speed layer in order to deliver to the user a unified and coherent view of the data. A serving layer can be composed of a persistence layer that integrates both historical aggregation and real-time updates. This component may be based on some kind of database, which can be relational or not, conveniently indexed in order to reduce latency and allow the fast retrieval of relevant data. The information is generally exposed to the user via either a direct connection to the database and is accessible using a specific domain query language, such as SQL, or via dedicated services, such as RESTful API servers (which in Python can be easily implemented using several frameworks, such as **flask**, **fastapi**, or **turbogear**), which provide the data via specifically designed endpoints.

The following diagram illustrates the Lambda architecture with these three layers:

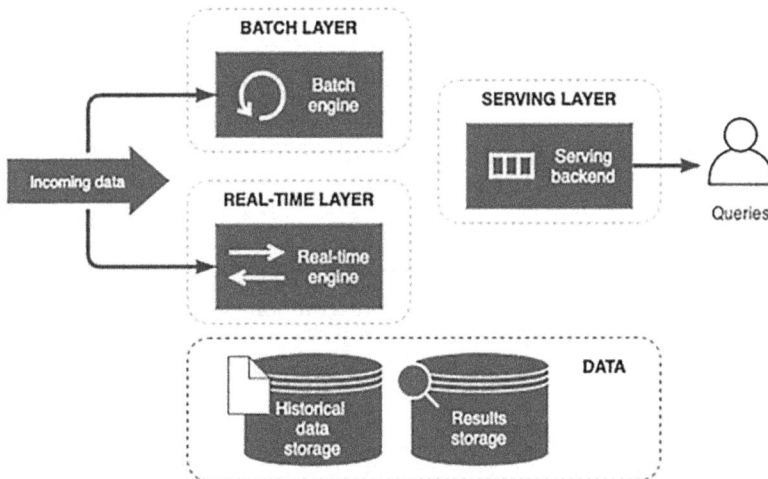

Figure 10.1: Functional diagram for an application based on Lambda architecture

All the service layers (batch, real-time, and serving) interact with the data layers, where all data – both operational and historical – are stored. Batch and real-time layers generally perform both read and write operations on the data, while the serving layers generally mostly read from the data layers.

Lambda architectures have several benefits that have motivated and promoted their use, especially in the context of *big data* applications. In the following bullet points, we list some of the main pros of Lambda architectures:

- **No server management**: The Lambda architectural design pattern typically abstracts the functional layers and does not require installing, maintaining, or administering any software/infrastructure
- **Flexible scaling**: The application can be either automatically scaled or scaled by controlling the number of processing units that are used in batch layers (for example, computing nodes) and/or in speed layers (for example, Kafka brokers) separately
- **Automated high availability**: It represents a serverless design for which we already have built-in availability and fault tolerance
- **Business agility**: Reacts in real time to changing business/market scenarios

Although very powerful and flexible, Lambda architectures come with some limitations mainly due to the presence of two interconnected processing flows: the **batch layer** and the **speed layer**. This may require developers to build and maintain separate code bases for batch and stream processes, resulting in more complexity and code overhead, which may lead to harder debugging, possible misalignment, and bug promotion.

So far, we have provided a short overview of Lambda architectures and their basic building blocks. For more details on how to design scalable architectures and the most commonly used architectural patterns, please refer to the book *Data Lake for Enterprises*, 2017, by Tomcy John and Pankaj Misra.

In the next section, we will show you how to implement a Lambda architecture for graph-powered applications. In particular, we will describe the main components and review the most common technologies.

Lambda architectures for graph-powered applications

When dealing with scalable, graph-powered, data-driven applications, the design of Lambda architectures is also reflected in the separation of functionalities between two crucial components of the analytical pipeline:

- The **graph processing engine** executes computations on the graph structure in order to extract features (such as embeddings), compute statistics (such as degree distributions, the number of edges, and cliques), compute metrics and **key performance indicators (KPIs)** (such as centrality measures and clustering coefficients), and identify relevant subgraphs (for example, communities) that often require OLAP.

- The **graph querying engine** allows us to persist network data (usually done via a graph database) and provides fast information retrieval and efficient querying and graph traversal (usually via graph querying languages). All of the information is already persisted in some data storage (that may or may not be in memory) and no further computation is required apart from (possibly) some final aggregation results, for which indexing is crucial to achieving high performance and low latency.

The following diagram illustrates this:

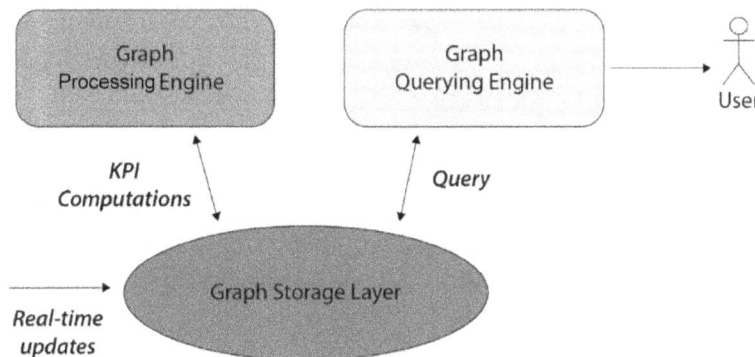

Figure 10.2: Graph-based architecture, with the main components also reflected in a Lambda architectural pattern

Graph processing engines sit on top of batch layers and produce outputs that may be stored and indexed in appropriate graph databases. These databases are the backend of graph querying engines, which allow relevant information to be easily and quickly retrieved, representing the operational views used by the serving layer.

Depending on the use cases and/or the size of the graph, it often makes sense to run both the graph processing engine and the graph query engine on top of the same infrastructure.

Instead of storing the graph on a low-level storage layer (for example, the filesystem, **Hadoop Distributed File System**, or S3), there are graph database options that could support both OLAP and OLTP. These provide, at the same time, a backend persistence layer where historical information processed by batch layers, together with real-time updates from the speed layer, is stored, and information that needs to be queried efficiently by the serving layer.

As compared to other use cases, this condition is indeed quite peculiar for graph-powered, data-driven applications. Historical data often provides a topology on top of which new, real-time updates and OLAP outputs (KPIs, data aggregations, embeddings, communities, and so on) can be stored. This data structure also represents the information that is later queried by the serving layer that traverses the enriched graph.

In the next sections, we will be discussing graph querying engines and the graph processing engine in more detail.

Graph querying engine

In the last decade, due to the large diffusion of non-structured data, NoSQL databases have started to gain considerable attention and importance. Among them, **graph databases** are indeed extremely powerful for storing information based on a relation between entities. Indeed, in many applications, data can naturally be seen as entities, associated with metadata in the form of node properties, connected by edges that also have properties that further describe the relationship between entities.

Examples of graph databases are libraries or tools such as Neo4j, OrientDB, ArangoDB, Amazon Neptune, Cassandra, JanusGraph (previously named TitanDB), Google Spanner Graph, and Microsoft Cosmos DB. In the following sections, we will briefly describe some of them, together with the languages that allow us to query and traverse the underlying graphs, which are called **graph querying languages**.

Neo4j

At the time of writing, **Neo4j** (https://neo4j.com/) is one of the most common graph databases around, with a large community supporting its use and adoption. It features two editions:

- *Community Edition*, released under a GPL v3 license, allows users/developers to openly include Neo4j in their applications

- *Enterprise Edition*, designed for commercial deployments where scale and availability are crucial

With its enterprise edition, Neo4j can scale out to fairly large datasets via **sharding**, that is, distributing data over multiple nodes and parallelizing queries and aggregation over multiple instances of the database. Additionally, the Neo4j federation also allows querying smaller separated graphs (sometimes even with a different schema) as if they were one large graph.

Some of Neo4j's strong points are its flexibility (which allows the schema to be evolved) and its user-friendliness. In particular, many operations in Neo4j can be done through its query language, which is very intuitive and easy to learn: **Cypher**. Cypher can just be seen as the counterpart of SQL for graph databases.

> It's noteworthy that **Graph Query Language (GQL)** has recently been published as a new ISO standard designed for property graphs. GQL is the first new ISO database language since SQL's introduction in 1987 and aims to standardize graph querying across different platforms. Cypher has significantly influenced GQL's development, providing a familiar foundation for those already acquainted with Cypher. However, GQL syntax and Cypher's are not that different, and, as noted in the GQL announcement, Cypher will be supported for many years to come.

Testing out Neo4j and Cypher is extremely easy. You could install the Community Edition (via Docker; see below) or play around with SaaS services, such as the online sandbox version (https://neo4j.com/sandbox/) or the free tier of Neo4j Aura (https://neo4j.com/free-graph-database/).

In the following, we will use the Docker installation. To start a single instance of the community edition of Neo4j, you can use the following Docker command:

```
docker run --rm --detach --name neo4j \
        --publish=7474:7474 --publish=7687:7687 \
        --user="$(id -u):$(id -g)" \
        --env NEO4J_AUTH=none \
        --env NEO4J_PLUGINS='["graph-data-science"]' \
        neo4j:5.26.0
```

Once the service is up and running, you can log in to the UI by opening the link http://localhost:7474 in one of your browsers and supplying the username neo4j and password neo5j. At this point, you are ready to play with your Neo4j database.

In the repository attached to the book, we provide a Cypher query to create the Movie dataset. The Movie dataset is made up of 38 movies and 133 people that acted in, directed, wrote, reviewed, and produced them. Both the on-premises version and the online version have a user-friendly UI that allows the user to query and visualize the data (see *Figure 10.3*). We start by listing 10 actors in the Movie dataset, by simply querying the following:

```
MATCH (p: Person) RETURN p LIMIT 10
```

But let's now use the information about relations between data points. We see that one of the actors that appears in the database is Keanu Reeves. We may wonder who all the actors that he has acted with in the listed movies are. This information can be easily retrieved using the following query:

```
MATCH (k: Person {name:"Keanu Reeves"})-[:ACTED_IN]-(m: Movie)-[:ACTED_
IN]-(a: Person) RETURN k, m, a
```

As shown in the following figure, the query intuitively and graphically indicates in its syntax how to traverse the graph by declaring the path we are interested in:

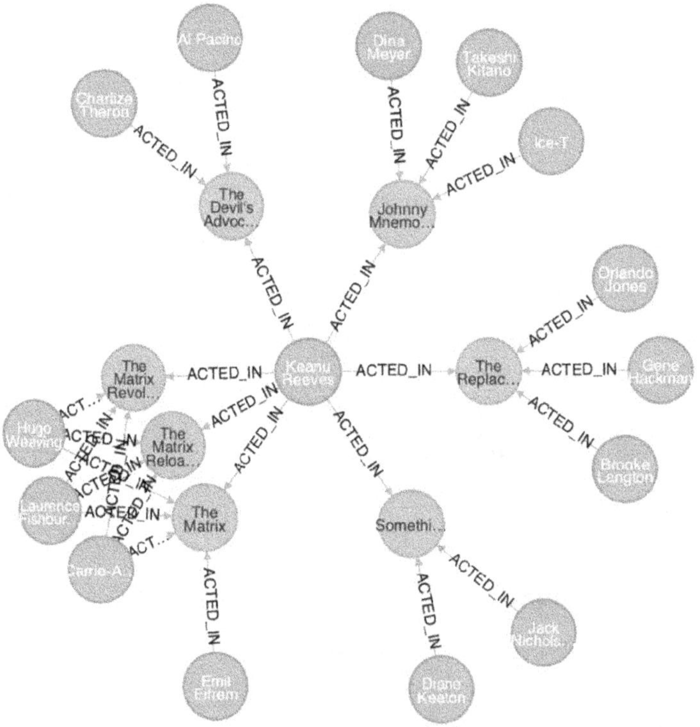

Figure 10.3: Example of Neo4j UI with the Cypher query to retrieve the co-actors of Keanu Reeves in the Movie dataset

Neo4j also provides bindings with several programming languages, such as Python, JavaScript, Java, Go, Spring, and .NET. For Python in particular, there are several libraries that implement connections with Neo4j, such as neo4j and graphdatascience, that are officially supported by Neo4j Inc. These libraries provide direct connections to the database via a binary protocol. When using the neo4j client library, creating a connection to the database and running a query is just a matter of a few lines of code:

```
from neo4j import GraphDatabase
driver = GraphDatabase("bolt://localhost:7687", "my-user", "my-password")
def run_query(tx, query):
    return tx.run(query)
with driver.session() as session:
    session.write_transaction(run_query, query)
```

A query could be any Cypher query, for instance, the one written previously to retrieve the co-actors of Keanu Reeves. In the notebook provided in the repository, we show how to create and query the Movie dataset programmatically using Python.

On the other hand, graphdatascience provides a higher-level API that abstracts away the need to instantiate the low-level sessions and transactions:

```
from graphdatascience import GraphDataScience

uri = f"bolt://localhost:7687"
gds = GraphDataScience(uri, auth=("neo4j", "neo5j"))

gds.run_cypher(query)
```

As we will see in the following, besides a higher-level API, the graphdatascience also provides intuitive APIs to interact with the Graph Data Science engine to perform various analytical computations on the graph.

JanusGraph – a graph database to scale out to very large datasets

Neo4j is an extremely great piece of software, unbeatable when you want to get things done quickly, thanks to its intuitive interface and query language. Neo4j is indeed a graph database that's suitable for production, particularly in its Enterprise edition, but it's especially good in MVPs when agility is crucial.

Indeed, when the volume of the data increases substantially, the Community Edition of Neo4j is likely to not be suitable anymore, and you should either consider switching to the Enterprise edition or other graph database options. Once again, this should be done only when the use case requirements start to hit the limitation of Neo4j Community Edition, as you need to evolve from the MVP initial requirements.

Besides other commercial alternatives, such as Amazon Neptune and Azure Cosmos DB, some open source options are also available. Among them, we believe it is worth mentioning **JanusGraph** (`https://janusgraph.org/`), which is a particularly interesting piece of software. JanusGraph is the evolution of a previously open source project that was called **TitanDB** and is now an official project under the Linux Foundation, also featuring support from top players in the tech landscape, such as IBM, Google, Hortonworks, Amazon, Expero, and Grakn Labs.

JanusGraph is a scalable graph database designed for storing and querying graphs distributed across a multi-machine cluster with hundreds of billions of vertices and edges. As a matter of fact, JanusGraph does not have a storage layer on its own, but it is rather a component, written in Java, that sits on top of other data storage layers, such as the following:

- **Google Cloud Bigtable** (`https://cloud.google.com/bigtable`), which is the cloud version of the proprietary data storage system built on Google File System, designed to scale a massive amount of data distributed across data centers (for more information, refer to *Bigtable: A Distributed Storage System for Structured Data*, Fay Chang et al., 2006)
- **Apache HBase** (`https://hbase.apache.org/`), which is a non-relational database that features Bigtable capabilities on top of Hadoop and HDFS, thus ensuring similar scalability and fault tolerance
- **Apache Cassandra** (`https://cassandra.apache.org/`), which is an open source distributed NoSQL database that allows handling a large amount of data, spanning multiple data centers
- **ScyllaDB** (`https://www.scylladb.com/`), which is specifically designed for real-time applications, and is compatible with Apache Cassandra while achieving significantly higher throughputs and lower latencies

Depending on its storage backend, JanusGraph inherits its various features, such as scalability, high availability, and fault tolerance, from scalable solutions, abstracting a graph view on top of them. Note that when running on top of eventually consistent databases, users should expect similar slack requirements on consistency for JanusGraph as well.

With its integration with ScyllaDB, JanusGraph handles extremely fast, scalable, and high-through-put applications. JanusGraph also integrates indexing layers that can be based on Apache Lucene, Apache Solr, and Elasticsearch in order to allow even faster information retrieval and search functionalities within the graph.

The usage of highly distributed backends together with indexing layers allows JanusGraph to scale to enormous graphs, with hundreds of billions of nodes and edges. This allows it to handle the so-called supernodes very efficiently. These are nodes that have an extremely large degree, which often arises in real-world applications (remember that a very famous model for real networks is the *Barabasi-Albert* model, based on preferential attachments, which makes hubs naturally emerge within the graph).

In large graphs, supernodes are often potential bottlenecks of the application, especially when the business logic requires traversing the graph passing through them. Having properties (such as timestamps or similarity metrics computed and updated regularly) that can help with rapidly filtering only the relevant edges during a graph traversal can speed up the process significantly and achieve better performance.

Similarly to what has been done for Neo4j, for JanusGraph we can easily start an instance of the database using Docker with the following command:

```
docker run --rm --detach --name janusgraph \
            --publish=8182:8182 \
            Janusgraph/janusgraph:1.1.0
```

Once the JanusGraph server is running you can start to interact and play with it. JanusGraph exposes a standard API to query and traverse the graph via the Apache TinkerPop library (https://tinkerpop.apache.org/), which is an open source, vendor-agnostic graph computing framework. Similarly to GQL described earlier, TinkerPop also provides a standard interface for querying and analyzing the underlying graph using the **Gremlin** graph traversal language. Using a standard querying engine can allow you to create application layers that can seamlessly integrate with the various compatible graph database systems. They allow you to build standard serving layers that do not depend on the backend technology, giving you the freedom to choose/change the appropriate graph technology for your application depending on your actual needs or different environments (e.g. customers having different technological stacks), also avoiding a vendor lock-in to some extent.

Besides Java connectors, Gremlin also has direct Python bindings thanks to the gremlinpython library, which allows Python applications to connect to and traverse graphs.

In the notebook provided in the repository, we show you how to use and interact with JanusGraph using Gremlin.

First of all, we need to connect to the database using the following code snippet:

```
from gremlin_python.driver.driver_remote_connection import
DriverRemoteConnection
from gremlin_python.driver.serializer import GraphSONSerializersV3d0

connection = DriverRemoteConnection(
    'ws://localhost:8182/gremlin', 'g',
    message_serializer=GraphSONSerializersV3d0())
)
```

Once the connection is created, we can then instantiate the GraphTraversalSource object, which is the basis for all Gremlin traversals, and bind it to the connection we just created:

```
from gremlin_python.structure.graph import Graph
from gremlin_python.process.graph_traversal import __
graph = Graph()
g = graph.traversal().withRemote(connection)
```

Once GraphTraversalSource is instantiated, we can reuse it across the application to query the graph database.

Using Gremlin, we can build a traversal that creates a node with a given label (here, 'student') and some properties (e.g. the name and their GPA):

```
g.addV('student')\
    .property('name', 'Jeffery')\
    .property('GPA', 4.0).next()
```

Note that at the end, we use next() since all Gremlin queries have lazy computation, which are first built and then executed. Multiple queries can also be chained together and different elements of the chain can be labeled:

```
g\
    .addV('student')\
        .property('name', 'Claire')\
        .property('GPA', 3.9).as_("n1")\
    .addV('student')\
        .property('name', 'Lisa')\
```

```
      .property('GPA', 3.6).as_("n2")\
   .addE("FRIEND_OF")\
     .from_("n1").to("n2")\
     .property("since", "2014")
   .iterate()
```

This query will both create the two nodes and the edge between them. Using these concepts, in the notebook attached we provide utility functions to create a graph starting from a list of nodes and edges. We then use these utilities to import in JanusGraph both the Karate Club network (discussed in *Chapter 1, Getting Started with Graphs*) and the Movie dataset used in the previous section.

Once the Movie dataset is imported, we can then re-write the Cypher query we used previously to find all the co-actors of Keanu Reeves using Gremlin:

```
co_actors = g.V()\
   .has('Person', 'name', 'KeanuReeves')\
   .out("ACTED_IN")\
   .in("ACTED_IN")\
   .values("name").dedup().to_list()
```

As can be seen in the preceding code, Gremlin is a functional language whereby operators are grouped together to form path-like expressions.

Now that we have stored the information in a graph database that can be retrieved using querying engines, it is time to process the data. The next subsection will present the options for processing and analyzing graphs by implementing a graph processing engine.

Graph processing engines

To select the right technology for a **graph processing engine**, it is crucial to estimate the size in memory of the network compared to the capacity of the target architecture. You can start by using simple frameworks that allow fast prototyping during the first phases of a project when the goal is to quickly build a **minimum viable product (MVP)**.

Such frameworks can then be substituted for more advanced tools later on when performance and scalability become more crucial. A microservice modular approach and proper structuring of these components will allow the switching of technologies/libraries independently from the rest of the application to target specific issues, which will also guide the choice of the backend stack.

Graph processing engines require information for whole graphs to be accessed quickly, such as having all of the graph in memory, and, depending on the context, you may need *distributed architectures*. As we saw in *Chapter 1, Getting Started with Graphs*, networkx is a great example of a library for building a graph processing engine when dealing with relatively small datasets. When datasets get larger but can still fit in single servers or shared memory machines, other libraries may help to reduce computational time. As seen in *Chapter 1, Getting Started with Graphs*, using libraries other than networkx where graph algorithms are implemented in more performant languages, such as C++ or Julia, may speed up the computation by more than two orders of magnitude.

However, there are cases where datasets grow so much that it is no longer technologically or economically viable to use shared memory machines of increasing capacity (fat nodes). In such cases, it is rather necessary to distribute the data on clusters of tens or hundreds of computing nodes, allowing horizontal scaling. The most popular frameworks that can support a graph processing engine in these cases are the following:

- **Apache Spark GraphX**, which is the module of the Spark library that deals with graph structures (https://spark.apache.org/graphx). It involves a distributed representation of the graph using **resilient distributed datasets (RDDs)** for both vertices and edges, or encoding the information using the Spark DataFrame API, resulting in the so-called GraphFrames that provide a more type-safe and structured interface. The graph repartition throughout the computing nodes can be done either with an *edge-cut* strategy, which logically corresponds to dividing the nodes among multiple machines, or a *vertex-cut* strategy, which logically corresponds to assigning edges to different machines and allowing vertices to span multiple machines. Although written in Scala, GraphX has wrappers that can be used with both R and Python. GraphX already comes with some algorithms implemented, such as *PageRank, connected components*, and *triangle counting*. There are also other libraries that can be used on top of GraphX for other algorithms, such as **SparklingGraph**, which implements more centrality measures.

- **Neo4j Graph Data Science** (**GDS**), a high-performance graph analytics engine that provides a broad set of **graph algorithms**, **ML pipelines**, and **enterprise support**. GDS is optimized for Neo4j's native graph storage, allowing scalable analysis of large networks in use cases such as **fraud detection, recommendations**, and **supply chain optimization**. In the attached notebooks, we show how the Neo4j Graph Data Science engine can be leveraged by using the graphdatascience Python library and perform an analysis (e.g., centrality computation) of the Cora dataset.

- **Amazon Neptune Analytics** is a powerful graph processing engine optimized for large-scale graph analysis for fast insights. It integrates well within the **AWS ecosystem**, making it easy for users already on AWS to get started without needing additional approvals. Neptune supports optimized graph algorithms and low-latency graph queries. The ease of integration makes it a great choice for enterprises using AWS services. In summary, Neptune Analytics offers a seamless and scalable graph analytics solution within the AWS cloud environment.

> **In the first version of this book, we mentioned Apache Giraph**, which is an iterative graph processing system built for high scalability (https://giraph.apache.org/). It was developed by Facebook to analyze the social graph formed by users and their connections and is built on top of the Hadoop ecosystem for unleashing the potential of structured datasets at a massive scale. Giraph is natively written in Java and, similarly to GraphX, also provides a scalable implementation for some basic graph algorithms, such as *PageRank* and *shortest path*. However, it is important to note that Apache Giraph has been retired due to inactivity as of 2023. Therefore, it is no longer recommended for current production applications.

When we consider scale-out to a distributed ecosystem, we should always keep in mind that the available choice for algorithms is significantly smaller than in a shared machine context. This is generally due to two reasons:

- First, implementing algorithms in a distributed way is a lot more complex than in a shared machine due to communication among nodes, which also reduces the overall efficiency
- Secondly, and more importantly, one fundamental mantra of big data analytics is that only algorithms that (nearly) scale linearly with the number of data points should be implemented in order to ensure the horizontal scalability of the solution, by increasing the computational nodes as the dataset increases

In this respect, GraphX and Neo4j GDS also allow you to define scalable, vertex-centric, iterative algorithms using standard interfaces based on `Pregel`, which can be seen as a sort of equivalent of iterative map-reduce operations for graphs (actually, iterative map-reduce operations applied to triplet node-edge-node instances). A Pregel computation is composed of a sequence of iterations, each called a `superstep`, each involving a node and its neighbors.

During the superstep, *S*, a user-defined function is applied for each vertex, *V*. This function takes the messages sent to *V* in superstep *S – 1* as input and modifies the state of *V* and its outgoing edges. This function represents the mapping stage, which can be easily parallelized. Besides computing the new states of *V*, the function also sends messages to other vertices connected to *V*, which will receive this information at superstep *S + 1*. Messages are typically sent along outgoing edges, but a message may be sent to any vertex whose identifier is known. In *Figure 10.4*, we show a sketch of what a Pregel algorithm would look like when computing the maximum value over a network:

Superstep S

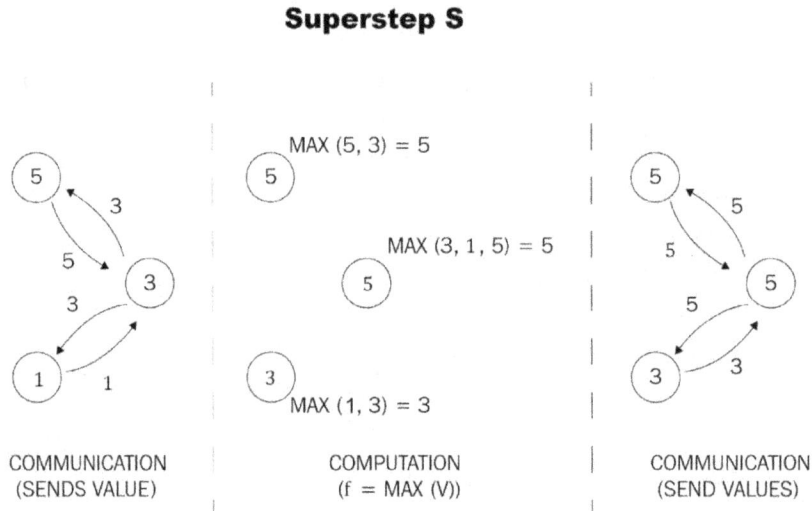

Figure 10.4: Example of calculating a maximum value over a node property using Pregel

For further details on this algorithm, please refer to the original paper, *Pregel: A System for Large-Scale Graph Processing*, written by Malewicz et al. in 2010.

By using Pregel, you can easily implement other algorithms, such as *PageRank* or *connected components*, in a very efficient and general way, or even implement node embeddings' parallel variants (for an example, see *Distributed-Memory Vertex-Centric Network Embedding for Large-Scale Graphs*, Riazi and Norris, 2020).

Selecting the right technology

Neo4j or GraphX? This is a question that often gets asked. However, as we have described briefly, the two pieces of software are not really competitors, but they rather target different needs. Neo4j, Neptune, JanusGraph, and CosmoDB allow us to store information in a graph-like structure and query the data, whereas GraphX, Neo4j Graph Data Science, and Amazon Neptune Analytics make it possible to analytically process a graph (especially for large graph dimensions). Although you could also use Neo4j as a processing engine and GraphX could also be used as an in-memory stored graph, this approach should be discouraged due to performance limitations, scalability concerns, and mismatched feature sets.

Graph processing engines usually compute KPIs that get stored in the graph database layers (potentially indexed such that querying and sorting become efficient) for later use. Thus, technologies such as GraphX are not competing with graph databases such as Neo4j, and they can very well co-exist within the same application to serve different purposes. As we stressed in the introduction, even in MVPs and at early stages, it is best to separate the two components, the *graph processing engine* and the *graph querying engine*, and use appropriate technologies for each of them.

Simple and easy-to-use libraries and tools do exist in both cases and we strongly encourage you to use them wisely in order to build a solid and reliable application that can be scaled out seamlessly.

Summary

In this section, we have provided you with the basic concepts of how to design, implement, and deploy data-driven applications that resort to graph modeling and leverage graph structures. We have highlighted the importance of a modular approach, which is usually the key to seamlessly scaling any data-driven use case from early-stage MVPs to production systems that can handle a large amount of data and large computational performances.

We have outlined the main architectural pattern, which should provide you with a guide when designing the backbone structure of your data-driven applications. We then continued by describing the main components that are the basis of graph-powered applications: *graph processing engines*, *graph databases*, and *graph querying languages*. For each component, we have provided an overview of the most common tools and libraries, with practical examples that will help you to build and implement your solutions. You should thus have by now a good overview of what the main technologies out there are and what they should be used for.

In the next chapter, we will explore recent advancements and emerging research trends in temporal graph machine learning. Specifically, we will discuss cutting-edge techniques (such as temporal graph neural networks) and applications. We will also highlight practical examples and potential use cases, drawing insights from the latest scientific literature.

Get This Book's PDF Version and Exclusive Extras

UNLOCK NOW

Scan the QR code (or go to packtpub.com/unlock).
Search for this book by name, confirm the edition,
and then follow the steps on the page.

*Note: Keep your invoice handy. Purchases made
directly from Packt don't require one.*

Part 4

Advanced topics in Graph Machine Learning

In this part, will learn about new trends in graph machine learning, beginning with the new trends and moving to dynamic temporal graph modeling. It ends with an analysis of the connection between graph methods and **large language models** (**LLMs**), highlighting cutting-edge research and future opportunities in structured data and deep learning.

This part comprises the following chapters:

- *Chapter 11, Temporal Graph Machine Learning*
- *Chapter 12, GraphML and LLMs*
- *Chapter 13, Novel Trends on Graphs*

11

Temporal Graph Machine Learning

In the ever-evolving landscape of data science and machine learning, the study of temporal graphs has emerged as a crucial field with widespread applications. **Temporal graphs** provide a dynamic representation of relationships and interactions between entities over time, offering a more realistic and nuanced perspective than traditional static graphs.

This chapter explores the fundamental concepts of temporal graphs, delving into their definitions, properties, and common applications in various domains. We will explore the definition of *dynamic* graphs and why they are needed. We will see common problems that can be modeled in the framework of dynamic graphs and we will explore several machine learning algorithms that have been developed for solving such problems, including **temporal graph neural networks**.

The following topics will be covered in this chapter:

- The definition of dynamic graphs
- Common problems that can be modeled with temporal graphs
- Embedding dynamic graphs
- A general taxonomy to navigate among temporal graph machine learning algorithms
- Hands-on temporal graphs

Technical requirements

All code files relevant to this chapter are available at https://github.com/PacktPublishing/ Graph-Machine-Learning/tree/main/Chapter11. Please refer to the *Practical exercises* section of *Chapter 1, Getting Started with Graphs,* for guidance on how to set up the environment to run the examples in this chapter, using either Poetry, pip, or Docker.

For more complex data visualization tasks provided in this chapter, Gephi (`https://gephi.org/`) may also be required. The installation manual is available here: `https://gephi.org/users/install/`.

What are dynamic graphs?

In the realm of graph theory, the conventional representation of relationships through *static* graphs has long been the cornerstone of various analytical approaches. As described in *Chapter 1, Getting Started with Graphs,* a static graph is denoted as G (V, E), where V is its set of vertices and E is its set of edges. However, the limitations inherent in static graphs have become increasingly evident, prompting the necessity to delve into the temporal dimension of dynamic graphs. Static graphs, while indeed efficient for capturing instantaneous relationships, fall short in encapsulating the *evolving nature of connections* over time, which is crucial in many real-world phenomena. To address this limitation, dynamic graphs extend the concept of static graphs to incorporate the temporal dimension. This concept can be used to solve several problems, as we will see in the next section.

Common problems with temporal graphs

The concept of temporal graphs is useful in all the real-world problems that can be represented as a graph, where the nodes and edges of the graph may change over time. For example, temporal graphs are extensively applied in modeling social networks. By capturing the evolving relationships between individuals, temporal graphs enable a more accurate representation of social dynamics. This is particularly useful for predicting changes in friendships, community structures, and the information diffusion over time.

Also, in transportation systems, such as road or airline networks, the dynamics of connection are time-dependent. Temporal graphs may help model the changing patterns of traffic, optimal routing, and impact of events such as rush hours or seasonal variations on the network's structure.

Temporal graphs are also useful for representing and analyzing communication networks, where the time of interactions is crucial. By considering the temporal order of messages or calls, analysts can gain insights into communication patterns, identify anomalies, and enhance the efficiency of network protocols.

Those are only a few examples of how temporal graphs can provide a powerful framework for modeling and understanding relationships over time. Another example is biology, where temporal graphs are employed to study dynamic processes, such as protein-protein interactions, gene regulatory networks, and ecological systems. In finance and stock markets, temporal graphs constitute a valuable tool for modeling transactions and movements to help explain market trends, detect anomalies, and improve predicting processes.

In epidemiology, temporal graphs can help to explain the dynamics between individuals, regions, and populations, aiding in predicting the course of the epidemic and implementing effective healthcare strategies.

As we have seen in *Chapter 6, Solving Common Graph-Based Machine Learning Problems*, all these problems can be reduced (and thus solved) to a specific task, such as node classification (or regression), graph classification (or regression), and link prediction. Similar consideration should also be made for problems involving a temporal component since tasks on static graphs have a natural translation into the temporal domain:

- **Node and graph classification over time**: Let V_t be a set of vertices of a graph G_t. Node classification at time t is the problem of classifying a vertex in V_t into one (or more) predefined classes, usually based on past or future information. If we aim to classify the whole graph at time t, the problem is called *graph classification over time*. A similar definition can be followed for the tasks of node regression and graph regression that we saw in *Chapter 6, Solving Common Graph-Based Machine Learning Problems*.

- **Link prediction over time**: This is the task of predicting when a new connection between two nodes will be created at a particular time point t, possibly exploiting past or future information on the graph.

It is worth mentioning that an alternative way of seeing these problems is under the setting of *time prediction*. In this scenario, we want to predict *when* an event happened or when it will happen. In other words, we want to predict at which time step t a new connection will be established, when a new node will appear in the graph, when a node will change its status, and so on.

But how can we formalize the concept of temporal graphs? Let's see in the next section.

Representing dynamic graphs

A critical aspect when modeling dynamic graphs is to define the granularity of the temporal dimension. In particular, two different methods may be adopted:

- **Discrete-time approaches**: Time is considered to be discrete, thus the evolution of the dynamic graph is described as a sequence of static graphs (snapshots) at a fixed timestamp

- **Continuous time**: Here, time is considered to be continuous, and specific *events* are recorded in real time

Based on the above considerations, the transition from static to dynamic involves embracing a spectrum of definitions. Beginning with static graphs of the form $G(V, E, X)$, where V is a set of nodes, E is a set of edges, and X is a set of features describing nodes, we progress to spatio-temporal graphs.

Spatio-temporal graphs are of the form $G(V, E, X_t)$, indicating that, while the topology remains the same, the features may change over time.

By further extending this concept, we encounter discrete-time dynamic graphs:

- **Discrete-time dynamic graphs** are of the form $G(V_t, E_t, X_t)$. Here, not only do the features change over time, but so does the topology of the graphs (connections appear and disappear, as well as nodes in the graph). In this scenario, time is considered to be discrete, meaning that we have *snapshots* of the changing graph over time at discrete time points.

  Finally, the concept of discrete-time dynamic graphs can be further generalized to continuous-time dynamic graphs, where each *event* (change in the graph) is observed and recorded individually, together with its timestamp.

- **Continuous-time dynamic graphs** are represented as a pair, (G, O), where $G(V_{t0}, E_{t0}, X_{t0})$ is the graph's initial state at time 0 and O is the set of events recorded for the graph through time. An event is a triple (event type, event, timestamp) where the event type can be any type of topology or feature update, including node addition/deletion, edge addition/deletion, feature update, and edge weight update. An example of an event is the tuple (node deletion, v4, 19-11-2023), meaning that on 19-11-2023 the node v4 was deleted.

In this section, we introduced temporal graphs and common ways of representing temporal graphs at different time granularities. In the next section, we will see how temporal graphs can be encoded to extract relevant features for downstream tasks!

Embedding dynamic graphs

As we saw in *Chapter 2, Graph Machine Learning*, most of the state-of-the-art machine learning algorithms on graphs can be modeled into an encoder-decoder framework. The same applies to dynamic graphs. More specifically:

- The **encoder** takes as input a dynamic graph and returns as output its embedded representation
- The **decoder** takes as input an embedded representation of the dynamic graph and, depending on the task, outputs a prediction (it can be a new line, a class, or even a reconstructed graph)

In fields where dynamic graphs can be used to describe various phenomena, accurately modeling the graph's evolution is often essential for precise predictions. Over time, various categories of machine learning models have been created to capture both the structure and evolution of dynamic graphs.

Notably, adaptations of **graph neural networks (GNNs)** tailored for dynamic graphs have recently demonstrated success in various domains, emerging as indispensable tools in the machine learning toolbox.

In this section, some of the most common methods for embedding dynamic graphs will be presented. However, for a thorough survey of representation learning methods for dynamic graphs, check out the scientific papers *Representation learning for dynamic graphs: A survey* by Kazemi et al. (2020), *A survey on embedding dynamic graphs* by Barros et al. (2021), and *Dynamic Graph Representation Learning With Neural Networks: A Survey* by Yang et al. (2024). For a more specialized exploration of GNN-based approaches to dynamic graphs, refer to the paper *Foundations and modelling of dynamic networks using dynamic graph neural networks: A survey* by Skarding et al. (2021).

Let's first introduce a mathematical formulation to define the problem of dynamic graph embedding. The task involves mapping a dynamic graph $G = (V_t, E_t, X_t)$ with evolving nodes and edges into a d-dimensional vector space over time. This mapping captures both the network's topological structure and temporal dependencies. The goal is to learn representations that can reconstruct the dynamic graph, predict its behavior beyond given timestamps, or address specific tasks like node classification. When the graph topology changes, there are two possible interpretations: either the vector representations move within the embedding space, allowing the tracking of node trajectories, or the embedding space evolves over time, enabling the learning of mappings between consecutive timestamps. Notice that the temporal granularity of the dynamic graph and the temporal embedding do not need to be the same. In fact, it is entirely reasonable for dynamics graphs to capture low-level interactions (e.g., daily events) mapped into a coarser granularity (e.g., months or years).

Following the paper *A survey on embedding dynamic graphs*, we propose a taxonomy of dynamic graph embedding methods (*Figure 11.1*). The taxonomy categorizes dynamic graph methods into the following:

- Factorization-based methods
- Random walk-based methods
- Graph kernel methods
- Temporal point process methods
- Deep learning-based methods
- Agnostic models

These are depicted here:

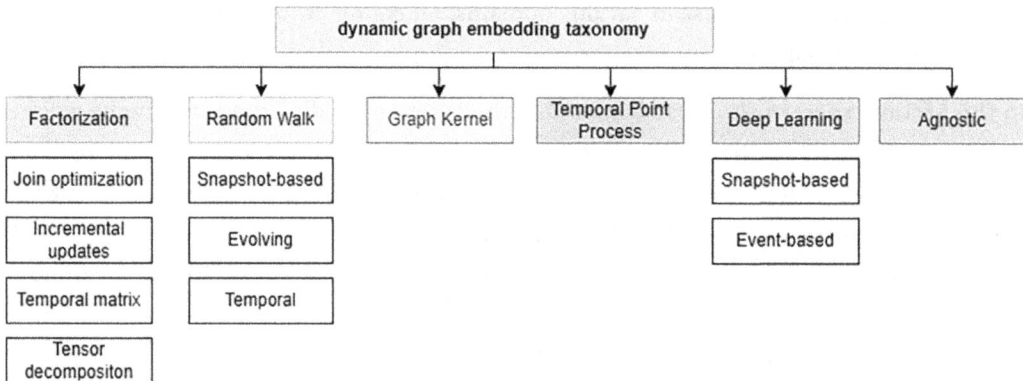

Figure 11.1: Dynamic graph embedding taxonomy

Let's discuss each of these methods in the following sections.

Factorization-based methods

Factorization-based techniques produce embeddings over time by decomposing low-rank representations of time-dependent similarity measures. As we saw in *Chapter 4, Unsupervised Graph Learning*, graphs can be represented either as a sequence of matrices or as three-way tensors linking nodes' similarity. This matrix (or tensor) can be factorized to emphasize specific properties of the input graph.

Dynamic graph embedding with matrix factorization follows static graph principles but introduces temporal dependence into the decomposition process. There are three main methods for inserting temporal dependence in the matrix factorization process:

- **Jointly optimizing reconstruction and temporal smoothing**: Here, the loss function consists of two terms. The first term measures the quality of the reconstruction (as in the static graph scenario) and the second term measures the similarity between the embeddings for each pair of consecutive time stamps, thus ensuring a smooth evolution of the embedding envelope.

- **Incremental updates on embeddings**: With this approach, the embeddings are computed by iteratively adjusting an initial representation based on the changes observed in the similarity matrices over time. For example, such an adjustment can be done using first-order matrix perturbation theory in symmetric matrices iteratively.

- **Temporal matrix factorization**: Here, each snapshot is usually decomposed into two components: a constant matrix U representing persistent properties between a pair of nodes and a time-dependent matrix V_t representing changes in topology over time. This temporally parameterized factorization (obtained, for example, by minimizing the sum of the squares between the adjacency matrix at time t and its reconstruction UV_t^T) can be used to reconstruct the structure of the network at any time t (past or future).

As per tensor-based approaches, it is natural to represent dynamic graphs as a three-way tensor (that is, a stack of adjacency matrices). Then, tensor factorization methods such as **CANDECOMP/PARAFAC (CP)** and Tucker decomposition can be used to learn both node embeddings and temporal embeddings.

Random walk-based methods

An alternative set of techniques for graph embedding revolves around random walks. As shown in *Chapter 4, Unsupervised Graph Learning*, multiple fixed-length random walks are considered as sentences, creating a context for each node and capturing higher-order dependencies without relying on adjacency matrices. The resulting node sequence matrix is then factorized, often employing a neural network architecture such as skip-gram. There are three main types of methods for embedding temporal graphs:

- **Random walks on snapshots**: This involves generating time-dependent node sequences by exploring the evolving graph structure at different timestamps.

- **Evolving random walks**: This method adapts to topological changes by incrementally updating representations. It ensures that the embeddings capture the evolving nature of the network as it changes over time.

- **Temporal random walks**: This defines time-dependent contexts by exploring random walks across consecutive timestamps. It captures evolving patterns and temporal dependencies in the graph, providing a comprehensive understanding of the network's dynamics.

Graph kernel-based methods

Another set of methods focuses on elementary substructures derived from an entire graph structure. These techniques integrate topological attributes, such as *graphlet transition count*, *graphlet frequencies over time*, and *adjacency matrix summation*, during network processing. The goal is to employ a shallow autoencoder to learn representations capable of reconstructing intricate attributes from these substructures. This enables a more nuanced understanding of the evolving nature of these topological building blocks in the network.

Temporal point process methods

Various techniques in dynamic graph embedding treat interactions between nodes as stochastic processes, where probabilities are influenced by the network's topological structure, node features, and historical context. In these approaches, events are assumed to impact a specific node, leading to potential interactions with other nodes susceptible to the influence of the current node.

Deep learning-based methods

Temporal graph neural networks (TGNNs) have seen significant progress in capturing spatial and temporal dependencies in graph-structured data. The first level of classification distinguishes between *snapshot-based and event-based* models:

- **Snapshot-based models:** These involve the use of two components: a suitable method to process the whole graph at each time point and a mechanism to learn the temporal dependencies. A further distinction is also made:

    - **Model evolution methods:** This kind of model, such as *EvolveGCN* (Pareya et al. 2020) uses a **recurrent neural network (RNN)** to adapt the parameters of a **graph convolutional network** over time

    - **Embedding evolution methods:** Instead of evolving the parameters of a static GNN, these methods use an RNN to evolve directly the embedding at the previous time point

- **Event-based models:** These models process event streams, updating node representations each time an event involving that node occurs. These models can be seen as an extension of the message-passing paradigm and can be further classified as follows:

    - **Temporal embedding methods:** These models (often based on *self-attention* mechanisms) process event streams, incorporating time into the model sequential information, node features, and graph topology interactions

    - **Temporal neighborhood methods:** These models use specialized modules to store and aggregate functions of events involving specific nodes at given times, updating node representations as time increases

Agnostic methods

All the approaches discussed so far are based on specific algorithms. For example, factorization-based approaches are based on factorization techniques, random walk-based methods are based on random walk algorithms, and so on.

There is a set of approaches that do not rely on any specific paradigm, namely *agnostic models*. These models focus on learning connections between representations at consecutive time points or within a time window. Two paradigms within this classification are as follows:

- **Retrofitted models:** These models leverage static network embeddings to learn an initial representation of the graph, and then they adapt and refine these representations (retrofitting) to capture the dynamic evolution of the graph in subsequent snapshots

- **Transformation methods:** These methods calculate representations for each graph snapshot independently using any static method and learn a transformation function connecting embeddings at different timestamps

You have now learned about several encoding methods for temporal graphs. It is now time to take a look at temporal graph machine learning! We will do this in the next section.

Hands-on temporal graphs

In this section, we will introduce representative examples of the machine learning approaches described in the previous sections for dealing with temporal graphs. We will offer a general understanding of how these approaches work and provide examples of their implementation using publicly available frameworks.

Temporal matrix factorization

Concerning the matrix factorization class of approaches, the **Temporal Matrix Factorization (TMF)** model by Yu et al. (2017) is a method used for temporal link prediction, particularly in dynamic network scenarios. This technique leverages matrix factorization with temporal dynamics to model the evolution of links in a dynamic network over time.

To exemplify this method, we adopted the implementation provided in the publicly available **OpenTLP** library (`https://github.com/KuroginQin/OpenTLP`). It integrates an encoder-decoder architecture, where the encoder learns model parameters through matrix factorization, and the decoder generates predictions based on these parameters. The optimization process involves minimizing a loss function that includes regularization and reconstruction error terms, capturing temporal dependencies in the network structure.

The core of this TMF implementation lies in the `TMF` class, which orchestrates the encoding-decoding process, as shown in the following code:

```
class TMF(Module):
    def __init__(self, num_nodes, hid_dim, win_size, num_epoch, alpha,
```

```
beta, theta, learn_rate, device):
        # ...
        self.enc = TMF_Enc(num_nodes, hid_dim, win_size, num_epoch, alpha,
beta, theta, learn_rate, device)
        self.dec = TMF_Dec()

    def TMF_fun(self, adj_list):
        self.enc.model_opt(adj_list)
        param_list, _ = self.enc()
        adj_estimated = self.dec(param_list, self.win_size+1)
        return adj_estimated
```

Let's break down the encoding and decoding steps.

The TMF_Enc class embodies the encoder, which is responsible for learning the model's parameters:

```
class TMF_Enc(Module):
    def __init__(self, num_nodes, hid_dim, win_size, num_epoch, alpha,
beta, theta, learn_rate, device):
        # ...
        self.dec_list = []  # List of decaying factor
        # ...

    def forward(self):
        adj_est_list = []
        for t in range(self.win_size):
            V = self.param[0] + self.param[1]*(t+1) + self.
param[2]*(t+1)*(t+1)
            U = self.param[3]
            adj_est = torch.mm(U, V.t())
            adj_est_list.append(adj_est)
        return self.param, adj_est_list

    def get_loss(self, adj_list, adj_est_list, dec_list, alpha, beta):
        def get_loss(self, adj_list, adj_est_list, dec_list, alpha, beta):
        ''''''
        Function to get training loss
        :param adj_list: sequence of historical adjacency matrix (ground-
truth)
```

```
        :param adj_est_list: sequence of estimated adjacency matrix
        :param dec_list: list of decay factors
        :param alpha, beta: hyper-parameters
        :return: loss function
    ''''''

        win_size = len(adj_list) # Window size (#historical snapshots)
        loss = 0.5*alpha*torch.norm(self.param[3],'p=''ro')**2
        loss += 0.5*beta*torch.norm(self.param[0],'p=''ro')**2
        loss += 0.5*beta*torch.norm(self.param[1],'p=''ro')**2
        loss += 0.5*beta*torch.norm(self.param[2],'p=''ro')**2
        for t in range(win_size):
            dec_t = dec_list[t] # Current decaying factor
            adj = adj_list[t]
            adj_est = adj_est_list[t]
            loss += 0.5*dec_t*torch.norm(a-j - adj_est,'p=''ro')**2
        return loss


    def model_opt(self, adj_list):
        ''''''

        Function to implement the model optimization
        :param adj_list: sequence of historical adjacency matrices
(ground-truth)
        :return:
    ''''''

        for epoch in range(self.num_epoch):
            _, adj_est_list = self.forward()
            loss = self.get_loss(adj_list, adj_est_list, self.dec_list,
self.alpha, self.beta)
            self.opt.zero_grad()
            loss.backward()
            self.opt.step()
```

The forward function reconstructs the adjacency matrices based on the learned parameters. The factorization is expressed as:

$$Adj_{est,t} = U \times (W_0 + W_1 \times (t + 1) + W_2 \times (t + 1)^2)^T$$

The get_loss function calculates the training loss, which includes terms for regularization and reconstruction errors:

$$Loss = \frac{1}{2}\alpha \|U\|_F^2 + \frac{1}{2}\beta (\|W_0\|_F^2 + \|W_1\|_F^2 + \|W_2\|_F^2) \sum_{t=1}^{win} \frac{1}{2} dec_t \|Adj_t - Adj_{est,t}\|_F^2$$

The model_opt function implements model optimization using the Adam optimizer.

Finally, the TMF_Dec class serves as the decoder, which generates predictions based on the learned parameters:

```
class TMF_Dec(Module):
    ''''''

    Class to define the decoder of TMF
    ''''''

    def __init__(self):
        super(TMF_Dec, self).__init__()

    def forward(self, param_list, pre_t):
        ''''''

        Rewrite forward function
        :param param_list: list of learned model parameters
        :param pre_t: time step of prediction result (e.g., win_size+1)
        :return: prediction result
        ''''''

        V = param_list[0] + param_list[1]*pre_t + param_
list[2]*pre_t*pre_t
        U = param_list[3]
        adj_est = torch.mm(U, V.t())

        return adj_est
```

The following code exemplifies the usage of the TMF method for temporal link prediction. An instance of the TMF model is created with specified parameters, and the TMF_fun function is applied to obtain predicted adjacency matrices based on a given sequence of historical adjacency matrices (adj_list):

```
TMF_model = TMF(num_nodes, hid_dim, win_size, num_epochs, alpha, beta,
theta, learn_rate, device)
adj_est = TMF_model.TMF_fun(adj_list)
```

Temporal random walk

As we saw in the previous section, random-walk-based approaches represent an important class of methods for the temporal domain. Among the various paradigms, we describe here a method based on **temporal random walk** called CTDNE, by Nguyen et al. (2018), for learning time-preserving embedding. Let's introduce the general idea.

Imagine selecting an initial edge $e_i = (u, v, t)$ at a certain time step t as the starting point of our temporal random walk. As in the static version described in *Chapter 4, Unsupervised Graph Learning*, at each step, we have to choose another edge to continue the walking. However, this time, the structure of the graph can change at each time step. Therefore, we need to define the set of temporal neighbors of a node v at a particular time t. One way is as follows:

$$N_t(v) = \{(w, t')|e = (v, w, t') \in E_T \wedge t' > t\}$$

That is, the set of all nodes that are connected to the node v after the time step t. Intuitively, starting from the node v at time t, we can *walk over time* by traversing all the edges that are present in a particular moment after t. Note that it is possible for the same neighbor w to appear in N_t multiple times since they can be connected multiple times over time (for example, v sends an email to w at time t, then w replies to v at time $t+1$, and so on).

It is worth noting that, according to the way we sample the next edge to traverse over time, the resulting random walk can be *biased*, allowing us to define effective sampling strategies for particular tasks. For example, in the context of temporal link prediction, if we aim to predict edges at time t, it is better to compute temporal walks from edges closer to time t rather than temporal walks sampled in the distant past, since the latter may offer lower predictive value.

The CTDNE method is available in `stellargraph`. Let's see an example of how to use it.

Let's first define the random walk parameters, including `walk_length` (the maximum length of each random walk), `context_window_size` (the size of the context window employed for training the Word2Vec model), and `num_cw` (the number of context windows we want to obtain):

```
walk_length = 80
context_window_size = 10
num_cw = 20
```

Let's then define the walks:

```
from stellargraph.data import TemporalRandomWalk

temporal_rw = TemporalRandomWalk(graph)
temporal_walks = temporal_rw.run(
    num_cw=num_cw,
    cw_size=context_window_size,
    max_walk_length=walk_length,
    walk_bias="exponential",
)
```

`graph` is the input graph. Notice the `walk_bias` parameter, which allows us to set the sampling strategy for the temporal walk.

Finally, we can train the Word2Vec model similarly to what we did in the static case but now using the obtained temporal walks generating the embeddings, which can then be used for downstream machine learning tasks (such as link prediction or node classification):

```
from gensim.models import Word2Vec

embedding_size = 128
temporal_model = Word2Vec(temporal_walks, size=embedding_size,
window=context_window_size, min_count=0, sg=1, workers=2, iter=1)
```

TGNNs

As GNNs have gained popularity, numerous architectures have emerged in recent years to address the temporal domain. From classical GNNs combined with recurrent modules (such as LSTM and GRU) to more specialized models, such as **DynGEM**, **EvolveGCN**, or generative models, these architectures have shown promising results in handling temporal dependencies.

In this section, we will introduce the **temporal graph network** (**TGN**) framework by Rossi et al. (2020), a generic and efficient approach for deep learning on dynamic graphs.

TGN is designed to handle continuous-time dynamic graphs represented as a sequence of time-stamped events. It produces as output a node-embedded representation. The core of the framework is composed of five different modules:

- **Memory**: The memory module of the TGN framework manages the model's state at a given time *t*. It consists of a state vector *si(t)* for each node *i* seen so far (that is, every time a new node is involved in an event a new vector is added to the memory).

- **Message function**: The message function module calculates the messages that are used to update the memory of a node *i* for each event it participates in. They are typically implemented using simple neural networks (for example, MLPs), taking as input the event details and the memory states of the nodes involved in the event.

- **Message aggregator**: Since multiple messages from the same node can occur in the same batch, an aggregator function can be used to aggregate messages (for example, by averaging all the messages for a given node).

- **Memory updater**: This is a learnable function that is typically implemented using recurrent units such as LSTM or GRU. It takes as input the current state *si(t)* of a node *i* involved in a certain event and the output computed by the message function based on the event and updates the memory of the node *i*.

- **Embedding**: This is a learnable function that computes the *temporal embedding* of a node *i* using the graph and the memory. The embeddings are then used for downstream tasks such as link prediction.

TGN and its modules are implemented in `pytorch_geometric` and can be easily used to solve simple and complicated tasks. Let's see how TGN can be used to embed temporal graphs. Below, we will be showing a general framework for computing embeddings using TGN. Notice that, to perform any downstream task such as link prediction and node classification, you should concatenate a proper decoder (a neural network that takes the embeddings and computes the predictions). You can find a complete example in the notebooks attached to this book.

The first step is to initialize the `TGNMemory` class, which will keep the model's state and handle the message computation and memory update. As shown in the following snippet of code, the `TGNMemory` class can be customized by using proper message modules (in our example we chose the `IdentityMessage` module) and a message aggregator (in our case the `LastAggregator` module, which keeps the most recent event for a node):

```
from torch_geometric.nn import TGNMemory
from torch_geometric.nn.models.tgn import (
    IdentityMessage,
    LastAggregator,
    LastNeighborLoader)

memory_dim = time_dim = embedding_dim = 100
```

```
memory = TGNMemory(
    data.num_nodes,
    data.msg.size(-1),
    memory_dim,
    time_dim,
    message_module=IdentityMessage(data.msg.size(-1), memory_dim, time_dim),
    aggregator_module=LastAggregator(),
).to(device)


# a data loader that performs neighbor sampling
neighbor_loader = LastNeighborLoader(data.num_nodes, size=10, device=device)
```

Together with the TGNMemory, we will also create a GNN for obtaining the embeddings. In this example, we will define a GraphAttentionEmbedding class, which uses the TransformerConv module (a message-passing module implemented in PyTorch):

```
from torch_geometric.nn import TransformerConv

class GraphAttentionEmbedding(torch.nn.Module):
    def __init__(self, in_channels, out_channels, msg_dim, time_enc):
        super().__init__()
        self.time_enc = time_enc
        edge_dim = msg_dim + time_enc.out_channels
        self.conv = TransformerConv(in_channels, out_channels // 2,
heads=2, dropout=0.1, edge_dim=edge_dim)

    def forward(self, x, last_update, edge_index, t, msg):
        rel_t = last_update[edge_index[0]] - t
        rel_t_enc = self.time_enc(rel_t.to(x.dtype))
        edge_attr = torch.cat([rel_t_enc, msg], dim=-1)
        return self.conv(x, edge_index, edge_attr)

gnn = GraphAttentionEmbedding(
    in_channels=memory_dim,
    out_channels=embedding_dim,
    msg_dim=data.msg.size(-1),
    time_enc=memory.time_enc,
).to(device)
```

Finally, we can insert the TGN components into a training loop, as shown in the following code:

```
def train():
    memory.train()
    gnn.train()

    memory.reset_state()  # Start with a fresh memory.
    neighbor_loader.reset_state()  # Start with an empty graph.

    total_loss = 0
    for batch in train_loader:
        optimizer.zero_grad()
        batch = batch.to(device)

        n_id, edge_index, e_id = neighbor_loader(batch.n_id)
        assoc[n_id] = torch.arange(n_id.size(0), device=device)

        # Get updated memory of all nodes involved in the computation.
        z, last_update = memory(n_id)
        z = gnn(z, last_update, edge_index, data.t[e_id].to(device),
                data.msg[e_id].to(device))

        # here we can compute the downstream task (for example link
        # prediction or node classification). This can be achieved by
        # concatenating a decoder (a neural network which takes the
        # embeddings and computes the predictions)

        # finally we can compute the loss and perform the update steps.
        # You can customize the criterion according to the downstream
        # task you want to perform
        loss = criterion(…)

        # Update memory and neighbor loader with ground-truth state.
        memory.update_state(batch.src, batch.dst, batch.t, batch.msg)
        neighbor_loader.insert(batch.src, batch.dst)

        loss.backward()
        optimizer.step()
```

```
memory.detach()
total_loss += float(loss) * batch.num_events
```

With this, we conclude the section where we have learned about the TGN framework, an advanced method for modeling dynamic graphs using memory-based representations. We also know how TGN's modular architecture enables effective temporal node embeddings for downstream tasks such as link prediction and node classification.

Summary

In this chapter, we introduced the concept of *temporal graph machine learning*. We discovered why it is needed and what the main problems that can be addressed using this paradigm are. We also learned a taxonomy for classifying temporal graph machine learning algorithms. Finally, we explored practical examples to understand how the theory can be applied to practical problems.

In the next chapter, we will explore the integration of language models with graphs, a rapidly evolving area at the intersection of natural language processing and graph-based learning. We will discuss recent advancements in leveraging graph structures to enhance language models, as well as techniques that incorporate textual data into graph-based representations.

Further reading

- Kazemi et al. *Representation learning for dynamic graphs: A survey.* The Journal of Machine Learning Research 21.1 (2020): 2648-2720.

- Barros et al. *A survey on embedding dynamic graphs.* ACM Computing Surveys (CSUR) 55.1 (2021): 1-37.

- Yang et al. (2024). *Dynamic graph representation learning with neural networks: A survey.* IEEE Access, 12, 43460-43484.

- Skarding et al. *Foundations and modeling of dynamic networks using dynamic graph neural networks: A survey.* IEEE Access 9 (2021): 79143-79168.

- Labonne, Maxime. *Hands-On Graph Neural Networks Using Python.* Packt Publishing Ltd. (2023)

- Nguyen et al. *Continuous-time dynamic network embeddings.* Companion proceedings of the web conference (2018)

12

GraphML and LLMs

In the rapidly evolving field of artificial intelligence, the convergence of **Graph Machine Learning (GraphML)** and **Large Language Models (LLMs)** represents a frontier rich with possibilities. This chapter explores how these two powerful technologies can be combined to unlock new insights and applications. From understanding context-rich relationships in text to enabling enhanced reasoning capabilities, this chapter aims to provide a comprehensive overview of state-of-the-art advancements in integrating GraphML and LLMs, before delving into theoretical insights and practical examples to illustrate their synergy.

In this chapter, we will:

- Provide an overview of the synergies between GraphML and LLMs.
- Illustrate the benefits of combining LLMs with graph-based approaches.
- Offer hands-on practical examples and code snippets to demonstrate these integrations.
- Highlight the challenges and opportunities in this emerging field.

Technical requirements

All code files relevant to this chapter are available at `https://github.com/PacktPublishing/ Graph-Machine-Learning/tree/main/Chapter12`. Please refer to the *Practical exercises* section of *Chapter 1, Getting Started with Graphs,* for guidance on how to set up the environment to run the examples in this chapter, either using Poetry, `pip`, or Docker.

LLMs are powerful but require significant computational resources, especially as their size increases (e.g., 3B, 7B, 12B, and 40B parameters). Not everyone has access to the necessary hardware to run these models locally. As a result, pay-per-use APIs (such as ChatGPT, Claude, and Gemini) are available to query remote LLMs. However, this chapter aims to provide model-agnostic examples and suggestions for running powerful LLMs locally on commonly available machines.

In our examples, we'll use the OpenAI library to interact with a server running the LLM. It can be used either with an API key (e.g., OpenAI) or with a local server. For those interested in running an LLM locally, we will deploy it using LM Studio (learn more at `https://lmstudio.ai/docs/api/server`). To run the LLM on your own machine, simply download the LM Studio software and follow the instructions on the website to download and set up the appropriate model.

When fine-tuning or training is required, we will be using the `transformer` Python module (`https://pypi.org/project/transformers/`), which provides APIs to quickly download, use, and fine-tune pretrained models, including LLMs.

Finally, we will be using Docker for running the Neo4j server.

LLMs: an overview

In the rapidly evolving field of artificial intelligence, LLMs have significantly advanced **natural language processing (NLP)** and understanding. These models, characterized by their extensive number of parameters and trained on large datasets, have demonstrated remarkable capabilities across a wide set of language-related tasks.

The journey of language models began with statistical approaches that relied on probabilistic methods to predict word sequences. These early models, while creating the foundations, were limited by their reliance on fixed-size context windows and the inability to capture long-range dependencies. However, as we also discussed in *Chapter 4, Unsupervised Graph Learning*, with the advent of neural networks, the field has undergone a significant shift, introducing models capable of learning **word embeddings**. In order to improve the ability to capture long-range dependencies, the initial neural network models were based on the **Long-Short Term Memory (LSTM)** and **Gate-Recurrent Unit (GRU)** architecture, which are forms of **Recurrent Neural Networks (RNNs)**. However, a pivotal moment occurred with the introduction of the Transformer architecture by Vaswani et al. in 2017. Unlike its predecessors, the Transformer model utilized *self-attention* mechanisms, enabling it to consider the entire context of a sentence without the sequential constraints inherent in RNNs. This innovation facilitated the development of models capable of processing and generating text in a more coherent and fluent way.

Building upon the Transformer architecture, researchers scaled models to unprecedented sizes, leading to the emergence of LLMs such as OpenAI's GPT series, Google's BERT and T5, and more recently, models such as GPT-3 and GPT-4.

In a nutshell, training LLMs involves optimizing a large number of parameters on very large datasets. This process, known as pretraining, typically employs unsupervised learning objectives, such as predicting missing words in a sentence (masked language modeling) or forecasting subsequent words (causal language modeling). As a side effect, the pre-training phase lets the model learn and "understand" a language, resulting in a remarkable ability to generalize across various tasks, often achieving state-of-the-art performance. LLMs have demonstrated proficiency in a diverse array of applications, reflecting their versatility and depth of language understanding. Key areas include text generation, language translation, question answering, and summarization, among many others.

Given the strengths of LLMs in unstructured text processing and generative tasks, an exciting frontier emerges when we consider their integration with graphs. While LLMs excel in understanding and generating natural language, graphs are particularly powerful for representing and analyzing structured relationships between entities. In the rest of the book, we will see examples of how we can take advantage of both.

Why combine GraphML with LLMs?

As we have learned throughout this book, GraphML excels at representing and analyzing structured data such as knowledge graphs, social networks, chemical structures, and so on. It is extremely useful for situations where exploiting relationships between entities is crucial for achieving good performances. However, LLMs are particularly good at interpreting unstructured text, offering generative skills, reasoning, and profound contextual awareness. When it comes to language-based activities such as content creation, question answering, and summarization, they excel.

Despite their impressive capabilities, LLMs are not without limitations. One of the most significant challenges is the problem of hallucination, where an LLM generates factually incorrect or misleading information that appears plausible. This is particularly problematic in domains requiring high factual accuracy, such as healthcare, finance, and legal applications. To mitigate hallucinations and enhance the reliability of LLM outputs, **Retrieval-Augmented Generation (RAG)** has emerged as a powerful technique. RAG works by dynamically retrieving relevant information from an external knowledge source (such as a knowledge graph) at inference time, rather than just relying on pre-trained knowledge. This approach ensures that the model has access to up-to-date and accurate data, grounding answers in verified information rather than generating content purely from its internal representations.

Recent advancements highlight how integrating GraphML with LLMs can drive significant innovation, enabling the development of applications that require both rich semantic understanding and relational analysis. For instance:

- **Graph-Augmented Question Answering**: LLMs can leverage knowledge graphs to answer domain-specific questions with factual accuracy.

- **Node Embedding Generation**: State-of-the-art frameworks such as GraphGPT use LLMs to generate node embeddings directly from textual data, enabling seamless integration with graph structures.

- **Knowledge Graph Construction and Enhancement**: Recent applications have shown how LLMs can be used to enrich knowledge graphs, where LLMs are used to extract semantic relationships and entities from text to enhance existing graph data.

Therefore, by bridging the gap between structured knowledge and natural language understanding, the synergy between GraphML and LLMs paves the way for more accurate, explainable, and intelligent systems.

In the next section, we will explore the state-of-the-art trends in combining GraphML and LLMs, as well as the current challenges.

State-of-the-art trends and challenges

Before diving into specific examples, it is crucial to understand the current landscape of GraphML and LLM integration. According to a recent survey by Jin et al. (`https://arxiv.org/abs/2312.02783`, 2024), the application scenario can be categorized into three main scenarios:

- **Pure Graphs**: These are graphs that lack associated textual information. Examples include social networks, traffic networks, and protein interaction networks. In such cases, the focus is on leveraging LLMs to process and analyze the structural aspects of the graph data.

- **Text-Attributed Graphs**: In these graphs, nodes or edges are enriched with textual attributes. For instance, in academic networks, papers (nodes) come with titles and abstracts, while authors (nodes) have profiles. E-commerce networks also fall into this category, where products (nodes) have descriptions, and user interactions (edges) may include reviews. The challenge here is to effectively combine the textual content with the graph's structural information.

- **Text-Paired Graphs:** This scenario involves graphs that are paired with separate textual descriptions or documents. Unlike text-attributed graphs, where text is embedded within the graph as attributes, text-paired graphs treat the graph and text as distinct but related entities. A pertinent example is molecular graphs accompanied by detailed textual descriptions of their properties. The objective is to align and integrate the information from both the graph structure and the associated text to enhance understanding and analysis.

To effectively utilize LLMs in these scenarios, three primary techniques can be used: LLMs as predictors, LLMs as encoders, and LLMs as aligners. Let's see these approaches one by one.

LLMs as predictors

The simplest and most direct approach is to use LLMs as predictors. In this paradigm, the LLM operates as a tool to infer outcomes directly from graph data. Imagine a scenario where textual information is either minimal or entirely absent (pure graphs). In this case, you can transform the graph data into a format that the LLM can process, such as converting graph structures into sequences or textual descriptions.

For instance, consider a simple social network graph where nodes represent people and edges indicate friendships (*Figure 12.1*). These features can be converted into a textual narrative, such as *Alice is linked with Bob*. An LLM can then process this narrative to predict new relationships or infer additional attributes about the nodes, such as professional interests or potential connections.

Bob   **Alice**   *"Alice is linked with Bob"*

Figure 12.1: Examples of how graphs can be converted to text narratives

Once the data is prepared, the LLM can be fine-tuned or prompted to perform specific tasks. These might include predicting node classifications, such as identifying the role of individuals in a social network, or link predictions, such as forecasting interactions between entities. In molecular research, LLMs as predictors can help determine the properties of chemical compounds based solely on their structural representations.

One advantage of this approach is its simplicity: LLMs can be applied directly to graph data without requiring extensive preprocessing or specialized models. However, this simplicity can also be a limitation. Purely structural information might not always be sufficient for complex tasks, particularly when additional contextual or textual data is available but not leveraged. Moreover, scalability and cost must be considered: encoding entire graphs as text can lead to an explosion of sentences, making inference expensive, potentially inefficient, and sometimes impossible (for example, if the maximum number of words an LLM can process at once is too small to contain the whole graph). Performance may also be limited, as this approach is similar to providing an LLM with a structured dataset and expecting accurate predictions without tailored adaptations. For this reason, more complex graph2text formalisms can be designed, incorporating node/edge descriptions into textual narratives while balancing efficiency and accuracy.

LLMs as encoders

When graphs are enriched with textual attributes, the LLM as encoder approach becomes particularly powerful. Here, the LLM is tasked with processing and encoding the textual information associated with nodes or edges, producing meaningful representations that can be integrated with the graph's structural features. These embeddings are then integrated into the graph through proper algorithms such as graph neural networks, which process the combined representation to perform downstream tasks.

This hybrid representation combines the strengths of both modalities, capturing the nuances of text alongside the relationships encoded in the graph. As depicted in *Figure 12.2*, each node could have attributes, such as a name and a brief bio for a node representing a person, while the edges might be annotated with information about the nature of the relation, e.g., *close friend* or *colleague* for a graph representing social networks. These features can be converted into a textual narrative, such as *Alice, a software engineer, is close friends with Bob, a data scientist.*

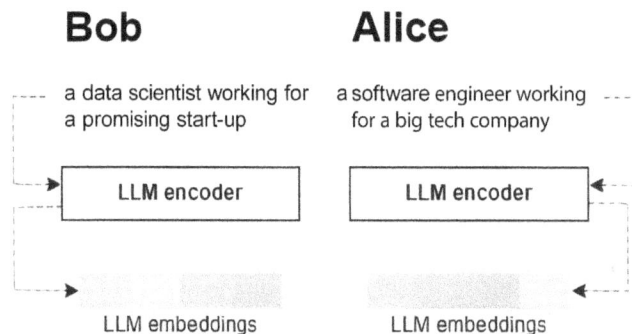

Figure 12.2: Examples of how LMMs can be used as encoders for node attributes

Other examples include academic citation networks, where papers (nodes) come with titles, abstracts, and keywords. An LLM can process these textual attributes to generate embeddings that encapsulate their semantic content. These embeddings are then combined with graph-specific features, such as the citation relationships between papers, to create a unified representation. Similarly, in e-commerce platforms, product descriptions and user reviews can be encoded by LLMs to enhance product similarity graphs or user behavior analysis.

It is worth noticing that the process of using LLMs as encoders typically involves fine-tuning the LLM on domain-specific textual data to ensure that the embeddings accurately reflect the requirements of the task.

This encoder approach offers significant benefits. By leveraging textual data, it captures context and nuances that purely structural methods might miss. It is particularly effective in scenarios where textual attributes can provide critical insights, such as identifying the themes of academic papers or understanding user preferences in recommendation systems.

LLMs as aligners

The goal here is to align and integrate the information from both structure and textual descriptions (or accompanying documents, in the case of text-paired graphs), enabling a comprehensive analysis that leverages the strengths of each. This can be achieved, for example, by finding a shared latent space or a semantic mapping that connects the two modalities. Such an approach might involve designing models that jointly optimize both modalities or using attention mechanisms to focus on the most relevant parts of each input.

In more detail, the synergy between textual encoding (handled by the LLM) and graph structure encoding (handled by, for example, a GNN), can be typically in two ways:

1. **Prediction Alignment**: Iterative training where LLMs and GNNs generate pseudo-labels to guide each other's learning

2. **Latent Space Alignment:** Contrastive learning to align the latent representation of the text and the graph structure in a shared space (e.g., *Figure 12.3*)

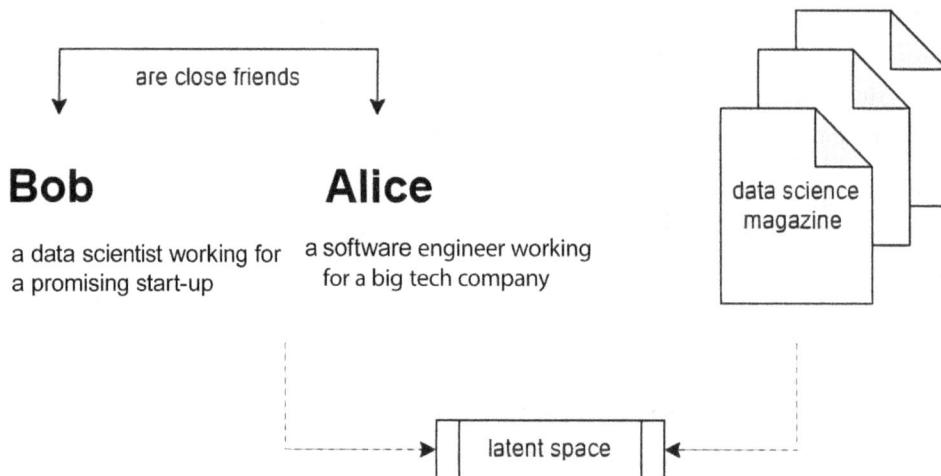

Figure 12.3: Graphs and associated texts can be embedded in a shared latent space

For example, in molecular research, a molecular graph might represent the structure of a compound, while a textual description provides information about its properties, synthesis, or applications. In this context, an LLM can be used to process the text to extract relevant features and align these with the structural characteristics of the graph, enabling tasks such as property prediction or drug discovery.

As you can imagine, this approach is particularly powerful in interdisciplinary fields where graphs and text provide complementary points of view. In computational social science, for instance, social graphs representing interactions between individuals can be aligned with news articles, social media posts, or other textual data to study the spread of information or public sentiment. Similarly, in e-commerce, user behavior graphs can be integrated with textual reviews to improve personalized recommendations.

Now that we have a clearer understanding of the LLM and graph landscape, let's dive into a practical example of how this integration works. We will explore this in the next section.

Hands-on GraphML with LLMs

Based on the previous characterization, the following sections present hands-on examples that showcase how GraphML and LLMs can be integrated.

LLM as predictor

First of all, let's define a simple social network using NetworkX. We have three nodes (Alice, Bob, and Carl), each coming with a short description of their job and what they like:

```python
import networkx as nx
# Create a directed graph
G = nx.DiGraph()
G.add_node(1, name="Alice", description="she is a software engineer and
she likes reading.")
G.add_node(2, name="Bob", description="he is a data scientist and he likes
writing books.")
G.add_node(3, name="Carl", description="he is a data scientist and he
likes swimming.")
G.add_edge(1, 3, relationship="is friend with")
```

Now that we have created the graph, let's define a function to encode it as a narrative text. We will be using a simple formalism in which we first declare each node and then we describe each connection. Notice that more complicated formalisms can be used to describe more complex scenarios, despite there being no standard way to do this:

```python
# Function to convert network to text
def graph_to_text(graph, edge_type):
    descriptions = []
    # 1. describe the graph structure
    descriptions.append(f"Num nodes: {graph.number_of_nodes()}.\n")
    for n in graph:
        descriptions.append(f"Node {n}: {graph.nodes[n]['name']}\n")

    for u, v, data in graph.edges(data=True):
        node_u = graph.nodes[u]
        node_v = graph.nodes[v]
        descriptions.append(f"The person named '{node_u['name']}'
({node_u['description']}) {edge_type} '{node_v['name']}'
({node_v['description']}).")
    return " ".join(descriptions)
text_input = graph_to_text(G)
print("Social Network as text:\n", text_input)
```

The output should be as follows:

```
Social Network as text:
Num nodes: 3.
Node 1: Alice
Node 2: Bob
Node 3: Carl
The person named 'Alice' (she is a software engineer and she likes
reading.) is friend with 'Carl' (he is a data scientist and he likes
swimming.).
```

We will now declare a *prompt*, which is a simple instruction describing the task to the LLM:

```
# Create a prompt
prompt = f"Here is a social network: {text_input}\nBased on the above,
suggest any missing link and explain why they might be relevant."
```

It is now time to send the prompt to the LLM server and wait for a response. To achieve this aim, we will be using the OpenAI API to create a client instance. The client will connect to the LLM server, and send and receive messages:

```
# Call the llm to generate a response
from openai import OpenAI
# Create a client for interacting with the LLM server. Here, we are
# running LM Studio locally, therefore we use the localhost address and
# "lm-studio" as api key. You can replace this line with a proper api key
# to a remote LLM service if you have one.
client = OpenAI(base_url="http://localhost:1234/v1/", api_key="lm-studio")
```

Let's use the client functionalities to create and send a message. Note that we are also specifying which language model to use (`minicpm-llama3-v-2_5`). If you are running LM Studio locally, you need to download the model first:

```
response = client.chat.completions.create(
    model="minicpm-llama3-v-2_5",
    messages=[{"role": "system", "content": "You are a helpful assistant."},
              {"role": "user", "content": prompt}],
    max_tokens=300,
)
```

Finally, let's check the answer:

```python
# Extract the generated text from the response
print(response.choices[0].message.content)
```

The output should be as follows:

```
A possible missing link in this citation network could be between Alice
and Bob. Since Alice is a software engineer who likes reading and Bob is
mentioned as a friend of Carl (a data scientist), it would make sense
for them to have some connection. Since Alice enjoys reading, she may
appreciate discussions or debates about literature with someone who shares
her interest.
```

The explanation is pretty clear: since Alice works in the tech industry and is a friend of Carl, it makes sense for Carl to introduce her to Bob, who shares similar interests.

Note that to add this link to the graph using NetworkX, you need to do some text processing. We leave this as an exercise for you. Furthermore, you may want to tune the prompt to let the model answer using a specific way to facilitate the parsing (e.g., it can answer something like "Alice -> Bob").

In the next example, we will see how to use the LLM as an encoder.

LLM as encoder

When graphs are enriched with textual attributes, the LLM as encoder approach becomes powerful. For example, in a recommendation system, products (nodes) might have textual descriptions, reviews, and other metadata. This textual data can be processed with LLMs to create meaningful embeddings, which can be further combined with graph-structured features such as user-product interactions.

Let's walk through an example of enhancing a movie recommendation graph using LLMs as encoders. Let's define our graph where nodes represent movies, and edges represent similarities between movies. Each node also contains a textual description of the movie:

```python
import networkx as nx
from openai import OpenAI

# Let's create a toy movie graph
G = nx.Graph()
G.add_node(1, title="Inception", description="A mind-bending thriller
about dreams within dreams.")
```

```
G.add_node(2, title="The Matrix", description="A hacker discovers the
shocking truth about reality.")
G.add_node(3, title="Interstellar", description="A team travels through a
wormhole to save humanity.")
G.add_edge(1, 2, similarity=0.8)
G.add_edge(1, 3, similarity=0.9)
```

As in the previous example, let's initialize the client to query the LLM server:

```
# Intialize the client
client = OpenAI(base_url="http://localhost:1234/v1/", api_key="lm-studio")
```

Let's write a function to compute the text embedding using the LLM. It takes as input a text and returns the embedding. For convenience, our function will exploit the `client.embeddings.create` method from the OpenAI API. Also, in this case, we have to specify an LLM. We have chosen the powerful *Nomic embedding model* (recall you have to download it in advance through LM Studio):

```
def encode_text(text):
    # Prepare the query for the LLM
    response = client.embeddings.create(
        input=text,
        model="text-embedding-nomic-embed-text-v1.5-embedding"
    )
    # Get 768-dimensional embedding
    embedding = response.data[0].embedding
    return embedding
```

For each node in the graph, let's compute the corresponding embedding and set it as a node attribute in the NetworkX graph:

```
# Encode movie descriptions and add embeddings to the graph
for node in G.nodes(data=True):
    description = node[1]['description']
    embedding = encode_text(description)
    node[1]['embedding'] = embedding
```

Once we have textual embeddings, we can integrate them with structural features such as node degrees or edge similarities. This hybrid representation is then fed into a downstream GraphML model (e.g., graph neural network):

```
import numpy as np
# Combine embeddings with structural features
for node in G.nodes(data=True):
    # We are using degree as a sample feature
    structural_features = np.array([G.degree[node[0]]])
    node[1]['combined_features'] = np.concatenate((node[1]['embedding'],
                                                   structural_features),
                                                   axis=None)
```

With the combined features, you can train a machine learning model to predict recommendations or similarities between nodes. For instance, we can build a simple transductive nearest-neighbor approach by computing the pairwise similarities between node features. This way, we can suggest "similar" movies to users:

```
from sklearn.metrics.pairwise import cosine_similarity
# Compute similarity between nodes based on combined features
node_features = [node[1]['combined_features'] for node in
G.nodes(data=True)]
similarity_matrix = cosine_similarity(node_features)
# Example: Find movies similar to 'Inception' (node 1)
movie_index = 0  # Index of the movie 'Inception'
# Let's take the top 2 similar
similar_movies = np.argsort(-similarity_matrix[movie_index])[1:3]
print("Movies similar to Inception:", similar_movies)
```

Of course, once you have extracted the node features, you can also use the various models we have described in previous chapters, such as GNNs seen in *Chapter 4, Unsupervised Graph Learning,* for unsupervised learning, and in *Chapter 5, Supervised Graph Learning,* for supervised learning. However, it is important to observe that, when combining textual embeddings with structural features, it's crucial to balance their influence. High-dimensional text embeddings can overshadow low-dimensional structural features, potentially distorting similarity computations. Proper scaling ensures both types of features contribute meaningfully as well as weighting the contribution of each feature. For example, you may want to assign different **weights** to text versus structure when concatenating, and using structural encoders such as GNNs may help balance dimensionalities.

LLM as aligner

As we have previously learned in this section, there are two typical approaches for achieving text-graph alignment: prediction alignment and latent space alignment. In the rest of this section, we will explore each in more detail with practical examples.

Prediction alignment

First, let's showcase how LLM-GNN *prediction alignment* can be achieved. Here's how we present an approach based on iterative training. The LLM learns from the text information in the graph (e.g., node descriptions), while the GNN learns from the graph structure. Each model generates pseudo-labels, which the other model uses to improve its training. Summarizing:

1. The LLM analyzes text data and generates node labels, which will serve as pseudo-labels for the GNN.

2. The GNN then processes the graph structure and produces node labels based on connectivity and relationships, which are then fed back to the LLM.

3. The process is repeated with each model refining its prediction based on insights from the other.

As we have previously said, LLMs are resource-intensive. As this example requires a bit of fine-tuning, it is difficult to showcase an example using very large models such as GPT. Therefore, we will be using a smaller but powerful model, BERT (https://arxiv.org/pdf/1810.04805). To access the model, we will use the transformer Python module.

Let's consider a toy citation network, where nodes represent research papers, edges represent citations between papers, and each node is described by title and abstract:

```python
from torch_geometric.data import Data
# Assume a toy dataset with 3 papers (nodes), edges, and labels
data = Data(
    x=torch.rand(3, 10),  # let's use random features for simplicity
    edge_index=torch.tensor([[0, 1], [1, 2]], dtype=torch.long),  # Edges
    y=torch.tensor([0, 1, 2], dtype=torch.long),  # True labels
    text=["Paper A abstract", "Paper B abstract", "Paper C abstract"],
    # Text data
)
```

Let's define a GNN module to encode structural information and a TextEncoder module, which uses the transformer API, to download and create the BERT model. Note that, since transformer is built in PyTorch, we will define our GNN using PyG:

```python
# 1. Define the Graph Neural Network (GNN)
class GNN(torch.nn.Module):
    def __init__(self, input_dim, hidden_dim, output_dim):
        super(GNN, self).__init__()
        self.conv1 = GCNConv(input_dim, hidden_dim)
        self.conv2 = GCNConv(hidden_dim, output_dim)

    def forward(self, x, edge_index):
        x = self.conv1(x, edge_index).relu()
        x = self.conv2(x, edge_index)
        return x


# 2. Define the LLM (e.g., BERT for text encoding)
class TextEncoder(torch.nn.Module):
    def __init__(self, model_name="bert-base-uncased", output_dim=128):
        super(TextEncoder, self).__init__()
        self.tokenizer = AutoTokenizer.from_pretrained(model_name)
        self.model = AutoModel.from_pretrained(model_name)
        self.fc = torch.nn.Linear(self.model.config.hidden_size, output_dim)

    def forward(self, texts):
        # Tokenize and encode text data
        inputs = self.tokenizer(texts, return_tensors="pt", padding=True,
truncation=True)
        outputs = self.model(**inputs)
        cls_embedding = outputs.last_hidden_state[:, 0, :]
        # [CLS] token embedding
        return self.fc(cls_embedding)
```

In the notebook attached to the repository, we have built a pretty standard GNN using two graph convolution layers. The TextEncoder, instead, is composed of the pretrained LLM model, followed by a trainable linear **fully connected (fc)** projection layer. The forward pass first converts the text into a format that is digestible by the LLM (tokenization). The resulting embeddings are then forwarded to the linear layer to make a prediction. Please refer to the notebook attached in the repository for the implementation details of the GNN and the text encoder.

Finally, using these analytical components (GNN and text encoder), we can define our training loop as follows:

```python
# 3. Training Loop with Pseudo-Label Exchange
def train_prediction_alignment(data, gnn, text_encoder, num_iterations=5):
    optimizer_gnn = torch.optim.Adam(gnn.parameters(), lr=0.01)
    optimizer_text = torch.optim.Adam(text_encoder.parameters(), lr=0.0001)

    # Initialize with true labels for first iteration
    gnn_pseudo_labels = data.y.clone()
    llm_pseudo_labels = data.y.clone()

    for iteration in range(num_iterations):
        # Train GNN using LLM pseudo-labels from previous iteration
        gnn.train()
        optimizer_gnn.zero_grad()
        gnn_logits = gnn(data.x, data.edge_index)
        gnn_loss = torch.nn.CrossEntropyLoss()(gnn_logits, llm_pseudo_labels)
        gnn_loss.backward()
        optimizer_gnn.step()

        # Generate new GNN pseudo-labels
        with torch.no_grad():
            gnn_pseudo_labels = torch.argmax(gnn_logits, dim=1)

        # Train Text Encoder using GNN pseudo-labels
        text_encoder.train()
        optimizer_text.zero_grad()
        text_logits = text_encoder(data.text)
        llm_loss = torch.nn.CrossEntropyLoss()(text_logits, gnn_pseudo_labels)
```

```
    llm_loss.backward()
    optimizer_text.step()

    # Generate new LLM pseudo-labels for next iteration
    with torch.no_grad():
        llm_pseudo_labels = torch.argmax(text_logits, dim=1)

    print(f"Iteration {iteration+1}: GNN Loss = {gnn_loss.item():.4f},
LLM Loss = {llm_loss.item():.4f}")
    print(f"  GNN predictions: {gnn_pseudo_labels.tolist()}")
    print(f"  LLM predictions: {llm_pseudo_labels.tolist()}")
```

In this supervised loop, the GNN model predicts pseudo-labels (the model is optimized using CrossEntropyLoss to minimize the difference between the prediction and the targets). The predicted labels are then used as targets to fine-tune the text encoder. This way, the final model benefits from both textual and structural insights, enabling more accurate classification of research papers.

> **Note**
>
> The "symmetric" approach can also be used in a unidirectional (asymmetric) manner, where pseudo-labels from only one model are used to train the other.

Of course, this is a toy example with random features, but we hope you grasp the principle to apply it in real-world cases and better understand related state-of-the-art approaches (check the end of this chapter for further reading!).

Latent space alignment

Instead of iteratively sharing labels, this method aligns the latent representations of text and graph data via *contrastive learning*. The goal is to force text and graph encodings for the same entity (e.g., a node) to be similar in a shared space while pushing encodings for unrelated entities far apart. Summarizing:

1. **Text Encoding**: Use an LLM to encode the node descriptions into a latent vector.
2. **Graph Encoding**: Use a GraphML model (e.g., GNN) to encode the graph structure around each node into latent vectors.

3. **Contrastive learning**: Use contrastive learning to maximize the similarity between the text and graph encoding for the same node or neighbor nodes, while minimizing the similarity between unrelated nodes.

Let's consider a toy knowledge graph, where nodes represent products, edges represent relationships such as "frequently bought together," and each node has a text description:

```
# Toy data with 3 products and their relationships
data = Data(
    x=torch.rand(3, 10),   # Node features
    edge_index=torch.tensor([[0, 1], [1, 2]], dtype=torch.long),   # Edges
    text=["Product A description", "Product B description", "Product C
description"],   # Text data
)
```

For simplicity, let's use the same GNN and text encoder as in the previous example. Therefore, we only need to define our contrastive loss and training loop.

As previously described, the contrastive loss will force the model to minimize differences between "similar" nodes, while maximizing the difference between unrelated nodes. In more detail, we compute a similarity matrix sim of shape (batch_size, batch_size), where sim[i, j] is the similarity between the i-th graph embedding and the j-th text embedding. Here, we assume a perfect one-to-one correspondence (labels), where the i-th graph embedding should match the *i*-th text embedding:

```
# Contrastive Learning Objective
def contrastive_loss(graph_emb, text_emb, tau=0.1):
    sim = F.cosine_similarity(graph_emb, text_emb)
    labels = torch.arange(sim.size(0)).to(sim.device)
    loss = F.cross_entropy(sim / tau, labels)
    return loss
```

The training loop simply optimizes the graph and text encoders using the contrastive loss:

```
# Training Loop for Latent Space Alignment
def train_latent_alignment(data, gnn, text_encoder, epochs=10):
    optimizer = torch.optim.Adam(list(gnn.parameters()) + list(text_
encoder.parameters()), lr=0.001)
    for epoch in range(epochs):
        optimizer.zero_grad()
```

```
# Encode graph and text
graph_emb = gnn(data.x, data.edge_index)  # Graph embeddings
text_emb = text_encoder(data.text)  # Text embeddings

# Compute contrastive loss
loss = contrastive_loss(graph_emb, text_emb)
loss.backward()
optimizer.step()
print(f"Epoch {epoch+1}: Loss = {loss.item()}")
```

This fusion may create richer node or entity embeddings, improving downstream tasks such as node classification and recommendation and retrieval systems (e.g., you may retrieve nodes from the graph based on their description). Interestingly, this unified representation can also support *zero-shot* and *few-shot* learning in graph-based tasks. Since LLMs process textual prompts, they can generalize to new, unseen categories within a graph without requiring extensive retraining. For example, if a graph-based dataset lacks labeled examples for a particular node class, an LLM can still classify nodes by leveraging semantic similarities and contextual cues from textual descriptions.

We have seen how text and graphs can be aligned to achieve a tight integration. In the next section, we will see another practical application of combining graphs and LLMs, which is how to build a knowledge graph from an unstructured text using an LLM.

Building knowledge graphs from text

In *Chapter 8, Text Analytics and Natural Language Processing Using Graphs,* we used spaCy to convert text into a graph. While spaCy is excellent for **named entity recognition** (**NER**) and dependency parsing, it has limitations when extracting complex relationships from text. Conventional NER performs well in domains with well-established taxonomies and is dependable for organized, predefined entity types. To extract entities and relationships from large, unstructured data sources, LLMs offer a more adaptable, context-aware alternative.

Here, we use an LLM-powered approach to build a **Knowledge Graph** (**KG**) from text. We will be using LangChain (https://www.langchain.com/), an open-source framework designed to help developers build applications powered by LLMs. It provides tools for prompt engineering, memory management, and data retrieval, amongst others.

In particular, we will make use of `LLMGraphTransformer` to extract entities and relationships from text, which features some nice properties:

- It may handle complex relationships better than rule-based NLP methods.
- It extracts a wider variety of entities beyond predefined spaCy models.
- It adapts dynamically to different domains without retraining a model.

Let's first consider a free text as follows:

```
text = """
Marie Curie, born in 1867, was a Polish and naturalized-French physicist
and chemist who conducted pioneering research on radioactivity.
She was the first woman to win a Nobel Prize, the first person to win
a Nobel Prize twice, and the only person to win a Nobel Prize in two
scientific fields.
"""
```

In the following snippet of code, we will use LangChain objects to convert the text into a set of nodes and relationships (KG) using an LLM. LangChain will handle all the operations to convert the text into a proper object (Document) and send it to our backend server to be parsed by our *MiniCPM* model in LM Studio:

```
from langchain_experimental.graph_transformers.llm import
LLMGraphTransformer
from langchain_openai import ChatOpenAI
from langchain_core.documents import Document

llm = ChatOpenAI(temperature=0, model_name="minicpm-llama3-v-2_5", base_
url="http://localhost:1234/v1", api_key="lm-studio")
llm_transformer = LLMGraphTransformer(llm=llm)

documents = [Document(page_content=text)]
graph_documents = llm_transformer.convert_to_graph_documents(documents)
print(f"Nodes: {graph_documents[0].nodes}")
print(f"Relationships: {graph_documents[0].relationships}")
```

The output should be as follows:

```
Nodes:[Node(id='Marie_Curie', type='Person', properties={}),
Node(id='Pierre_Curie', type='Person', properties={})]
Relationships:[Relationship(source=Node(id='Marie_Curie',
```

```
type='Person', properties={}), target=Node(id='Pierre_Curie',
type='Person', properties={}), type='MARRIED_TO', properties={}),
Relationship(source=Node(id='Marie_Curie', type='Person', properties={}),
target=Node(id='Nobel_Prize', type='Award', properties={}), type='WON_
NOBEL_PRIZE', properties={}), Relationship(source=Node(id='Pierre_
Curie', type='Person', properties={}), target=Node(id='Marie_Curie',
type='Person', properties={}), type='MARRIED_TO', properties={})]
```

That's it! We can parse the nodes and relationships to encode the graph in a proper framework such as NetworkX or Neo4j.

Note that different results can be achieved using longer text (e.g., a custom PDF) and different models. For example, it is expected that fine-tuned models or larger models (more than 100B parameters) can achieve better results. However, they would need more resources.

In the next section, we will see how can we load the extracted graph into Neo4j to perform GraphRAG, that is, how we can augment an LLM using a knowledge graph for more precise question answering.

Real-world scenarios: GraphRAG

In the context of integrating LLMs with graph-based data, the term "LLM as aligner" refers to the role of LLMs in aligning or integrating textual information with graph structures to enhance understanding and retrieval. This approach is exemplified by techniques such as RAG and its extension, GraphRAG.

RAG is a framework that combines the strengths of traditional information retrieval systems with the generative capabilities of LLMs. In this setup, an LLM is augmented with a retrieval component that fetches relevant information from external data sources, such as knowledge bases or databases, to produce more accurate and contextually relevant responses. This method enhances the LLM's output by grounding it in authoritative, up-to-date information.

GraphRAG builds upon the RAG framework by incorporating KG into the retrieval process. In this approach, the retrieval component utilizes a knowledge graph—a structured representation of entities and their relationships—to provide contextually relevant information to the LLM. This integration allows for a richer understanding of complex data by combining text extraction, network analysis, and LLM prompting into a cohesive system.

By acting as an aligner, the LLM effectively bridges the gap between unstructured textual data and structured graph representations, facilitating more comprehensive data analysis and retrieval. This alignment enhances the LLM's ability to generate responses that are both contextually relevant and grounded in structured knowledge, leading to improved performance in tasks such as question answering, recommendation systems, and data summarization.

In recent years, several tools have risen for GraphRAG. Here, we will be using Neo4j and LangChain to achieve a fully local, efficient GraphRAG.

First of all, we need to start our Neo4j server. It will act as a backend for storing the KG and performing the RAG operations. As you can read in *Chapter 10, Building a Data-Driven Graph-Powered Application*, Neo4j is one of the most common graph databases, which can scale to fairly large datasets and can be distributed over multiple nodes.

Similarly to *Chapter 10, Building a Data-Driven GraphPowered Application,* we can locally deploy a Neo4j server instance using Docker. However, since here we need the apoc plugins, we need to feed to the docker commands some extra environment variables with respect to the command seen in the previous chapter:

```
docker run --rm --detach --name neo4j \
            --publish=7474:7474 --publish=7687:7687 \
            --env NEO4J_AUTH=neo4j/defaultpass \
            --env NEO4J_PLUGINS='["apoc-extended"]' \
            --env NEO4J_apoc_export_file_enabled=true \
            --env NEO4J_apoc_import_file_enabled=true \
            --env NEO4J_apoc_import_file_use__neo4j_config=true \
            neo4j:5.26.0
```

It's now time to implement our GraphRAG system! First of all, let's connect to Neo4j (this part can be adapted if you are using a different edition):

```
from neo4j import GraphDatabase
from langchain_neo4j import Neo4jGraph


NEO4J_URI = "bolt://localhost:7687"
NEO4J_USER = "neo4j"
NEO4J_PASSWORD = "your_password"


driver = GraphDatabase.driver(NEO4J_URI, auth=(NEO4J_USER, NEO4J_PASSWORD))
graph = Neo4jGraph(url=NEO4J_URI, username=NEO4J_USER, password=NEO4J_
PASSWORD)
```

Once we have our graph connection instance, we can store the KG created above in Neo4j, as follows:

```
graph.add_graph_documents(graph_documents)
```

At this point, our extracted KG is stored in Neo4j and ready for querying. You can visualize it in your browser at localhost:7474, and it should look as follows:

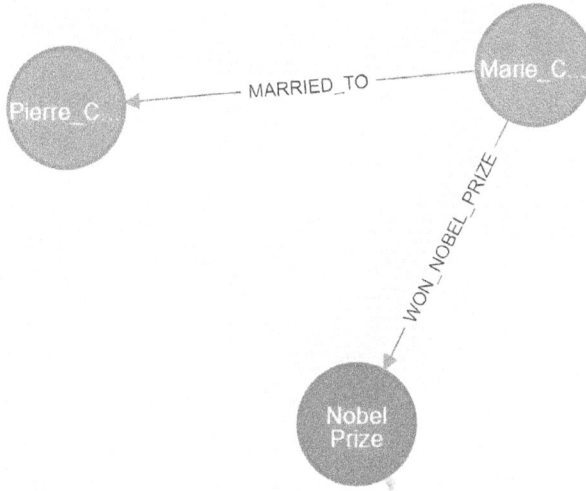

Figure 12.4: Visualization of extracted KG in Neo4j

Our GraphRAG method is fairly simple: we provide an LLM with our KG and a textual query, and we ask it to generate a Cypher query representing the textual query. Then, Neo4j performs the query and provides the LLM with the extracted results. Finally, the LLM uses this information to generate an answer.

To help the LLM generate a correct query, we first define a Cypher template to guide query generation:

```
CYPHER_GENERATION_TEMPLATE = """You are a Neo4j expert. Generate a Cypher
query to answer the given question.

Database Schema:
- Nodes:
  * Person (properties: id)
  * Award (properties: id)
- Relationships:
  * (Person)-[:MARRIED_TO]-(Person)
```

```
  * (Person)-[:WON_NOBEL_PRIZE]->(Award)

Rules:
1. Always use explicit `MATCH` for relationships.
2. Never use `WHERE` for relationship matching.
3. Use `RETURN DISTINCT` when appropriate.

Example Queries:
1. Question: "Who won the Nobel Prize?"
   Cypher: MATCH (p:Person)-[:WON_NOBEL_PRIZE]->(:Award) RETURN p.id AS
winner

Question: {query}
Return only the Cypher query without any explanation or additional text.
Cypher:"""
```

Let's use the GraphCypherQAChain object in LangChain. This object passes context to the Q&A prompt in the _call method. It retrieves the results from the graph database using the generated Cypher query and passes these results as the context to the LLM. The LLM then uses this context along with the question to generate an answer:

```
from langchain_neo4j import GraphCypherQAChain
from langchain_core.prompts import PromptTemplate

chain = GraphCypherQAChain.from_llm(
    llm=llm,
    graph=graph,
    verbose=True,
    cypher_prompt=PromptTemplate(
        input_variables=["query"],
        template=CYPHER_GENERATION_TEMPLATE
    ),
    allow_dangerous_requests=True
)
```

Let's ask something simple but not trivial, for example, "Who married a Nobel Prize winner?":

```
question = "Who married a Nobel Prize winner?"

print(f"\nQuestion: {question}")
response = chain.invoke(question)
print("Response:", response['result'])

# Close the driver
driver.close()
```

The output should be as follows:

```
Response: Pierre Curie married a Nobel Prize winner.
```

The answer is correct! Note that, to answer the question, the engine had to build a specific Cypher query that looks like the following:

```
MATCH (p1:Person)-[:MARRIED_TO]-(p2:Person)-[:WON_NOBEL_PRIZE]->(:Award)
RETURN p1.id AS winner
```

Therefore, the LLM received grounded information to prepare the answer, which makes the response more trustable compared to a pure generative method. However, while this method effectively retrieves structured data, alternative approaches (such as embedding-based retrieval) could further enhance results. GraphRAG can be implemented in multiple ways:

- **Cypher Query-Based Retrieval (our approach)**: Uses an LLM to translate queries into **Cypher queries**, which fetch structured data from a Neo4j graph
- **Graph Embedding Retrieval**: Stores knowledge graph entities as vector embeddings and retrieves relevant nodes using similarity search
- **Hybrid Approach**: Combines structured graph querying with embedding-based retrieval for greater flexibility

Neo4j and LangChain offer a well-curated and robust environment to implement and try new approaches, which can result in very powerful applications.

This section concludes our overview of practical examples of combining graphs with LLMs. However, in these rapidly evolving fields, it is important to stay updated on new discoveries that may overcome current challenges, some of which we will outline in the next section.

Challenges and future directions

The integration of GraphML and LLMs opens up a lot of possibilities, but it also presents significant challenges that must be addressed for widespread adoption. One of the foremost concerns is scalability since dealing with large-scale graphs alongside computationally intensive LLMs requires expensive resources in terms of memory, processing power, and efficient data pipelines.

As with many other deep learning-based approaches, another major challenge is in interpretability. While knowledge graphs provide a structured and transparent way to store relationships, LLMs operate as a black box, making it difficult to understand how specific outputs are generated. For example, considering the presented GraphRAG approach, it is not guaranteed that the generated query will be semantically correct, and correcting the result is not an easy task.

Data alignment is also a key issue, as structured knowledge graphs and unstructured text data must be carefully preprocessed to ensure consistency. Differences in data formats, ontology mismatches, and information redundancy can create inefficiencies when integrating these two paradigms. Developing robust pipelines that seamlessly connect graph-based insights with LLM-generated text remains an open challenge.

By tackling these challenges, the synergy between GraphML and LLMs has the potential to bridge the gap between structured knowledge and flexible, natural language reasoning.

Being such a recent and rapidly evolving field, we encourage you to dive deeper into GraphML and LLMs. Reading scientific papers is probably the best way to stay updated. Packt has also a nice collection of books to further expand your knowledge. However, living in the AI era, you can easily find online blogs, materials, and resources to stay updated with the current trends!

Summary

This chapter has provided an introduction to combining GraphML and LLMs with practical examples. By leveraging the strengths of both technologies, researchers and practitioners can push the boundaries of what's possible in AI-driven applications.

We have learned what LLMs are and how they can work with graphs using state-of-the-art techniques. We also explored the current trends and challenges in the landscape of GraphML and LLM integration. Finally, we saw how to start developing useful tools such as knowledge graph builders and GraphRAG systems.

In the next chapter, we will turn to some recent developments and the latest research and trends in machine learning that have been applied to graphs. In particular, we will describe some of the latest techniques (such as generative neural networks) and applications (such as graph theory applied in neuroscience) available in the scientific literature, providing some practical examples and possible applications.

Further reading

There are some excellent books and papers that can help you further. Please have a look at the following:

- *Applied Deep Learning on Graphs* by *Lakshya Khandelwal and Subhajoy Das*; https://www. amazon.com/Applied-Deep-Learning-Graphs-Architectures/dp/1835885969/ref=s r_1_1?crid=10ETRBSQUJFUV&dib=eyJ2IjoiMSJ9.WJ7uHlfdD3FBONCaE1YNb2GqKTx1PUU bbCBoOHHU3-Q.Z0RYDj_sTRNkealzx8eBCAszlWghNWziW0lSCOv2A6k&dib_tag=se&keywo rds=Applied+Deep+Learning+on+Graphs%3A+Leverage+graph+data+for+business+ap plications+using+specialized+deep+learning+architectures&qid=1743499437&sp refix=applied+deep+learning+on+graphs+leverage+graph+data+for+business+app lications+using+specialized+deep+learning+architectures%2Caps%2C720&sr=8-1

- There is a great tutorial here: https://www.packtpub.com/en-us/learning/how-to-tutorials/large-language-models-llms-and-knowledge-graphs?srsltid=AfmBOoq ieAeasuGZ1y7i_uVtFo-5AsUZkl1qvoDhRmz2pTwONBcGzu23

- You can also get more information from *Chapter 1* of *Building LLM Powered Applications*, by *Valentina Alto*: https://www.packtpub.com/en-us/product/building-llm-powered-applications-9781835462317/chapter/introduction-to-large-language-models-1/ section/most-popular-llm-transformers-based-architectures-ch01lvl1sec03?sr sltid=AfmBOopgI8gtJ6sH-vTJgaLUDn1zTJ8WHAkVC3G

- Li, Q., Zhao, T., Chen, L., Xu, J., & Wang, S. (2024, December). *Enhancing Graph Neural Networks with Limited Labeled Data by Actively Distilling Knowledge from Large Language Models*. In *2024 IEEE International Conference on Big Data (BigData)* (pp. 741-746). IEEE.

- Wang, D., Zuo, Y., Li, F., & Wu, J. (2024). *LLms as zero-shot graph learners: Alignment of GNN representations with llm token embeddings. Advances in Neural Information Processing Systems, 37*, 5950-5973.

- Li, Y., Li, Z., Wang, P., Li, J., Sun, X., Cheng, H., & Yu, J. X. (2023). *A survey of graph meets large language model: Progress and future directions. arXiv preprint arXiv:2311.12399.*

Get This Book's PDF Version and Exclusive Extras

13

Novel Trends on Graphs

In the previous chapters, we described different supervised and unsupervised algorithms that can be used in a wide range of problems concerning graph data structures. However, the scientific literature on graph machine learning is vast and constantly evolving and, every month, new algorithms are published. In this chapter, we will provide a high-level description of some new techniques and applications concerning graph machine learning.

This chapter will be divided into two main parts – advanced algorithms and applications. The first part is mainly devoted to describing some interesting new techniques in the graph machine learning domain. You will learn about some data sampling and data augmentation techniques for graphs based on random walk and generative neural networks. Then, you will learn about topological data analysis, a relatively novel tool for analyzing high-dimensional data. In the second part, we will provide you with some interesting applications of graph machine learning in different domains, ranging from biology to geometrical analysis. After reading this chapter, you will be aware of how looking at the relationships between data opened the door to intriguing novel solutions.

Specifically, we will cover the following topics in this chapter:

- Data augmentation for graphs
- Topological data analysis
- Applying graph theory in new domains

Technical requirements

All code files relevant to this chapter are available at https://github.com/PacktPublishing/
Graph-Machine-Learning/tree/main/Chapter13. Please refer to the *Practical exercises* section of
Chapter 1, Getting Started with Graphs, for guidance on how to set up the environment to run the
examples in this chapter, either using Poetry, pip, or Docker.

Data augmentation for graphs

In *Chapter 9, Graph Analysis for Credit Card Transactions*, we described how graph machine
learning can be used to study and automatically detect fraudulent credit card transactions. While
describing the use case, we faced two main obstacles:

- There were too many nodes in the original dataset to handle. As a consequence, the
  computational cost was too high to be computed. This is why we selected only 20% of
  the dataset.

- From the original dataset, we saw that less than 1% of the data had been labeled as
  fraudulent transactions, while the other 99% of the dataset contained genuine
  transactions. This is why, during the edge classification task, we randomly subsampled
  the dataset.

The techniques we used to solve these two obstacles, in general, are not optimal. For graph data,
more complex and innovative techniques are needed to solve the task. Moreover, when datasets
are highly unbalanced, as we mentioned in *Chapter 9, Graph Analysis for Credit Card Transactions*,
we can solve this using anomaly detection algorithms.

In this section, we will provide a description of some techniques and algorithms we can use to
solve the aforementioned problems. We will start by describing the graph sampling problem,
and we will finish by describing some graph data augmentation techniques. We will share some
useful references and Python libraries for both of these topics.

Sampling strategies

In *Chapter 9, Graph Analysis for Credit Card Transactions*, to perform the edge classification task, we
started our analysis by sampling only 20% of the whole dataset. Unfortunately, this strategy, in
general, is not optimal. Indeed, the subset of nodes selected with this simple strategy may form
a subgraph that does not accurately represent the overall topology of the graph. Due to this, we
need to define a strategy for building a subgraph of a given graph by sampling the right nodes.
The process of building a (small) subgraph from a given (large) graph by minimizing the loss of
topological information is known as **graph sampling**.

A good starting point so that we have a full overview of the graph sampling algorithm is available in the *Little Ball of Fur: A Python Library for Graph Sampling* paper, which can be downloaded from `https://arxiv.org/pdf/2006.04311.pdf`. The Python implementation of using the `networkx` library is available at the following URL: `https://github.com/benedekrozemberczki/littleballoffur`. The algorithms that are available in this library can be divided into node and edge sampling algorithms. These algorithms sample the nodes and edges in the graph bundling, respectively. As a result, we get a node- or edge-induced subgraph from the original graph. We will leave you to perform the analysis proposed in *Chapter 9, Graph Analysis for Credit Card Transactions*, using the different graph sampling strategies available in the `littleballoffur` Python package.

Exploring data augmentation techniques

Data augmentation is a common technique when we're dealing with unbalanced data. In unbalanced problems, we usually have labeled data from two or more classes. Only a few samples are available for one or more classes in the dataset. A class that contains a few samples is also known as a *minority* class, while a class that contains a large number of samples is known as a *majority* class. For instance, in the use case described in *Chapter 9, Graph Analysis for Credit Card Transactions*, we had a clear example of an unbalanced dataset. In the input dataset, only 1% of all the available transactions were marked as fraudulent (the minority class), while the other 99% were genuine transactions (the majority class). When dealing with *classical* datasets, the problem is usually solved using random down- or up-sampling or using data generation algorithms such as *SMOTE*. However, for graph data, this process may not be as easy since generating new nodes or graphs is not a straightforward process. This is due to the presence of complex topological relations. In the last decade, a large range of data augmentation graph algorithms have been made. Here, we will introduce two of the latest available algorithms, namely *GAug* and *GRAN*.

The GAug algorithm is a node-based data augmentation algorithm. It is described in the paper *Data Augmentation for Graph Neural Networks*, which is available at `https://arxiv.org/pdf/2006.06830.pdf`. The Python code for this library is available at `https://github.com/zhao-tong/GAug`. This algorithm can be useful for use cases where edge or node classification is needed, as in the use case provided in *Chapter 9, Graph Analysis for Credit Card Transactions*, where the nodes belonging to the minority class can be augmented using the algorithm. As an exercise, you can extend the analysis we proposed in *Chapter 9, Graph Analysis for Credit Card Transactions*, using the GAug algorithm.

The GRAN algorithm is a graph-based data augmentation algorithm. It is described in the *Efficient Graph Generation with Graph Recurrent Attention Networks* paper, which is available at `https://arxiv.org/pdf/1910.00760.pdf`. The Python code for the library is available at `https://github.com/lrjconan/GRAN`. This algorithm is useful for generating new graphs when we're dealing with graph classification/clustering problems. For example, if we're dealing with an unbalanced graph classification problem, it could be useful to create a balance step for the dataset using the GRAN algorithm and then perform the classification task.

More graph sampling and data augmentation techniques have been developed in recent years, and here we have provided a brief introduction to help you explore this rapidly evolving and fascinating field.

In the next section, we will introduce a novel approach—a new technique for learning about and analyzing graph features, rooted in topological data analysis.

Learning about topological data analysis

Topological Data Analysis (**TDA**) is a rather novel technique that's used to extract features that quantify the *shape of the data*. The idea of this approach is that by observing how data points are organized in a certain space, we can reveal some important information about the process that generated it.

The main tool for applying TDA is **persistent homology**. The math behind this method is quite advanced, so let's introduce this concept through an example. Suppose you have a set of data points distributed in a space, and let's suppose you are "observing" them over time. Points are static (they do not move across the space); thus, you will observe those independent points forever. However, let's imagine we can create associations between these data points by connecting them together through some well-defined rules. In particular, let's imagine a sphere expanding from these points through time. Each point will have its own expanding sphere and, once two spheres collide, an "edge" can be placed by these two points. This is exemplified in the following diagram:

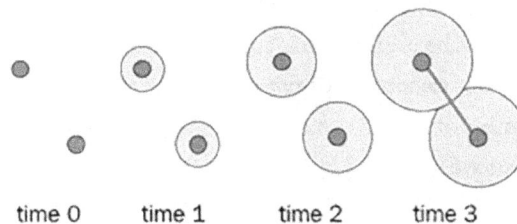

Figure 13.1: Example of how relationships between points can be created

The more spheres that collide, the more associations will be created, and the more edges will be placed. This happens when multiple spheres intersect more complex geometrical structures such as triangles, tetrahedrons, and so on appear:

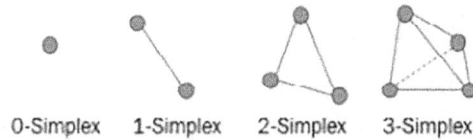

Figure 13.2: Example of how connections among points generate geometrical structures

When a new geometrical structure appears, we can note its *birth* time. On the other hand, when an existing geometrical structure disappears (for example, it becomes part of a more complex geometrical structure), we can note its *death* time. The survival time (time between birth and death) of each geometrical structure that's observed during the simulation can be used as a new feature for analyzing the original dataset.

We can also define the so-called **persistent diagram** by placing each structure's corresponding pair (birth, death) on a two-axis system. Points closer to the diagonal normally reflect noise, whereas points distant from the diagonal represent persisting features. An example of a persistence diagram is as follows. Notice that we described the whole process by using expanding *spheres* as an example. In practice, we can change the dimension of this expanding shape (for instance, using 2D circles), thus producing a set of features for each dimension (commonly indicated using the letter H):

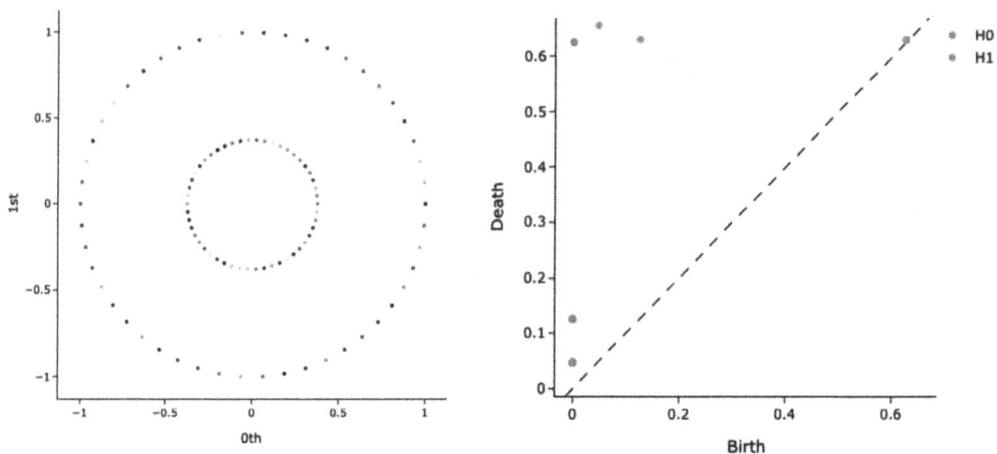

Figure 13.3: Example of a 2D point cloud (right) and its corresponding persistence diagram (left)

A good Python library for performing topological data analysis is **giotto-tda**, which is available at https://github.com/giotto-ai/giotto-tda. Using the giotto-tda library, it is easy to build the simplicial complex and its relative persistence diagram, as shown in the preceding figure.

Topological machine learning

Now that we know the fundamentals behind TDA, let's see how it can be used for machine learning. By providing machine learning algorithms with topological data (such as persistent features), we can capture patterns that might be missed by other traditional approaches.

In the previous section, we saw that persistence diagrams are useful for describing data. Nevertheless, using them to feed machine learning algorithms (such as **RandomForest**) is not a good choice. For instance, different persistent diagrams may have different numbers of points, and basic algebraic operations would not be well defined.

One common way to overcome such a limitation is to transform diagrams into more suitable representations. Embeddings or kernel methods can be used to obtain a *vectorized* representation of the diagrams. Moreover, advanced representation methods such as *persistence images*, *persistence landscapes*, and *Betti curves*, among others, have been shown to be very useful in practical applications such as shape analysis, biomolecular structure prediction, network science, material science, and machine learning for graph-based data. Persistent images (*Figure 13.4*), for instance, are bi-dimensional representations of persistence diagrams that can easily be fed into convolutional neural networks. An example of a persistent image is shown here:

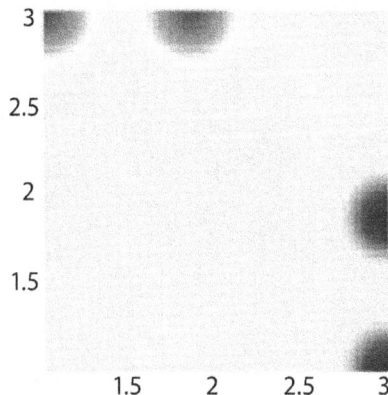

Figure 13.4: Example of a persistent image

Several possibilities arise out of this theory, and there is still a connection between the findings and deep learning. Several new ideas are being proposed, making the subject both hot and fascinating.

Topological data analysis is a rapidly growing field, especially since it can be combined with machine learning techniques. Several scientific papers are published on this topic every year and we expect novel exciting applications in the near future.

In this book, we have explored several examples of how graph machine learning can help solve practical problems. In the next section, we will provide an overview of its applications, aiming to expand your perspective on how the world can be understood through nodes and links!

Applying graph theory in new domains

In recent years, due to there being a more solid theoretical understanding of graph machine learning, as well as an increase in available storage space and computational power, we can identify a number of domains in which such learning theories are spreading. With a bit of imagination, you can start looking at the surrounding world as a set of *nodes* and *links*. Our work or study place, the technological devices we use every day, and even our brains can be represented as networks.

In this section, we will look at some examples of how graph theory (and graph machine learning) has been applied to, apparently, unrelated domains.

Graph machine learning and neuroscience

The study of the brain by means of graph theory is a prosperous and expanding field. Several ways of representing the *brain as a network* have been investigated, with the aim of understanding how different parts of the brain (nodes) are *functionally* or *structurally* connected to each other.

By means of medical techniques such as **Magnetic Resonance Imaging** (**MRI**), a three-dimensional representation of the brain can be obtained. Such an image can be processed by different kinds of algorithms to obtain distinct partitions of the brain (parcellation).

There are different ways in which we can define connections between those regions, depending on whether we are interested in analyzing their functional or structural connectivity:

- **Functional Magnetic Resonance Imaging** (**fMRI**) is a technique that's used to measure whether a part of the brain is "active" or not. Specifically, it measures the **blood-oxygen-level-dependent** (**BOLD**) signal of each region (a signal indicating the variation of the level of blood and oxygen at a certain time). Then, the *Pearson correlation* between the BOLD series of two brain regions of interest can be computed. High correlation means that the two parts are "functionally connected," and an edge can be placed between them. An interesting paper on graphically analyzing fMRI data is *Graph-based network analysis of resting-state functional MRI*, which is available at https://www.frontiersin.org/articles/10.3389/fnsys.2010.00016/full.

- On the other hand, by using advanced MRI techniques such as **Diffusion Tensor Imaging (DTI)**, we can also measure the strength of the white matter fiber bundles physically connecting two brain regions of interest. Thus, we can obtain a graph representing the structural connectivity of the brain. A paper where graphs neural networks are used in combination with graphs generated from DTI data is called *Multiple Sclerosis Clinical Profiles via Graph Convolutional Neural Networks* and is available at `https://www.frontiersin.org/articles/10.3389/fnins.2019.00594/full`.

- Functional and structural connectivity can be analyzed using graph theory. There are several studies that enhance significant alterations of such networks related to neurodegenerative diseases, such as Alzheimer's, multiple sclerosis, and Parkinson's, among others.

The final result is a graph describing the connection between the different brain regions, as shown here:

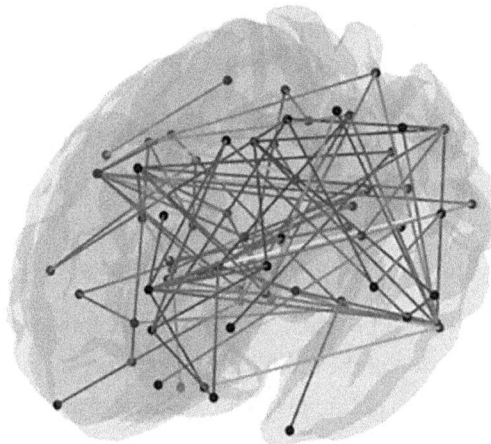

Figure 13.5: Connections between brain regions as a graph

Here, we can see how different brain regions can be seen as nodes of a graph, while the connections between those regions are edges.

Graph machine learning has been shown to be very useful for this kind of analysis. Different studies have been conducted to automatically diagnose a particular pathology based on the brain network, thus predicting the evolution of the network (for example, identifying potentially vulnerable regions that are likely to be affected by the pathology in the future).

Network neuroscience is a promising field, and, in the future, more and more insight will be collected from those networks so that we can understand pathological alterations and predict a disease's evolution.

Graph theory and chemistry and biology

Graph machine learning can be applied to chemistry. For example, graphs provide a natural method for describing **molecular structures** by treating atoms as the nodes of a graph and bonds as their connections. Such methods have been used to investigate different aspects of chemical systems, including representing reactions, and learning chemical fingerprints (indicating the presence or absence of chemical features or substructures), among others.

Several applications can also be found in biology, where many different elements can be represented as a graph. **Protein-protein interactions** (**PPI**), for example, is one of the most widely studied topics. Here, a graph is constructed, where nodes represent protein and edges represent their interaction. Such a method allows us to exploit the structural information of PPI networks, which has proved to be informative in PPI prediction.

Graph machine learning and computer vision

The rise of deep learning, especially **convolutional neural network** (**CNN**) techniques, has achieved amazing results in computer vision research. For a wide range of tasks, such as image classification, object detection, and semantic segmentation, CNNs can be considered the state of the art. However, recently, central challenges in computer vision have started to be addressed using graph machine learning techniques – **geometric deep learning** in particular. As we have learned throughout this book, there are fundamental differences between the 2D Euclidean domain in which images are represented and more complex objects such as 3D shapes and point clouds. Restoring the world's 3D geometry from 2D and 3D visual data, scene understanding, stereo matching, and depth estimation are only a few examples of what can be done. Let's see some tasks in the next sections.

Image classification and scene understanding

Image classification, one of the most widely studied tasks in computer vision, nowadays dominated by CNN-based algorithms, has started to be addressed from a different perspective. Graph neural network models have shown attractive results, especially when huge amounts of labeled data are not available. In particular, there is a trend in combining these models with *zero-shot and few-shot learning techniques*. Here, the goal is to classify classes that the model has never seen during training. For instance, this can be achieved by exploiting the knowledge of how the unseen object is semantically related to the seen ones.

Similar approaches have also been used for scene understanding. Using a relational graph between detected objects in a scene provides an interpretable structured representation of the image. This can be used to support high-level reasoning for various tasks, including captioning and visual question answering, among others.

Shape analysis

In contrast with images, which are represented by a bi-dimensional grid of pixels, there are several methods for representing 3D shapes, such as *multi-view images*, *depth maps*, *voxels*, *point clouds*, *meshes*, and *implicit surfaces*, among others. Nevertheless, when applying machine and deep learning algorithms, such representations can be exploited to learn specific geometric features, which can be useful for designing a better analysis.

In this context, geometric deep learning techniques have shown promising results. For instance, GNN techniques have been successfully used to find correspondence between deformable shapes, a classical problem that leads to several applications, including texture animation and mapping, as well as scene understanding.

For those of you who are interested, some good resources to help you understand this application of graph machine learning are available at `https://arxiv.org/pdf/1611.08097.pdf` and `http://geometricdeeplearning.com/`.

Recommendation systems

Another interesting application of graph machine learning is in recommendation systems, which we can use to predict the rating or the preference that a user would assign to an item. In *Chapter 7, Social Network Graphs*, we provided an example of how link prediction can be used to build automatic algorithms that provide recommendations to a given user and/or customer. In the paper *Graph Neural Networks in Recommender Systems: A Survey*, available at `https://arxiv.org/pdf/2011.02260.pdf`, the authors provide an extensive survey of graph machine learning that's been used to build recommendation systems. More specifically, the authors describe different graph machine learning algorithms and their applications.

Graph machine learning and NLP

Graph machine learning has become increasingly relevant in **natural language processing** (NLP) as it enables the modeling of complex relationships between entities, concepts, or words in a structured manner. One common application is the use of knowledge graphs to represent semantic relationships, where nodes represent entities or concepts, and edges capture the interactions or relationships between them.

These knowledge graphs can be used in tasks such as question answering, where the model learns to reason over the graph to provide accurate answers based on the relationships between entities, similar to what we did in *Chapter 8, Text Analytics and Natural Language Processing Using Graphs*.

Another trend in NLP is the representation of text as graphs. For instance, words in a document can be treated as nodes, and their relationships (such as syntactic dependencies or semantic proximity) can form the edges. Graph neural networks can be applied to these graphs to capture higher-order dependencies that traditional sequence models might miss, improving performance on tasks such as document classification, sentiment analysis, or summarization. Graph-based approaches can model not just word-to-word relations, but also long-range dependencies between distant words, offering a more robust understanding of text.

In addition to these structural applications, as we have seen in the previous chapters, **large language models (LLMs)** are also benefiting from graph-based methods. **Retrieval-augmented generation (RAG)** is an exciting development where LLMs retrieve information from an external knowledge base, often structured as a graph, to enhance their responses. In RAG, the model can first query a knowledge graph to retrieve relevant information and then use this knowledge to generate more accurate and contextually relevant outputs. This is particularly useful in tasks requiring factual accuracy, such as summarizing or generating responses in open-domain question-answering systems.

By integrating graph-based techniques with LLMs, it becomes possible to leverage external knowledge for more reliable and interpretable text generation, thus advancing the field of NLP in areas such as knowledge-based QA, dialogue systems, and even semantic search.

Summary

In this chapter, we provided a high-level overview of some emerging graph machine learning algorithms and their applications for new domains. At the beginning of the chapter, we described, using the example provided in *Chapter 9, Graph Analysis for Credit Card Transactions*, some sampling and augmentation algorithms for graph data. We provided some Python libraries that can be used to deal with graph sampling and graph data augmentation tasks.

We continued by providing a general description of topological data analysis and how this technique has recently been used in different domains.

Finally, we provided several descriptions of new application domains, such as neuroscience, chemistry, and biology. We also described how machine learning algorithms can also be used to solve other tasks, such as image classification, shape analysis, and recommendation systems.

This is it! In this book, we provided an overview of the most important graph machine learning techniques and algorithms. You should now be able to deal with graph data and build machine learning algorithms. We hope that you are now in possession of more tools in your toolkit and that you will use them to develop exciting applications. We also invite you to check the references we provided in this book and to address the challenges we proposed in the different chapters.

The world of graph machine learning is fascinating and rapidly evolving. New research papers are published every day with incredible findings. As usual, a continuous review of the scientific literature is the best way to discover new algorithms, and arXiv (`https://arxiv.org/`) is the best place to search for freely available scientific papers.

Get This Book's PDF Version and Exclusive Extras

UNLOCK NOW

Scan the QR code (or go to `packtpub.com/unlock`).
Search for this book by name, confirm the edition,
and then follow the steps on the page.

*Note: Keep your invoice handy. Purchases made
directly from Packt don't require one.*

Index

‹packt›

Other Books You May Enjoy

If you enjoyed this book, you may be interested in these other books by Packt:

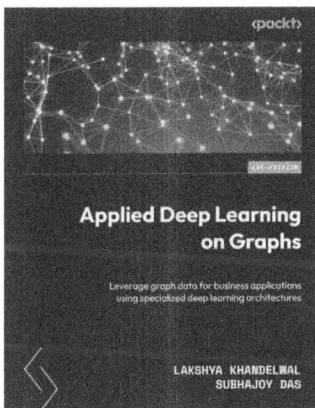

Applied Deep Learning on Graphs

Lakshya Khandelwal, Subhajoy Das

ISBN: 978-1-83588-596-3

- Discover how to extract business value through a graph-centric approach
- Develop a basic understanding of learning graph attributes using machine learning
- Identify the limitations of traditional deep learning with graph data and explore specialized graph-based architectures
- Understand industry applications of graph deep learning, including recommender systems and NLP
- Identify and overcome challenges in production such as scalability and interpretability
- Perform node classification and link prediction using PyTorch Geometric

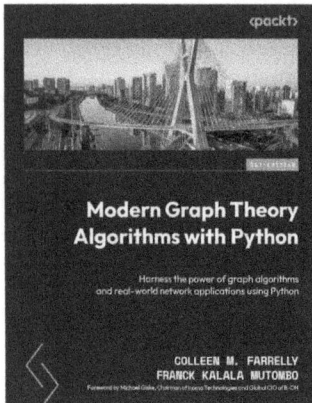

Modern Graph Theory Algorithms with Python

Colleen M. Farrelly, Franck Kalala Mutombo

ISBN: 978-1-80512-789-5

- Transform different data types, such as spatial data, into network formats
- Explore common network science tools in Python
- Discover how geometry impacts spreading processes on networks
- Implement machine learning algorithms on network data features
- Build and query graph databases
- Explore new frontiers in network science such as quantum algorithms

Packt is searching for authors like you

If you're interested in becoming an author for Packt, please visit authors.packtpub.com and apply today. We have worked with thousands of developers and tech professionals, just like you, to help them share their insight with the global tech community. You can make a general application, apply for a specific hot topic that we are recruiting an author for, or submit your own idea.

Share Your Thoughts

Now you've finished *Graph Machine Learning, Second Edition,* we'd love to hear your thoughts! Scan the QR code below to go straight to the Amazon review page for this book and share your feedback or leave a review on the site that you purchased it from.

https://packt.link/r/1803248068

Your review is important to us and the tech community and will help us make sure we're delivering excellent quality content.

* 9 7 8 1 8 0 3 2 4 8 0 6 6 *